MW00760954

L
STELET
HERZEN   JEANS   RUTHERFORD
WIEN   EINSTEIN   LANGEVIN
  Madame CURIE   POINCARÉ   KAMERLINGH ONNES

THE THEORY OF THE
QUANTUM WORLD

Proceedings of the 25th Solvay Conference on Physics

THE THEORY OF THE
QUANTUM
WORLD

Brussels, Belgium   19 – 22 October 2011

Editors

David Gross
Director, Kavli Institute for Theoretical Physics
University of California, Santa Barbara, USA

Marc Henneaux
Université Libre de Bruxelles
Director, International Solvay Institutes, Belgium

Alexander Sevrin
Vrije Universiteit Brussel
Also at the Universiteit Antwerpen and the KULeuven
Deputy-director, International Solvay Institutes, Belgium

 **World Scientific**

NEW JERSEY · LONDON · SINGAPORE · BEIJING · SHANGHAI · HONG KONG · TAIPEI · CHENNAI

Published by

World Scientific Publishing Co. Pte. Ltd.
5 Toh Tuck Link, Singapore 596224
USA office: 27 Warren Street, Suite 401-402, Hackensack, NJ 07601
UK office: 57 Shelton Street, Covent Garden, London WC2H 9HE

British Library Cataloguing-in-Publication Data
A catalogue record for this book is available from the British Library.

THE THEORY OF THE QUANTUM WORLD
Proceedings of the 25th Solvay Conference on Physics

Copyright © 2013 by World Scientific Publishing Co. Pte. Ltd.

All rights reserved. This book, or parts thereof, may not be reproduced in any form or by any means, electronic or mechanical, including photocopying, recording or any information storage and retrieval system now known or to be invented, without written permission from the Publisher.

For photocopying of material in this volume, please pay a copying fee through the Copyright Clearance Center, Inc., 222 Rosewood Drive, Danvers, MA 01923, USA. In this case permission to photocopy is not required from the publisher.

ISBN 978-981-4440-61-5
ISBN 978-981-4518-84-0 (pbk)

Printed in Singapore by Mainland Press Pte Ltd.

The International Solvay Institutes

Board of Directors

Members

Mr. Jean-Marie Solvay
President

Prof. Franz Bingen
Vice-President and Emeritus-Professor at the VUB

Prof. Lode Wyns
Secretary and Vice-Rector for Research at the VUB

Mr. Philippe Busquin
Minister of State and Former European Commissioner

Prof. Eric De Keuleneer
Solvay Brussels School of Economics & Management

Mr. Bernard de Laguiche
Chief Financial Officer of Solvay S.A.

Mr. Alain Delchambre
President of the Administrative Board of the ULB

Baron Daniel Janssen
Honorary Chairman of the Board of Directors of Solvay S.A.

Mr. Eddy Van Gelder
President of the Administrative Board of the VUB

Honorary Members

Baron André Jaumotte
Honorary Director of the Solvay Institutes, Honorary Rector,
Honorary President of the ULB

Mr. Jean-Marie Piret
Emeritus Attorney General of the Supreme Court of Appeal
and Honorary Principal Private Secretary to the King

Prof. Jean-Louis Vanherweghem
Former President of the Administrative Board of the ULB

Prof. Irina Veretennicoff
Emeritus Professor at the VUB

Guest Members

Prof. Marc Henneaux
Director and Professor at the ULB

Prof. Alexander Sevrin
Deputy Director, Professor at the VUB and Scientific Secretary of the
Committee for Physics

Prof. Franklin Lambert
Emeritus Professor at the VUB

Prof. Anne De Wit
Professor at the ULB and Scientific Secretary of the Committee for Chemistry

Prof. Hervé Hasquin
Permanent Secretary of the Royal Academy of Sciences, Letters and Fine Arts
of Belgium

Prof. Géry van Outryve d'Ydewalle
Permanent Secretary of the Royal Flemish Academy of Belgium for Sciences
and the Arts

Director

Prof. Marc Henneaux
Professor at the ULB

Solvay Scientific Committee for Physics

Prof. David Gross (Chair)
Kavli Institute for Theoretical Physics (Santa Barbara, USA)

Prof. Roger Blandford
Stanford University (USA)

Prof. Steven Chu
Stanford University (USA)

Prof. Robbert Dijkgraaf
Universiteit van Amsterdam (The Netherlands)

Prof. Bert Halperin
Harvard University (Cambridge, USA)

Prof. Giorgio Parisi
Università la Sapienza (Roma, Italy)

Prof. Pierre Ramond
University of Florida (Gainesville, USA)

Prof. Gerard 't Hooft
Spinoza Instituut (Utrecht, The Netherlands)

Prof. Klaus von Klitzing
Max-Planck-Institut (Stuttgart, Germany)

Prof. Peter Zoller
Universität Innsbruck (Austria)

Prof. Alexander Sevrin (Scientific Secretary)
Vrije Universiteit Brussel (Belgium)

In Memoriam Jacques Solvay (1920-2010)

It is with great sadness that the Board of Directors of the International Solvay Institutes for Physics and Chemistry, founded by Ernest Solvay, learned of the passing away of its President, Jacques Solvay on April 29, 2010.

Throughout his life, Jacques Solvay demonstrated a keen interest for scientific research and supported and encouraged the development of science at the highest level.

Jacques Solvay was the inaugural President of the Board of Directors of the International Solvay Institutes upon their creation in 1970. From as early as the fifties, he expressed his strong intention to continue the work initiated by his great-grandfather, Ernest Solvay, by becoming an active member of the administrative committees of the Institute for Physics and the Institute for Chemistry prior to their merger.

Under his direction and that of Ilya Prigogine – with whom he shared a common purpose – the Institutes broadened their activities and developed new research directions. At the same time, Jacques Solvay maintained a strong focus on the Solvay Conferences for Physics and Chemistry, an activity which he kept close to his heart. The hospitality which Mr. and Mrs. Solvay extended to the participants contributed greatly to the prestige and the success of these events.

With the passing of Jacques Solvay, the international scientific community loses a great friend of science. The Institutes will keep a vivid and grateful memory of Jacques Solvay and offer their condolences to his widow and the whole Solvay Family.

***

To honor the memory of Jacques Solvay, the Board of Directors created the "International Jacques Solvay Chair for Physics". The Board of Directors expresses their gratitude to the Solvay Family which supported this initiative.

Hotel Métropole (Brussels), 19-22 October 2011

The Theory of the Quantum World
Chair: Professor David Gross

The 25th Solvay Conference on Physics took place in Brussels from October 19 through October 22, 2011 celebrating one century of tradition and scientific excellence initiated by Lorentz at the 1st Solvay Conference on Physics in 1911 (*Premier Conseil de Physique Solvay*). The conference was preceded by an Academic Session *Why Curiosity Driven Science?* in the presence of His Royal Highness King Albert II of Belgium. The conference was followed by a public event entitled *The Future of Physics*. William Phillips and Frank Wilczek each delivered a lecture and a panel of scientists – led by David Gross and consisting of Alain Aspect, Gerard 't Hooft, William Phillips, Subir Sachdev, Frank Wilczek and Peter Zoller – answered questions from the audience.

The organization of the 25th Solvay Conference has been made possible thanks to the generous support of the *Solvay Family*, the *Solvay Company*, the *Université Libre de Bruxelles*, the *Vrije Universiteit Brussel*, the *Belgian National Lottery*, the *Foundation David and Alice Van Buuren*, the *Belgian Science Policy Office*, the *Communauté française de Belgique*, de *Actieplan Wetenschapscommunicatie* of the *Vlaamse Regering*, the *City of Brussels*, the *Max-Planck-Gesellschaft* and the *Hôtel Métropole*.

Participants

Yakir	**Aharonov**	Chapman University
Boris	**Altshuler**	Columbia University
Ignatios	**Antoniadis**	CERN
Nima	**Arkani-Hamed**	Institute for Advanced Study
Alain	**Aspect**	Institut d'Optique
David	**Awschalom**	University of California Santa Barbara
Leon	**Balents**	University of California Santa Barbara
Niklas	**Beisert**	ETH Zürich
Michael	**Berry**	University of Bristol
Roger	**Blandford**	Stanford University
Lars	**Brink**	Chalmers University of Technology
Claudio	**Bunster**	Centro de Estudios Científicos
Ignacio	**Cirac**	Max-Planck-Institut für Quantenoptik
Sankar	**Das Sarma**	University of Maryland
Seamus	**Davis**	Cornell University
Robbert	**Dijkgraaf**	University of Amsterdam
Savas	**Dimopoulos**	Stanford University
Michael R.	**Douglas**	Stony Brook University
Georgi	**Dvali**	New York University
François	**Englert**	Université Libre de Bruxelles
Matthew	**Fisher**	Microsoft Station Q
Murray	**Gell-Mann**	Santa Fe Institute
Howard	**Georgi**	Harvard University
Gary	**Gibbons**	Cambridge University
Steven	**Girvin**	Yale University
Gian Francesco	**Giudice**	CERN
Michael B.	**Green**	Cambridge University
David	**Gross**	University of California Santa Barbara
Alan	**Guth**	MIT
F. Duncan M.	**Haldane**	Princeton University
Bertrand	**Halperin**	Harvard University
Jim	**Hartle**	University of California Santa Barbara
Stephen	**Hawking**	Cambridge University
Alan J.	**Heeger**	University of California Santa Barbara
Marc	**Henneaux**	Université Libre de Bruxelles
Gary	**Horowitz**	University of California Santa Barbara
Shamit	**Kachru**	Stanford University
Wolfgang	**Ketterle**	MIT
Igor	**Klebanov**	Princeton University
Daniel	**Kleppner**	MIT

Anthony J.	**Leggett**	University of Illinois at Urbana-Champaign
Juan	**Maldacena**	Institute for Advanced Study
Viatcheslav	**Mukhanov**	Ludwig-Maximilians-Universität
Nikita	**Nekrasov**	Institut des Hautes Etudes Scientifiques
Hermann	**Nicolai**	Max-Planck-Institut für Gravitationsphysik
Hirosi	**Ooguri**	California Institute of Technology
Giorgio	**Parisi**	University La Sapienza
William D.	**Phillips**	National Institute of Standards and Technology; University of Maryland
Joseph	**Polchinski**	University of California Santa Barbara
Alexander	**Polyakov**	Princeton University
John	**Preskill**	California Institute of Technology
Eliezer	**Rabinovici**	Hebrew University
Lisa	**Randall**	Harvard University
Valery	**Rubakov**	Insitute for Nuclear Research of the Russian Academy of Sciences
Subir	**Sachdev**	Harvard University
Nathan	**Seiberg**	Institute for Advanced Study
Ashoke	**Sen**	Harish-Chandra Research Institute
Mikhail	**Shifman**	University of Minnesota
Eva	**Silverstein**	Stanford University
Gerard	**'t Hooft**	Utrecht University
Gabriele	**Veneziano**	Collège de France
Erik	**Verlinde**	University of Amsterdam
Klaus	**von Klitzing**	Max-Planck-Institut für Festkörperforschung
Spenta	**Wadia**	Tata Institute of Fundamental Research
Xiao-Gang	**Wen**	MIT
Frank	**Wilczek**	MIT
David J.	**Wineland**	National Institute of Standards and Technology
Edward	**Witten**	Institute for Advanced Study
Anton	**Zeilinger**	University of Vienna
Peter	**Zoller**	University of Innsbruck
Wojciech H.	**Zurek**	Los Alamos National Laboratory

Auditors

Riccardo	**Argurio**	Université Libre de Bruxelles
Glenn	**Barnich**	Université Libre de Bruxelles
Nicolas	**Cerf**	Université Libre de Bruxelles
Ben	**Craps**	Vrije Universiteit Brussel
Jan	**Danckaert**	Vrije Universiteit Brussel
Frederik	**Denef**	Katholieke Universiteit Leuven
Thomas	**Durt**	Institut Fresnel
Frank	**Ferrari**	Université Libre de Bruxelles
Jean-Marie	**Frère**	Université Libre de Bruxelles
Pierre	**Gaspard**	Université Libre de Bruxelles
Jean-Marc	**Gérard**	Université Catholique de Louvain
Nathan	**Goldman**	Université Libre de Bruxelles
Thomas	**Hertog**	Katholieke Universiteit Leuven
Axel	**Kleinschmidt**	Université Libre de Bruxelles
Fabio	**Maltoni**	Université Catholique de Louvain
Serge	**Massar**	Université Libre de Bruxelles
Stefano	**Pironio**	Université Libre de Bruxelles
Alexander	**Sevrin**	Vrije Universiteit Brussel
Philippe	**Spindel**	Université de Mons
Jacques	**Tempere**	Universiteit Antwerpen
Peter	**Tinyakov**	Université Libre de Bruxelles
Michel	**Tytgat**	Université Libre de Bruxelles
Christian	**Van Den Broeck**	Universiteit Hasselt
Antoine	**Van Proeyen**	Katholieke Universiteit Leuven
Irina	**Veretennicoff**	Vrije Universiteit Brussel

Opening Session

The inaugural session took place on October 19, 2011 in the Brussels City Hall, in the presence of the Solvay Family and Mrs Sabine Laruelle, Minister for Belgian Science Policy. We are grateful to Mr. Freddy Thielemans, Mayor of Brussels, for hosting the session in this prestigious and historical place and for greeting the participants.

Welcoming Address by Jean-Marie Solvay, President of the International Solvay Institutes

Dear Friends,

In the name of the Solvay Family, I cannot tell you how proud I am to be here welcoming you today at the opening session of the 25th Solvay Conference on Physics. One hundred years have gone by since the first council and so much has changed in the world that we live in, thanks in a large part, to the work of the curious people, who like you, gathered here together in order to solve complex and exciting problems.

The program of today's opening session is outstanding and I am looking forward to hearing the extraordinary lectures that will be held this morning in this beautiful hall.

I wish to thank Marc Henneaux, the Director of our Institutes, and David Gross, the Chair of the Scientific Committee, for making this gathering possible. I am intensely curious to see what will unfold in the next coming days as you all discuss your views on the "Theory of the Quantum World".

Thank you for being here. Have a fruitful meeting!

Opening Address by Marc Henneaux,
Director of the International Solvay Institutes

Dear Colleagues, Dear Friends,

Today starts the 25th Solvay Conference on Physics, entitled "The Theory of the Quantum World". In the name of the International Solvay Institutes, I would like to welcome all of you to the first session of the conference, which is mostly devoted to more historical aspects — as it is natural since we are celebrating in 2011 one hundred years of Solvay Conferences.

The history of the Solvay Conferences is intimately connected with the development of quantum mechanics. It was at the first Solvay Conference on Physics that the conceptual rupture between the old "classical" physics and the new theory of quanta was clearly realized to be inevitable. It was at the 5th Solvay Conference in 1927 that the formulation of quantum mechanics still used today was definitely established. Most of the subsequent Solvay Conferences dealt with quantum mechanics in one form or the other (to list some in the last fifty years: 1961: "Quantum Field Theory"; 1967: "Fundamental Problems in Elementary Particle Physics"; 1982: "Higher Energy Physics"; 1991: "Quantum Optics"; 2005: "The Quantum Structure of Space and Time"; 2008: "Quantum Theory of Condensed Matter").

It was thus quite logical that quantum mechanics was chosen to be the central theme of the 25th Solvay Conference on Physics, the conference of the centenary.

Before we start the conference, I would like to address special thanks to the conference chair, David Gross, who is also the chair of the Scientific Committee for Physics. He has been an exceptional driving force in the scientific revival of the Solvay Conferences, which are now organized back again along the elitist format set up by Lorentz, the first chair of the Solvay Scientific Commitee for Physics.

It is well documented that Lorentz spent days of meticulous preparation of the Solvay conferences when he was their chair. It was that exceptional preparation that was key to their success, where discussions and exchanges of ideas played the central role.

I can assure you that the same meticulous preparation went into this conference, which started about two years ago. David is a very busy man, and it is for us of exceptional significance that he devotes a great amount of his time to the Solvay activities. His help has been a tremendous support.

In fact, he helped us enormously not just for the 25th scientific conference itself, but for the organisation of all the activities of the centenary celebration. You saw him two days ago in the staged reading of Copenhagen and he helped me establish the program of yesterday's academic session on the importance of fundamental research.

On the occasion of the hundredth anniversary of the first Solvay Conference on Physics, the International Solvay Institutes have created a special Solvay centenary chair. This chair has been granted to David not only for his outstanding scientific merits, but also as a way to express the gratitude of the International Solvay Institutes for the exceptional role he is playing since 2004 in giving new vigor to their activities.

We are very pleased and honoured that he accepted.

We are also most grateful to the Solvay Family for their support to the Institutes for a century — now I can say it, as a century has indeed elapsed since 1911. And finally, I would like to thank all the participants in the 25th Solvay Conference. It is your positive response to our invitation and your presence here that gives sense to the efforts of the International Solvay Institutes.

I thank you very much for your attention. Let us move on to the next introductory speech.

Opening Address by David Gross,
Chair of the 25th Solvay Conference on Physics
and Chair of the Solvay Scientific Committee for Physics

A Century of Quantum Mechanics

One hundred years ago, twenty-four physicists met at the Hotel Métropole in Brussels; they were invited by Ernest Solvay to participate in a new kind of scientific congress. One of the first international scientific meetings, the Solvay Conferences were characterized by a highly restricted invitation list and an unusual mixture of short talks and long discussions. Solvay played a unique and important role in the development of twentieth century physics — most notably in the quantum revolution whose birth overlapped the initiation of these meetings. The Solvay tradition has continued with a physics conference every three years, except for unfortunate lapses due to war. Solvay, one of the few traditions that remain in the rapidly changing scientific landscape, represents excellence, internationalism, free discussion and lively debate — a tradition worth preserving, but a tradition that is hard to maintain as the number of physicists has increased in the last century by two orders of magnitude, whereas the number of invitees can only increase by a factor of two or three! I thank the members of the Solvay Physics Committee for their help in the difficult task of compiling the invitation list to this conference. Many of our colleagues that should be here are absent. We will miss them. We must regard ourselves as representatives of a much larger community. I also thank the chairs of the individual sessions, the members of the Solvay physics committee (R. Blandford, S. Chu, R. Dijkgraaf, B. Halperin, G. 't Hooft, G. Parisi, P. Ramond, K. Von Klitzing, P. Zoller) and the scientific leaders of the Solvay Institutes, Marc Henneaux (Director) and Alexander Sevrin (Scientific Secretary) for their invaluable contributions to making this conference possible.

For this centenary year and for the 25th Solvay Conference in Physics, we decided to cover physics broadly, but to focus on quantum mechanics, whose early birth pangs and later applications were most often the subjects of the Solvay meetings. Quantum mechanics emerged in the period between 1900, when Planck first quantized the energy of radiating oscillators, and 1925-26 with Heisenberg, Schrodinger, Born and Dirac's formulation of the principles of quantum mechanics, and thus is approximately one century old. The development of quantum mechanics and its application to atomic theory and the structure of matter dominated the first five Solvay Conferences, culminating in the most famous 1927 Solvay meeting, where the meaning of quantum reality was heatedly debated between the pioneers and the revolutionaries of quantum mechanics.

The first Solvay conference, one hundred years ago to the month, addressed the central problem of physics at that time: *Was the quantum structure of nature truly unavoidable?* Lorentz's opening address at the first Solvay conference reverberates with the anguish that this master of classical physics felt at the first glimpses of the quantum world:

Modern research has encountered more and more serious difficulties when attempting to represent the movement of smaller particles of matter and the connection between these particles and phenomena that occur in the ether. At the moment, we are far from being completely satisfied that, with the kinetic theory of gases gradually extended to fluids and electron systems, physicists could give an answer in ten or twenty years. Instead, we now feel that we reached an impasse; the old theories have been shown to be powerless to pierce the darkness surrounding us on all sides.

We face no such crisis today.

Quantum mechanics is the most successful of all the frameworks that we have discovered to describe physical reality. It works, it makes sense, and it is hard to modify. The order of this list of successes is in the order of importance that most physicists demand of a physical theory: It works, it makes sense, and it is hard to modify. I shall start with the second point.

Quantum mechanics does make sense, although the transition, a hundred years ago, from classical to quantum reality was not easy. It took time to learn how to get out of phase space and to live in Hilbert space. Some of the boldest pioneers of quantum theory (notably Einstein) resisted the replacement of classical determinism with a theory that often can only make probabilistic predictions. Even harder to get used to was the idea that in quantum mechanics one can describe a system in many different and incompatible ways, and that there is no unique exhaustive description. The freedom one has to choose among different, incompatible, frameworks does not influence reality — one gets the same answers for the same questions, no matter which framework one uses. That is why one can simply "shut up and calculate". Most of us do that most of the time. Different, incompatible aspects cannot both enter a single description. If one errs by mixing incompatible descriptions or histories, we produce paradoxes.

By now, especially with the consistent (or decoherent) histories approach, initiated by R. Griffiths, and further developed by Gell-Mann, Hartle, Omnes, Zurek and others, we have a completely coherent and consistent formulation of quantum mechanics that corresponds to what we actually do in predicting and describing experiments and observations in the real world. For most of us there are no problems.

Nonetheless, there are dissenting views. Experimentalists continue to test the predictions of quantum theory, and some theorists continue to question the foundations. We will hear from them, and we will debate them in our second session. Most interesting to me is the growing understanding as to how the classical framework emerges from quantum mechanics, especially interesting as our experimental friends continue to astonish us with their ability to control and manipulate quantum systems while preserving their quantum coherence. How can we explain measurements without invoking the absurd collapse of the wave function? How does classical physics emerge from quantum reality? The mathematics of the classical limit and the growing understanding of decoherence will also be discussed in our second session.

Quantum mechanics is more powerful and richer than classical mechanics, for, after all, classical physics is just a limiting, special case of quantum physics. In recent years we have also become aware of the increased computational power of quantum mechanical states. Entanglement, the strange new feature of quantum states, can be efficiently used to amplify computation, and has motivated an intensive effort to develop a quantum computer. This goal might take many decades to realize, but meanwhile the effort has provided enormous stimulation to atomic and condensed matter physics. Quantum information theory will be discussed in our second session.

The dream of a quantum computer is only conceivable because of the enormous advances made in recent years towards greater control and understanding of matter, down to the scale of individual atoms: mesoscopics, atomic traps, quantum optics, and spintronics. A new field is developing that might be called quantum engineering, with enormous potential for both technological innovation and for use as a marvelous tool for the experimental exploration, and the simulation, of fascinating states of quantum matter. These tools enable not only the study of the static phases of complicated many body systems, but also of their dynamics and non-equilibrium behavior. This development is the subject of our third session.

Quantum mechanics works.

It works not just for simple systems such as single atoms and molecules, but also for collections of 10^{23} atoms, sometimes strongly interacting, over an enormous range of energies. It explains not just the anomalies in the classical description of blackbody radiation and the specific heat of solids at low temperatures (that stimulated early developments), but also the detailed properties of ordinary matter, such as conductors, insulators, semiconductors as well as more exotic materials.

The quantum theory of matter (many-body theory) and the quantum theory of fields share many common features; indeed they are essentially the same thing. Thus, critical developments in condensed matter physics and in elementary particle physics towards the end of the twentieth century often occurred in parallel. In these developments, symmetry principles played a fundamental role. But if the secret of nature is symmetry, much of the texture of the world is due to mechanisms of symmetry breaking. Magnetism and chiral symmetry breaking are two important examples of the spontaneous breaking of a global symmetry.

One of the most important quantum phenomenon — that of superconductivity — was discovered by Onnes 100 years ago and discussed at the first Solvay conference. Parenthetically, Rutherford's discovery of the nucleus of atoms, made also in 1911, was not discussed, although Rutherford attended! It took almost half a century (until 1957) for Bardeen, Cooper and Schrieffer to come up with a full understanding of this first example of the spontaneous breaking of a local symmetry, which later played a fundamental role in the understanding of the weak nuclear force — the so-called Higgs mechanism — with the final confirmation coming from the LHC. Even today, unconventional superconductors are still a great mystery at the frontiers of the understanding of quantum states of matter. It now appears that that there are new forms of matter — labeled not by symmetry but by topology. The important question, "What are the possible quantum phases of matter?" remains wide open, and will be discussed in our fourth session.

Quantum mechanics works.

It works at distances that are a billion times smaller than the size of the atom, well within the nucleus and its constituent quarks. It works for energies that are a trillion times larger than atomic energies. From the beginning it was clear that quantum mechanics fit together seamlessly with special relativity and with Maxwells theory of the electromagnetic field, despite a few technical difficulties that took some time to resolve. The resulting edifice, the quantum theory of fields, resolved the perplexing duality of particles and waves, and, in what I regard as one of the most amazing successes of theoretical physics, predicted anti-matter, the first examples of which were soon discovered.

Quantum field theory has been tested with extraordinary precision. Much of the incredible precision that physics is able occasionally to achieve rests on quantum features of nature, such as the identity of indistinguishable particles and the existence of discrete sharp states. I cannot refrain from noting one of the most amazing of these precision tests, that of the measurement of the anomalous magnetic moment of the electron:

$$a_e = \frac{g_e - 2}{2} = 0.00115965218085 \pm .00000000000076 \,,$$

a test of Quantum Electrodynamics (QED) to almost one part in 10^{12}, sensitive to all the components of the standard model, but especially QED (the comparison involves 5 loop quantum effects). Quantum field theory works and has been tested over an incredible range of physical phenomena, from the edge of the galaxy (10^{27} *cm.*) to the nano-nano centimeter scale, over forty-five orders of magnitude. In fact, we know of no reason why the framework of quantum field theory could not continue to be adequate until we reach the Planck scale (10^{-33} *cm*), where quantum effects of gravity become important.

Quantum Mechanics works.

It provides the explanation, not only of the structure of atoms and molecules, but also of the structure of the nucleus, and the nature of the strong and weak nuclear forces. In a reductionist sense, the standard model of elementary particles (with 3 families of quarks and leptons, charged under 3 gauge groups that generate three forces) is an amazing theory, powerful enough to encompass almost all of the known forces that act on the known particles of nature (with the exception of dark matter and the right-handed partner of the neutrino). The standard model is so extraordinarily successful that we currently strain, so far unsuccessfully, to find deviations. The successes and failures of the standard model will be discussed in the fifth session.

Finally, Quantum Mechanics is hard to modify.

Our present fundamental framework, quantum field theory, appears under no threat from observation or experiment, and seems to be completely adequate for the understanding of macroscopic and microscopic physics, from the edge of the universe to the nano-nano meter scale. It is very difficult to construct consistent alternatives to this framework that agree with observation. But no framework, no theory, is likely to survive untouched forever. Where might our present quantum mechanical framework breakdown and how?

Hints from observation and from experiment point to physics beyond the standard model. The existence of dark matter, the non-vanishing neutrino masses and the many unanswered questions regarding quark and lepton masses and their mixing require non-standard-model physics; but the necessary modifications do not necessarily force us to abandon the framework of quantum field theory. More hints come from trying to extend our standard theory to new regimes of energy and distance and from challenging our concepts with thought experiments.

The extrapolation of the standard model to high energy, or equivalently short distance, suggests that the atomic and nuclear forces are unified at very high energy. Such unification does not necessarily suggest a breakdown of the framework of quantum field theory; we can construct grand unified gauge theories. However, the fact that the implied unification scale is so close to the Planck scale, where the quantum nature of gravity becomes essential, is an important hint that the grand synthesis must include quantum gravity. Traditional quantum field theory appears to be at a loss to consistently describe gravity, due to the uncontrollable quantum fluctuations of the metric at the Planck scale. In the search for a unified theory of standard model forces, we have been led to string theory, which also automatically includes gravity and yields a consistent extension and quantization of classical Einstein gravity.

String theory was originally thought to break with traditional quantum field theory in important ways, but recently we have realized that string theory and quantum field theory are not mutually exclusive. Quantum field theory, in the old fashioned sense, is not sufficient to contain gravity. But it is part of a bigger framework that includes extended objects, strings, membranes, and higher dimensional "branes". The formulation in terms of strings is often best understood — thus "string theory". String theory always describes dynamical space-time — gravity. On the other hand, some string theory quantum states can be usefully described in terms of quantum field theory. This insight has been inspired by the remarkable duality between supersymmetric gauge theory in four dimensions (or more generally conformal field theory) and string theory in an AdS background. Even the theoretical framework we use for the standard model, consisting of quantum gauge theory with fundamental fermions and a few scalars has (many of us believe) a dual description in terms of a string theory with highly curved extra dimensions. A close cousin of Quantum Chromodynamics, endowed with extra (super) symmetry, is undoubtedly identical to string theory in AdS space. So string theory and quantum field theory are part of a larger quantum mechanical framework, whose structure and extent are still being explored.

Finally, there are indications that once again we might be forced to modify our most fundamental of physical concepts, that of space and time. Many of us are more and more convinced that space is an emergent, not fundamental, concept. We have many examples of interesting quantum mechanical states, for which we can think of some (or all) of the spatial dimensions as emergent. Together with emergent space, we have the emergent dynamics of space and thus emergent gravity. But it is hard to imagine how time could be emergent? How would we formulate quantum mechanics without time as a primary concept? Were time to be emergent, our understanding of quantum mechanics would have to change.

To describe nature and to make predictions, we need more than just the framework of quantum mechanics, or of quantum field theory, or of quantum string theory. We need a particular dynamical principle, a Hamiltonian that determines the time development, and we also need an initial state. So what picks the dynamics? Quantum field theory offers little guide, except symmetry. String theory, in which all parameters are dynamical, appeared at first to offer the hope of providing a unique answer. But this hope appears to be a mirage. String "theory" does not provide such a principle; rather it consists of a set of tricks to find consistent quantum states, often constructed in a perturbative semiclassical expansion. And there are many such quantum states, an infinite number in fact, perhaps 10^{500} that resemble our universe. Some believe that this is the complete story, and that all of these universes might exist somewhere in a multiverse, and that to make predictions we must resort to arguing that our patch of the multiverse is particularly suited for our existence.

Since a theory of quantum gravity is a dynamical theory of space-time, we must finally come to grips with quantum cosmology. Here it makes no sense to separate the observer and the observed, and we are faced with many puzzling conceptual issues. What picks the initial condition? the final condition? In addition, we are challenged by astrophysics. In the last hundred years, we have learned much about the universe, including a detailed description of most of its history. The outstanding mysteries that remain — the dynamics of inflation, the mystery of the big bang and the accelerated expansion — represent serious challenges to our theoretical framework.

String theory and quantum cosmology will be discussed in the sixth session.

So what is the whole picture? We are faced today not with a crisis but with confusion at the frontiers of knowledge. Fundamental physics today is in a state more analogous to the one that prevailed in 1891, rather than in 1911. In 1891, with all the successes of classical physics — mechanics, electrodynamics, kinetic theory and statistical mechanics — physics appeared in fine shape. Who could have dreamed of the conceptual revolutions that lay in store?

Many of the issues I have alluded to will be discussed towards the end of our meeting, in the final session. We are unlikely to come to a resolution during this meeting. The most we can hope for is that our discussions will clarify the issues and most importantly stimulate the advances that are necessary. In any case it should be lots of fun.

Contents

Session 1

History and Reflections

Chair: *Marc Henneaux*, Université Libre de Bruxelles, Belgium
Rapporteurs: *John L. Heilbron*, University of California at Berkeley, USA and
Murray Gell-Mann, Santa Fe Institute, USA
Scientific secretary: *Philippe Spindel*, Université de Mons, Belgium

John L. Heilbron: The First Solvay Council "A sort of private conference"[a]

1. Introduction

It is a great an honor for me to address this opening session of the 25th Solvay Conference on Physics, and also a great challenge. For the charge the organizers laid upon me was to define the historical importance of the initial Solvay Conference, or Council as it was called for a reason I'll mention presently. This first Council, more famous now, perhaps, for the eminence of its members than for the significance of its deliberations, convened here, a century ago, as a result of an appeal by the physical chemist Walther Nernst to the philanthropic industrialist Ernest Solvay.

---

[a]Professor of History, Emeritus, University of California, Berkeley. I am grateful to the editor for allowing me to maintain the informal style of the lecture from which this text derives, and to Franklin Lambert for generous references to archival documents. The following abbreviations are used: Erreygers, Guido Erreygers, "The economic and social reform programs of Ernest Solvay," in Warren J. Samuels, ed., European economists of the early 20th century, Vol. 1. Studies of neglected thinkers of Belgium, France, The Netherlands and Scandinavia (Chattenham: E. Elgar, 1998), 220-62; Solvay I, La théorie du rayonnement et les quanta. Rapports et discussions de la réunion tenue à Bruxelles du 30 octobre au 3 novembre 1911, ed. Paul Langevin and Maurice de Broglie (Paris: Gauthier-Villars, 1911); Warnotte, Daniel Warnotte, Ernest Solvay et l'Institut de Sociologie (2 vols., Brussels: E. Bruylant, 1946).

Nernst asked Solvay to defray the expenses of bringing together the few scientists who then cared about quantum theory to talk about the problems it presented for received physics.

The Council did not reach agreement about the nature of the quantum or its place in physics. However, most of its participants took away the conviction that the foundations of physics had to be enlarged, if not built anew, to accommodate the quantum of energy unintentionally introduced into the theory of radiation by Max Planck a decade earlier.[1] The place of the first Solvay Council in the history of physics – which is not at all the same as its place in history – was at the watershed between classical and modern physics. Perhaps, on this celebratory occasion, it might not be too great an exaggeration to say that the first Solvay Council defined the watershed; for in identifying the difficulties that beset received theories, it had to specify the elements of what we now refer to as classical physics.[2]

This watershed was a feature in a much wider landscape molded by forces not recognized by physics. These forces – let's call them social – created the extraordinary figure of Ernest Solvay, who responded favorably to Nenst's request for deeply held reasons of his own. Hence, perhaps, the label conseil for the gathering. Since the use of the term "council" for a scientific conference was unusual, the editors of its proceedings felt obliged to explain that it meant "a sort of private conference." In fact, from Solvay's point of view, it was a sort of council, a panel advisory to himself. Though Nernst dominated the arrangements, Solvay retained overall authority, which extended to the choice of participants. That agreed with his practice with the scientific enterprises he had established over the previous two decades. Experience with the Council was to cause him to relax his control and give the physicists, who were no more docile then than now, their head.

2. Soda and Energy

The so-called second industrial revolution, to which Solvay's soda was an important contributor, celebrated itself in a series of international fairs during the later nineteenth century. These were made possible and occasionally profitable by a revolution in transportation. To fix ideas, the first pubic railroad, an uncomfortable line between Manchester and Liverpool, opened in 1830 with about 60 kilometers of track. At the time of the Solvay Council, the world's railroads extended over a million kilometers and ran first-class carriages more opulent than Eurostar's.

They also carried freight, which, by 1910, included two million tons of ammonia soda a year, all of it produced directly by, or under license from, Ernest Solvay et Compagnie. This family-owned business with its licensees then constituted the world's largest chemical enterprise.[3] Soda was used in all sorts of industries, soap, glass, ceramics, food, medicine, artificial silk, fine chemicals. A multicolor "soda tree" drawn up for the Universal Exhibition in Paris in 1900 has roots labeled "brine," "ammonia," and "limestone," and more branches, twigs, and leaves than the generations of David on a Jesse tree.[4]

The Solvay enterprises must have earned close to fifty million francs a year in 1912, when Solvay endowed his Institut de Physique. He could have taken the million he gave it from his current income. As this benefaction may suggest, Solvay's generous gifts, grants, and endowments favored science and technology. In this he fell in line with the practice of most 19th-century industrialists. New money tended to go to practical things, whereas old money tended to support high culture. To take a clear-cut example, inherited wealth founded and supported the Metropolitan Museum of Art while new riches did the same for the American Museum of Natural History – at the same time, the 1860s, and in the same place, New York City.[5] If Solvay had had his way, there would have been no old money. One of his more remarkable proposals was to fund the state entirely on confiscatory taxes on bequests and inheritances.[6]

Solvay did not invent the ammonia-soda process. Unknown to him, many patents existed on it and attempts to work it commercially had driven more than one firm close to bankruptcy. The secret of his success, besides ignorance of previous failures, was supreme self-confidence and know-how gained from handling ammonia and other gases at his uncle's small-scale chemical works. It owed nothing to his university training in chemistry, for the good reason that he had none. Like Alfred Nobel, who also converted a small family business into a large industrial enterprise, and many other 19th-century self-made entrepreneurs, Solvay had great confidence in his own judgment and a complementary skepticism about the wisdom of academics.[7]

Solvay was not one of those industrial giants who climbed to riches by extracting the last drop of energy from his workers. At a time when unions were beginning to feel their power and the first socialist deputies entered European parliaments, Solvay was in the van of enlightened industrialists who understood that they would do better with a cheerful than with an oppressed workforce. He introduced several improvements in working conditions such as the eight-hour day (to be sure, six days a week), paid vacations, and welfare schemes; and he concerned himself with the education of the classes from which he drew his employees.[8] With proper education, proper training, proper capacitation (to use Solvay's term), workers would keep factories prosperous, and society would reap an increasingly bountiful harvest. The key to prosperity therefore was also the keystone of social morality: equality of opportunity. With equality of opportunity, every member of society can have the education needed for maximum productivity. The punitive tax on inheritance was one of the ways to eliminate inequalities in birth.[9]

So, maximization of profit was not incompatible with humanitarian impulses. With a little experiment to decide the right proportion of discipline and indulgence, humankind, led by enlightened industrialists, could gain access to the laws of its own development. That there must be such laws was clear to Solvay from Quetelet's statistics on the behavior of humans in bulk.[10] The fact that the apparently uncoordinated and random paths pursued by individuals summed to regular and predictable group behavior indicated that individuals in society must act in accordance with some elementary law of nature.

This inference gave rise to the characteristic 19th-century pseudo-science of neo-classical economics, which operated with equations adapted from physics. Solvay occasionally illustrated his economic thought with a gentle thermodynamic equation, but he refrained, possibly from ignorance of the necessary mathematics, from following neoclassical economists into commodity space where vectors indicate the direction of money flow.[11] Instead, he proposed to abolish money of exchange altogether. The individual would make and receive payments solely via a personal account kept at the National Bank. That does not sound very attractive today. The economists to whom Solvay sent the proposal did not like it either, which, of course, only increased his attachment to it.[12]

Solvay's success with the ammonia-soda process depended on astute measures to conserve energy as well as material. Of course, energy conserves itself: that was the grand conclusion reached by all sorts of scientists, from physicists to physiologists, during the decade when Solvay began to labor in his family's gas works. But as Solvay found from practice, and physicists were proving from theory, all the energy locked up in hot water or chemical bonds is never available to the entrepreneur – or to anybody else. What counts for the chemist is a calculable fraction he can dispose of freely.

The job of the chemical engineer is to direct as much of this free energy as possible into productive channels. Since all natural processes involve the transfer of energy, every individual is, or should be, a chemical engineer in his or her interest. In keeping with neo-classical economics, Solvay identified this interest with the selfish pursuit of the best existence possible achievable with the resources available. This was the expression in the social sphere of the "physico-chemical law of maximum energy".[13] It is the fundamental law of what Solvay called "social energetics."

The increasing prominence of energy considerations in science in the second half of the 19th century culminated in a program to reduce all of physics and chemistry, including the fundamental laws of motion and material entities like atoms, to energy. This program, known as Energetics, reached its acme in the 1890s. One of its main leaders was the German chemist Wilhelm Ostwald. In a famous speech delivered before the Deutsche Naturforscherversamlung in 1895, entitled "On overcoming scientific materialism," Ostwald appropriated Scripture to exorcize the atoms, ethers, and mechanical models to which his errant colleagues were addicted. In the words of Moses he thundered, "Thou shalt not take unto thee any graven images, or any likeness of anything".[14] For in the Beginning God created Energy, and nothing more.

Ostwald eventually retreated from this theology, but not before he discovered Solvay's writings. They were a revelation. Ostwald jumped on the train to Brussels to track down the sage of soda. Although his French was weak and Solvay knew no German, they immediately understood one another. They were brothers in arms. "Solvay too was an energeticist,"[15] Ostwald wrote, a master energeticist, who had stretched the program far beyond a reformulation of physical science. Ostwald later made many visits to the man he acknowledged as "The Founder of Social Energetics".[16]

3. Positivism and Progress

In the beginning came theology. Then thinking men created metaphysics. By the time of Solvay's birth in 1838, they had taken their first steps into the positivist age, which would see the sciences perfected, one after the other, according to their hierarchy when Auguste Comte worked out his account of the three stages of humankind's ascent to truth. The perfection of mathematics was already in sight; the physical sciences were advancing steadily; physiology had begun to find its principles; and it was only a matter of time until the capstone of the system, sociology, would be hoisted into place.[17] Solvay accepted this scheme and applied his wealth to hasten the day.

He began by setting up two institutes of physiology, one for fostering anatomy at the Université Libre de Bruxelles, the other for hastening the day. He made no secret of the purpose of his benefaction: *The physical-chemical sciences have triumphed everywhere, they dominate all modern industries, and soon they will take possession of life itself....[Physiology has to] begin from the profound conviction that the phenomena of life must be explained only by the play of the physical forces that direct the universe.*[18]

After institutionalizing physiology, Solvay moved on to sociology, which he entrusted to an engineer and statistician named Emile Waxweiler. At the inauguration of its institute in 1902, Solvay declared that it would work with its predecessors in physiology to "reduce the realm of biological and sociological phenomena to physico-chemical actions with the help of the principles [of social energetics]".[19] On these principles, Solvay continued, we must consider the individual as "a purely energetic apparatus capable of acting externally on its environment." Social energetics teaches this apparatus how to maximize its output.[20]

In maximizing social output, the energetic engineer must include in his calculations every factor that can increase or decease the activity of the individual at different times and places. These include not only food, clothing, shelter, state of health, but also recreation, education, exercise, inheritance, fads and fashions, everything. Intellectual activity, although not disaggregated in the calculations, is no different from other social functions. Individuals excrete books and ideas as naturally as other bodily waste. Ideas useful for science, technology, or recreation can increase output far beyond the resources required to maintain writers, scientists, and engineers. Even the most artificial mental products can have their uses. Solvay gives theology and metaphysics as examples. Although unproductive in themselves, they had the merit, in Comte's system, of paving the way for their destruction and the advent of positive science.[21]

Despite the cogency of this reasoning, theology and religion did not recognize that positive science had killed them. On the contrary. In 1870, the Catholic Church defined and promulgated the doctrine of papal infallibility. Solvay directed his first extended essay on social problems, published anonymously in 1879, against this doctrine. It showed, he said, that Rome was as opposed as it had ever been to people "who dare to think and study freely".[22] Falling in line with the views of a fellow chemist, John William Draper, whose History of the conflict of science and religion (1875) was then a best seller, Solvay wrote that religion had always stood in the way of science.

It certainly stood in his way. Not only Catholics, but all true Christians, had to oppose Solvay's social science, which modeled human beings as physico-chemical appliances without free will. "A mind that is truly scientific and entirely free," Solvay wrote, "knows perfectly well that it is not free at all".[23] But like the Monsignor in Benjamin Disraeli's novel Lothair, which came out in the same year as papal infallibility, most people had not yet advanced to the social-physical idea that "thought is phosphorus, the soul complex nerves, and [the] moral sense a secretion of sugar".[24]

In 1909, on Nernst's proposal, the Akademie der Wissenschaften in Berlin awarded Solvay its gold Leibniz Medal – not for reducing morality to sugar, but for supporting scientific research in Prussia.[25] He took the occasion to alert German scientists to three interrelated problems in the "new ways of science." They were the constitution of matter in space and time, the mechanism of life, and the evolution of the individual and social groups. The list indicates a return on Solvay's part to the physical sciences. The time was ripe for him, if not for physics, for a summit. And so he was prepared to go much further than Nernst could have anticipated when he asked him to defray the expenses of an ad-hoc meeting of physicists. After the Council met, Solvay made it semi-permanent as the main activity, along with the giving of grants and fellowships, of the Institute of Physics he set up in 1912. He did the same thing for chemistry a year later. That completed his empire of thought and action from fundamental physics and chemistry through sociology and beyond, to commerce, whose institute he had already established, and, of course, to science-based industry. Informed by this vertical cartel, humankind would be able to harness cosmic energy efficiently in its endeavor to secure the maximum output of social energy.[26]

The inexorable laws determining the development of society decree that in a brief period science and industry will arrive at their pinnacles, which will be the higher, and reached the sooner, if society organized itself in the manner worked out by Solvay's intellectual cartel. There was no need, therefore, to endow the institutes in perpetuity. Solvay thought that 25 or 28 or 30 years would be enough, and directed that the institutes spend down their capital accordingly. These numbers were not the product of laborious calculations. Solvay was a practical man. They were the durations of the leases he obtained from the city of Brussels for the land on which the institutes stood.[27]

4. International Connections

"In Belgium internationalism was regarded and cultivated as a sort of domestic industry".[28] Or so Ostwald wrote, in connection with Solvay's institutes, after World War I. In fact, despite his benefactions around Europe, Solvay did not organize his institutes internationally before adding physics to his cartel in 1912. Perhaps the domestic cultivation of internationalism as well as the nature of physical science prompted this innovation. Belgium then was the seat of more international bodies than any other European country, even though it missed out on some big ones, like the International Union of Vegetarian Speakers of Esperanto, which chose Hamburg instead. In 1910, with the foundation of the Union of International Associations, Brussels became the clearinghouse as well as the meeting place for cosmopolitan organizations.

My own line of work, the academic study of the history of science, originated in a dribble from the same cauldron of internationalism, positivism, and scientism that prompted Solvay's Councils. I say this in gratitude and not presumption, and in reference to George Sarton, who, in 1912, floated a plan for the first professional periodical devoted entirely to the history of science. He named it Isis after the Egyptian goddess identified with all knowledge. It survives as the polymath journal of the History of Science Society of the US. Sarton was a Belgian mathematician inspired by the ideas of Comte and Quetelet. He believed that knowledge of the global history of the sciences was essential for acting wisely in the present. This unusual view of his subject fitted it for a place in Solvay's Institut de Sociologie, which already had sections or cabinets for anthropology, history, statistics, and technology. Sarton worked for a time in Waxweiler's institute and had the benefit of his advice about a career: choose something sensible, he said, something with an institutional locus, not the history of science. Despite Sarton's rejection of this sound advice, Waxweiler directed some Solvay money to support Isis and allowed his name to appear among the first patrons of Sarton's enterprise.

The international movement of Solvay's time owed much to the second industrial revolution. For example, pacifism, one of the movement's leading sectors, prospered as knowledge of the horrors of mechanized warfare spread during the 1860s. Cheap newspapers illustrated by new processes and distributed by new means of communication brought the Crimean War, the American Civil War, and the battles for the unification of Italy and of Germany to the breakfast table. Improved steels and methods of manufacture promised faster machine guns and bigger artillery. Chemistry created powerful new explosives to maximize the effectiveness of the guns. Factories built to meet and stimulate peacetime demands stood ready for conversion to instrumentalities of war. Nobel's dynamite factories were so many plowshares beatable into spears; so too were Solvay's soda plants, as Britain showed during the first world war. No one knew this better than Nobel, who made dynamite for building railroads and smokeless powder for firing guns, and gave a portion of his fortune to endow a prize for peace.

Another, and more benign technological push to internationalism came from the new means of mingling and distributing goods, people, and services. International conventions were required for post and telegraph, police and security, transportation, and, with respect to manufacture, standardization of measures, quantities, and qualities. Scientists and engineers representing their governments met regularly to argue over the definition and realization of units and standards. And these meetings in turn were made possible by the revolution in transportation and communication whose consequences the experts assembled to discuss. Without the railroad and the steamship, frequent, short, international conferences among specialists could not have occurred.

Solvay Councils combined features of these specialist meetings as expressed particularly in the first international congress of chemistry, held in Karlsruhe in 1860, and the first international congress of physics, held in Paris in connection with the International Exhibition of 1900. The purpose of the Karlsruhe conference, to which Nernst referred as a precedent in his approach to Solvay, was to discuss, and if possible to dissolve, the fog that then enveloped the concepts of atom, molecule, and chemical equivalent. Differences over definitions and methods, and a general rejection of Avogadro's hypothesis of the diatomic nature of simple gases, had resulted in many different formulas for the same chemical compound. At the meeting Stanislao Cannizzaro argued for his countryman Avogadro, but did not prevail. As the meeting broke up, delegates received copies of the lecture on the subject that Cannizzaro had prepared for his beginning students. Reading it at their leisure, the chemists who had not been persuaded when together were conquered individually. Within a decade, chemistry had its first periodic table of the elements.

The physicists came together in 1900 in a jubilant mood to review their progress during what they called the century of science and to celebrate their spectacular recent discoveries: x rays, radioactivity, alpha rays, beta rays, the electron. The quantum had not yet made its appearance. Despite uncertainty over the nature of x rays and radioactivity, they could celebrate an unprecedented synthesis, or Grand Unifying Theory, created almost single-handedly by Joseph John Thomson, who had identified the electron in cathode rays and had measured its charge and mass. To Thomson it was, as it still is, an elementary particle. He showed that it tied together the concepts of electrolytic ion, gaseous ion, spectral radiator, radioactivity, conductivity, and a few other things; and he deduced, no doubt precipitously, that it was a, or rather the, constituent of all matter. Thomson had given a preliminary version of his grand vision to a joint meeting of the British and French Associations for the Advancement of Science in 1899. As one of the participants recorded its reaction, "the scientific world seemed to awake to the fact that their fundamental conceptions had been revolutionized." This participant, Arthur Schuster, could have been a member of the first Solvay Council. But he declined the invitation, perhaps because he could not face another transformation of his worldview.

Solvay had kept himself informed about progress in physics, especially as related to energy, and probably knew that Planck's theory of radiation, as developed by

Einstein, challenged basic principles. That agreed perfectly with Solvay's expectations, for he had sketched out a system of space and matter that deviated from received physics more than the quantum theory. When Nernst informed him in 1910 that the quantum rot had spread from radiation to matter, Solvay thought the time had come to try his system with the acknowledged masters of the subject. As he told its members, had Nernst not had the idea of convoking them, "I would have been able, perhaps, to think of it myself."

5. The Council of 1911

In a press release just before the Council opened, Solvay's associate Edouard Herzen explained that it concerned the foundations of physics, in which Solvay had been interested for half a century, and to which he had contributed the notions that energy is interconvertible with matter and is exchanged in jumps or steps. The release thus insinuated that the quantum physicists were meeting to help Sovay work out his theory of everything. To help the Council counsel him, Solvay gave each of its members a hundred printed pages outlining his system of gravito-matérialitique. He hoped that with his advisors' help his system might "lead to an exact and therefore definitive knowledge of the fundamental finite elements of the active universe."

When Nernst learned that Solvay intended to issue his theory of everything at the Council, he naturally worried that its participants would regard its convener as a crank. He asked Planck to read the proofs of Solvay's 100-page handout. Planck's evaluation must have surprised him. "I became more and more interested in it [so he wrote Nernst] not only because the author shows that he knows the laws of physics, especially those of planetary motion, in a way that would do credit to a professional theoretical physicist, but also because of his entirely independent and original [approach]." Lorentz also studied Solvay's physics and later published a note in the Comptes rendus of the Académie des sciences in Paris praising the independence of Solvay's mind and his deduction of the interconvertibility of mass and energy before Einstein thought of it. With good will and good hindsight, one can also see some features of Bohr's stationary states in the opening sections of the brochure Solvay gave his councilors.

After Solvay's opening address, his private conference considered a report by Lorentz on reasons to believe that Planck's radiation formula, which had withstood every experimental test, was incompatible with ordinary physics. Then came separate contributions by two mathematical physicists trained at Cambridge, Lord Rayleigh and James Jeans, who hoped to evade Lorentz' conclusion by filling the theory with a great many undetermined parameters whose purpose was to impede the flow of energy into the radiation field. To this the mathematician and methodologist Henri Poincaré, who was no friend of the quantum theory, replied, in effect, that any fool could save the phenomena with an inexhaustible supply of disposable parameters.

Having put down the English counter-revolution, the Council heard the experimental evidence in favor of Planck's radiation law and Planck's new version of his theory, which, with no comfort to counter-revolutionaries, replaced his quantum of energy with one of action. The meeting then evaluated evidence for the existence of atoms and molecules that had recently persuaded Ostwald and other energetic holdouts to accept graven images. The long report on this evidence by Jean Perrin ended with a connection with the quantum theory. He showed that the magnitudes of molecular constants obtained from his investigations of Brownian motion as analyzed by ordinary physics agreed well with values calculated via the radiation formula and quantum theory.

This encouraging agreement preceded deliberation of two papers describing results apparently inexplicable on ordinary physics. One, by Nernst, presented the confirmation of Planck's formula as applied to the specific heats of solids in theory by Einstein and to experiment by himself. The other, by Arnold Sommerfeld, developed a theory of the photo-effect and other non-periodic phenomena based on a quantum of action.

The upshot of the Council's discussions was ambivalence about the place of the quantum, whether of energy or of action, in physics; agreement that, in its straightforward construal, ordinary physics could not lead to Planck's formula; doubt whether it might be made to do so with a few patches and disposable constants; and, among those who saw no acceptable loophole, disagreement over whether the needed changes could be limited to repairs or would require entirely new foundations. In short, although the Council aired everything it settled nothing; and, as one of its secretaries said, left the puzzles of the quantum more puzzling than ever. Or, as Einstein put it in his pungent way, "the h-disease looks increasingly hopeless," "nobody really knows anything."

This was also Solvay's view. At the end of the Council, after declaring his satisfaction at the amount of energy it had expended, he observed that it had not managed to identify "the very simple primordial elements of this active universe." It would be necessary for the Council to reconvene to judge a new version of the gravito-materialistic system. Meanwhile, Solvay added, it would advance physics if his dispersed councilors could contrive experiments to show that Brownian particles do not obtain their energy from the medium in which they swim and that radioactivity does not get its energy from disintegrating atoms. This was a hard assignment. But not a necessary one. In Solvay's opinion, calculations from a general theory are always closer to the truth than the results of experiments.

He expected that his general theory eventually would provide the basis of the reform in physics that his councilors agreed was necessary but could not formulate. We know this expectation from a letter Solvay wrote in March 1912, some five months after his Council convened. "I'll try to commission researches by specialists from everywhere that will verify my principles, and I still firmly believe that... I shall be the theorist with the correct solution to the great remaining problems."

Nonetheless – and this is a measure of the man – the statutes of the Institut International de Physique Solvay, which were drawn up early in 1912, assigned its scientific program to an international committee independent of Solvay. His earlier policy, which required that the work of his institutes be guided by his ideas, would not work internationally if he wanted to bring people of the caliber of Lorentz, Planck, Einstein and the rest under his banner. To engage them, Solvay had to advance from directed to curiosity-driven philanthropy.

Solvay's councilors did not manage to confirm his physical theory. Therefore, from his point of view, their immediate social output did not match the resources consumed in convening them. Soon, however, the indirect pay-off provided for in his calculus of the social utility of intellectual work began to show itself. First, a Karlsruhe effect occurred: meditating on the deliberations of the Council and other information, the most hesitant of its members, notably Poincaré and Jeans, came to see the necessity for significant change in the foundations of physics. Then, just as the scientific committee of the new Solvay Institute was planning a second council to consider Solvay's pet subject, the structure of matter, a prediction Lorentz had made at the start of the first council was about to become true. "It is very likely [Lorentz had said] that while we are discussing a problem an isolated researcher, in some remote corner of the world, is finding the solution."

This corner was Copenhagen. Although remote, it was not isolated from the latest news from Brussels. The professor of physics at its university, Martin Knudsen, who had attended the Council as a representative of Denmark and molecular physics, was the secretary of the International Scientific Committee of the Institut de Physique Solvay. In the fall of 1912 his new assistant, Niels Bohr, came from Manchester, where he was completing a postdoctoral year that had begun in Cambridge. Each of these centers had provided a councilor for Solvay. Bohr was in contact with both of them: at Cambridge, with James Jeans, whose courses he attended; at Manchester, with Ernest Rutherford, whose atomic model he perfected. Bohr may have been the only young physicist in the world to learn about the Solvay Council from three of its participants.

A lesser man would have been discouraged by the news that Einstein, Planck, Lorentz, and Poincaré had not been able to solve problems with which he too was engaged. Bohr would have found it exhilarating. Three months before the second Solvay Council convened in October 1913, he published the first of his revolutionary papers on the constitution of atoms and molecules. He received strong endorsements from Knudsen, Jeans, and Rutherford. Support for innovation, as institutionalized in Solvay's International Institute of Physics, was the immediate payoff for basic science of the enlightened industrial philanthropy, the scientism and positivism, the social physics and real physics, behind the famous Council that met in Brussels a hundred years ago.

It has become a byword for a watershed in physics. In 1947 quantum mechanics had reached a state resembling that of quantum theory in 1911. Once again a physical chemist, Duncan MacInnes, took the lead. He appealed to the US National Academy of Sciences to sponsor an elite specialist conference at a comfortable inn on Shelter Island. The conference set the direction of quantum field theory. A participant who knew something abut the history of physics made the inevitable comparison. "[The] Shelter Island meeting... has proved more important than it seemed even at the time, and [will] be remembered as the 1911 Solvay Congress is remembered, for having been the starting-point of remarkable new developments."

Works Cited

- Adam, Thomas. "Philanthropy in a transatlantic world." In Thomas Adam, ed. Philanthropy, patronage, and civil society. Experiences from Germany, Great Britain, and North America. Bloomington: University of Indiana Press, 2004. Pp. 15-33.
- Adam, Thomas. Buying respectability. Philanthropy and urban society in transnational perspective. 1840s to 1930s. Bloomington: University of Indiana Press, 2009.
- Barkan, Diana Kormos. Walther Nernst and the transition to modern physical science. Cambridge: Cambridge University Press, 1999.
- Cohen, J.M. The life of Ludwig Mond. London: Methuen, 1956.
- Crombois, Jean-François. "Energétisme et productivisme: La pensée morale, sociale et politique d'Ernest Solvay." In Despry-Meyer and Devriese, Solvay (1997), 209-20.
- Despry-Meyer, Andrée, and Valérie Montens. "Le mécénat des frères Ernest et Alfred Solvay." In Despry-Meyer and Devriese, Solvay (1997), 221-45.
- Despry-Meyer, Andrée, and Didier Devriese, eds. Ernest Solvay et son temps. Brussels: Archives de l'Université Libre, 1997.
- Devriese, Didier. "Du premier Conseil à l'Institut de Physique." In Pierre Marage and Grégoire Wallenborn, eds. Les conseils Solvay et les débuts de la physique moderne. Brussels: Université Libre, 1995. Pp. 43-56.
- Eckert, Michael. "From x-rays to the h-hypothesis: Sommerfeld and the early quantum theory 1909-1913." In Lambert, Early Solvay councils (to appear).
- Eijkman, Pieter Hendrik. L'internationalisme scientifique (sciences pures et lettres). The Hague: Bureau préliminaire de la fondation pour l'internationalisme, 1911.
- Grau, Conrad. Die Berliner Akademie der Wissenschaften. 3 vols. Berlin: Akademie-Verlag, 1975-79.
- Haber, L.F. [I] The chemical industry during the nineteenth century. A study of the economic aspect of applied chemistry in Europe and North America. Oxford: Oxford University Press, 1958.
- Haber, L.F. [II] The chemical industry 1900-1930. International growth and technological change. Oxford: Oxford University press, 1971.
- Héger, Paul, and Charles Lefebure. Vie d'Ernest Solvay. Brussels: Lamertin, 1929.
- Heilbron, J.L. "Lectures on the history of atomic physics." In Charles

Weiner, ed. History of twentieth-century physics. New York: Academic Press, 1977. Pp. 40-108.

- Heilbron, J.L. "British participation in the first Solvay councils on physics." In Lambert, Early Solvay councils (to appear).
- Herren, Madeleine. Hintertüren zur Macht. Internationalismus und modernisierungs-orientierte Aussenpolitik in Belgien, der Schweiz und den USA, 1865-1914. Munich: R. Oldenbourg, 2000.
- Ihde, Aaron. The development of modern chemistry. New York: Harper & Row, 1964.
- Institut International de Physique Solvay. La structure de la matière. Rapports et discussions du Conseil de Physique tenu à Bruxelles du 27 au 31 octobre 1913. Paris: Gauthier-Villars, 1921.
- Kuhn, Thomas S. Black-body theory and the quantum discontinuity, 1894-1912. Oxford: Oxford University Press, 1978.
- Lambert, Franklin, ed. The early Solvay councils and the advent of the quantum era (to appear).
- Lorentz, H.A., and Edouard Herzen. "Les rapports de l'énergie et de la masse d'après Ernest Solvay." Académie des Sciences, Paris. Comptes rendus, 177:20 (12 Nov 1923), 925-9.
- Lyons, F.S.L. Internationalism in Europe, 1815-1914. Leiden: Sijthoff, 1963.
- Mirowski, Philip. More heat than light. Economics as social physics, physics as nature's economics. Cambridge: Cambridge University Press, 1989.
- Nye, Mary Jo. Molecular reality. A perspective on the work of Jean Perrin. London: MacDonald, 1972.
- Or, Louis d', and Anne-Marie Wirtz-Cordier. Ernest Solvay. Brussels: Palais des Académies, 1981.
- Ostwald, Wilhelm. Lebenslinien. 3 vols. Berlin: Klasing, 1926-27.
- Overbergh, C. van, et al., eds. Le mouvement scientifique en Belgique, 1830-1905. 2 vols. Brussels: C. Bulens, 1907-08.
- Pickering, Mary. Auguste Comte. An intellectual biography, Vol. 1. Cambridge: Cambridge University Press, 1993.
- Pyenson, Lewis. The passion of George Sarton. Philadelphia: American Philosophical Society, 2006. (APS, Memoires, Vol. 260.)
- Pyenson, Lewis, and Christophe Verbruggen. "Ego and the International. The modernist circle of George Sarton." Isis, 100 (2009), 60-78.
- Quinn, Terry. From artefacts to atoms. The BIPM and the search for ultimate measurement standards. Oxford: Oxford University Press, 2011.
- Schuster, Arthur. The progress of physics during 33 years (1875-1908). Cambridge: Cambridge University Press, 1911.
- Schweber, Silvan S. QED and the men who made it. Princeton: Princeton University Press, 1994.
- Seth, Suman. Crafting the quantum. Arnold Sommerfeld and the practice of theory, 1890-1926. Cambridge: MIT Press, 2010.

- Solvay, Ernest. Science contre religion au point de vue social, ou Faut-il avancer or reculer? Brussels, 1879.
- Solvay, Ernest. Principes d'orientation sociale. Résumé des études de M. Ernest Solvay sur le productivisme et le comptabilisme. 2nd edn. Brussels: Misch and Throu, 1904. (Instituts Solvay. Travaux de l'Institut de Sociologie. Actualités sociales.)
- Solvay, Ernest. Note sur les formules d'introduction à l'énergétique physio- et psycho-sociologique. Brussels: Misch and Throu, 1906. (Instituts Solvay. Travaux de l'Institut de Sociologie. Notes et mémoires, 1.)
- Solvay, Ernest. Sur l'établissement des principes fondamentaux de la gravito-matérialitique. Brussels: Bothy, 1911.
- Staley, Richard. Einstein's generation: The origins of the relativity revolution. Chicago: University of Chicago Press, 2008.
- Williams, Trevor I. "Heavy chemicals." In Charles Singer, et al., eds. A history of technology, Vol. 5. The late nineteenth century. Oxford: Oxford University Press, 1958. Pp. 235-56.

References

1. Kuhn, Black-body theory (1978), 102-10, 115-20.
2. Cf. Staley, Einstein's generation (2008), 410-22.
3. Williams, in Singer, History, 5 (1958), 244; Cohen, Mond (1956), 134; Haber, Chem. ind. [I] (1958), 157-8, and [II] (1971), 160-1.
4. Picture in Héger and Lefebure, Vie (1929), 36-7.
5. Adam, in Adam, Philanthropy (2004), 27-9, and Buying respectability (2009), 96-7, 105.
6. Erreygers, 223-4, 231-4.
7. Haber, Chem. inst. [I] (1958), 42, 87-9; Héger and Lefebure, Vie (1929), 22-31.
8. Haber, Chem. ind. [II] (1971), 377, 383, 385-6, 400-1; D'Or and Wurtz-Cordier, Solvay (1981), 34-5, give several instances in which the working practices in the Solvay enterprises were several decades in advance of the law in Belgium and France.
9. Solvay, Principes (1904), 14-15, 18, 36-7, 60-3, 88-9; Erreygers, 226-7, 245-7.
10. Warnottte, 58.
11. Solvay, Formules (1906), 4-9; Mirowski, More heat than light (1989), 217-31, 252-70, including a caricature of Solvay (p. 269).
12. Solvay, Principes (1904), 17-18, 72-81; Erreygers, 234-43, 252-4. Solvay called his scheme "comptabilisme social."
13. Solvay, Principes (1904), 30-2, and Formules (1906), 3-7; Warnottte, 59.
14. Heilbron, in Weiner, History (1977), 44-5.
15. Ostwald, Lebenslinien (1926), 3, 284 (quote), 322-4.
16. Ibid., 327.
17. Pickering, Comte (1993), 201-3, 335-7.
18. Talk of 14 Dec 1893, in Warnotte, 75.
19. Speech of 1902, in Warnotte, 76.
20. Quoted in Warnotte, 56 (first quote), and in Crombois, in Despy-Meyer and Devriese, Solvay (1997), 222.
21. Warnotte, 42-6.
22. D'Or and Wirtz-Cordier, Solvay (1981), 29-30; Héger and Lefebure, Vie (1929), 42-3.

23. Solvay, Science contre religion (1879), in Warnotte, 18-19, 58. Brookes, Science and religion (1991), 34-5.

24. Disraeli, Lothair (3 vols., 1870).

25. Grau, Die Berliner Akademie (1975), 1, 105. Cf. Haber, Chem. ind. [II] (1971), 49 (gift for a Chemische Reichsanstalt); Overbergh, Mouvement (1907), 1, 331 (seismic observatory in Belgium), and 2, 89 (laboratory for bacteriology and parasitology in the Congo).

26. Warnotte, 14, 19, 52-3.

27. Warnotte, 20-1; Despy-Meyer and Montens, in Despy-Meyer and Devriese, Solvay (1997), 235, 238.

28. Ostwald, Lebenslinien (1926), 3, 278; Herren, Hintertüren (2000), 141-85.

29. Lyons, Internationalism (1963), 17-18, 205-8; Eijkman, Internationalisme (1911), 88, and App., no. 531.

30. Eijkman, Internationalisme (1911), 90; Pyenson and Verbruggen, Isis, 100 (2009), 61-4, 67-70.

31. Waxweiler, in Overbergh, Mouvement (1907), 2, 539-40.

32. Sarton, "Le but d'Isis," reprinted in Isis, 54 (1963), 7-10, and list of patrons; Pyenson, Passion (2006), 95, 202, 209, 287, 384-5, 417; and information from Franklin Lambert.

33. Haber, Chem. ind. [II], (1971), 207.

34. E.g., Quinn, Artefacts (2011), chapts. 1-3.

35. Ihde, Development (1964), 228-30.

36. Heilbron, in Weiner, History (1977), 40-1, 50-1, 100n43; Staley, Einstein's generations (2008), chapt. 5.

37. Schuster, Progress (1911), 71; Heilbron, in Lambert, Earliest Solvay councils (to appear).

38. Solvay I, 2.

39. Press release, 26 Oct 1911, in Devriese, in Marage and Wellenborn, Conseils (1995), 48; cf. Solvay, Gravito-matérialitique (1911), 62.

40. Planck to Nernst, 22 Oct 1911, Solvay Archives, Brussels, courtesy of F. Lambert.

41. Lorentz and Herzen, Académie des Sciences, Paris, CR, 177:20 (1923), 925-9.

42. Solvay, Gravito-matérialitique (1911), 5-18, proving that the energy of an elliptical orbit, which depends only on its major axis, defines a stable molecular "state," and that an increment of energy drives the molecule to a higher "state" of similar stability.

43. Solvay, in Solvay I, 1-5; Lorentz, ibid., 12-39.

44. Rayleigh, in Solvay I, 49-50; Jeans, ibid., 53-73; Poincaré, ibid., 77.

45. Planck, in Solvay I, 93-114; Kuhn, Black-body theory (1978), 236-54, 340-56.

46. Perrin, in Solvay I, 153-250, esp. 246-50; Nye, Molecular reality (1972), 151-6.

47. Nernst, in Solvay I, 254-90; Barkan, Nernst (1999), 147-80.

48. Sommerfeld, in Solvay I, 313-72; Eckert, in Lambert, Early Solvay councils (to appear); Seth, Crafting the quantum (2010), 149-56.

49. Solvay I, 451-4.

50. F.A. Lindemann to his father, 4 Nov 11, in Lindemann Papers, Nuffield College, Oxford.

51. Resp., Einstein to Lorentz, 23 Nov 1911, and to Heinrich Zangger, 15 Nov 11, quoted from Barkan, Nernst (1999), 103, 199.

52. Solvay I, 456.

53. Solvay to Héger, 26 Mar 1912, in Devriese, in Marage and Wellenborn, Conseils (1995), 54-5. Cf. Solvay to Lorentz, 1912, ibid., 53: long before the Physics Institute's thirty years had elapsed, "physics will have said its last say."

54. Institut, Structure (1921), vii-xii.

55. Solvay I, 6-8.

56. I.I. Rabi, as quoted in K.K. Darrow to MacInnes, Jan 1948, in Schweber, QED (1994), 156.

Murray Gell-Mann: From Solvay 1961 to Solvay 2011

At the 1961 Solvay meeting on the Quantum Theory of Fields, those present included Heisenberg and Dirac and even Niels Bohr, along with much younger theoretical physicists such as Feynman, Chew, Pais, Goldberger, and Mandelstam. I was probably the youngest. Those younger ones set most of the agenda, including a review of QED by Feynman and a review by Pais of some developments in thinking about fundamental particles and their interactions.

Now what changes took place during those fifty years from 1961 to 2011? One important change was connected to the emergence of cosmology as more of an empirical science. There are still some mysteries, however. Today we are still puzzled by the nature of dark matter. There is also the cosmological constant, which is probably what underlies the recently discovered accelerating expansion of the universe. Why is that constant some 118 orders of magnitude smaller than one might expect from the values of other natural constants? To a particle physicist the cosmological constant is just the average energy density of the vacuum. It would be zero if supersymmetry were exact, but of course that symmetry is badly broken, if it is there at all.

Over those fifty years, the main change, as far as elementary particle physics is concerned, was the development of the standard model. In 1961 we were just beginning to construct it and it was not until 1972 or 1973 that it was largely finished.

In 1961 we were still lacking quarks, exactly conserved color, charm, the "Higgs boson", and many other features. We had the idea of a generalized Yang-Mills theory based on $SU(3)$ for the strong interaction, but we didn't know that there were two separate spaces for flavor and color and some of us tried to squeeze the strong and the electro-weak interactions into the same space, which doesn't work, of course.

But we need not dwell on these now familiar topics. Let us go on to another one, which has had a curious history during the last half-century and played an important role in the discussions at the Solvay Congress in 1961. It has to do with how we treat the relation between quantum field theory and the S-matrix of Breit and Wheeler. I hope you will forgive my discussing this topic partly from my own point of view, based on my own experience.

For renormalizable field theories, perturbation theory was perfected in the late 1940's, especially with the use of the elegant method of Feynman and Stueckelberg, so successfully applied to quantum electrodynamics by Feynman and others . There is, however, an alternative way of arriving at the same series for the S-matrix. I suggested that at the 1956 Rochester Conference, drawing on my work on dispersion theory with Murph Goldberger. In the alternative approach, one utilizes only amplitudes on the mass shell. I called it field theory on the mass shell. In many cases the four-momenta have to be imaginary or complex, of course, in order to fit on the mass shell.

The method utilizes analyticity properties of the amplitudes, leading to dispersion relations. It also utilizes unitarity relations generalized to the momenta involved. Finally, it utilizes the "crossing" symmetry of the amplitudes. In this way the entire perturbation series for QED or other renormalizable field theories can be reproduced. Although the integrals look different from the ones coming out of Feynman diagrams, they yield the same S-matrix.

An important feature of the mass-shell approach is that only renormalized quantities appear. (The unrenormalized ones are replaced by the high energy behavior of certain functions, such as the square of the effective electron charge including vacuum polarization.) An important question is the following: since the dispersion theory, extended unitarity, and crossing properties are universal, how do we know which field theory we are considering? In the perturbation expansion, we start with free elementary particles and the way is clear after that. But in a non-perturbative treatment, what should one do?

I tried for years, between 1956 and 1961, to convince Geoffrey Chew of the value of the mass-shell approach, but he kept resisting. I think his reluctance was connected with this question of how to specify the theory. Finally, in 1961, at a conference in La Jolla, California, Geoff gave a remarkable speech in which he came forward with what he called S-matrix theory (presumably in homage to Heisenberg, who had created so much interest in the S-matrix of Breit and Wheeler and had suggested that it might be possible to make a theory of the S-matrix directly rather than calculating it from a field theory). Geoff argued that field theory had failed in describing the hadrons and the strong interaction and that it needed to be replaced by what he called S-matrix theory, based on analyticity, extended unitarity, and crossing. He did not stress that these were the same three features as in field theory on the mass shell, as discussed in 1956. The implication was that, for the hadrons at least, one should throw away field theory and adopt something that looked like field theory on the mass shell but presumably wasn't the same.

Then Geoffrey Chew and Steven Frautschi came up with the ingenious proposal that there are no elementary hadrons. Instead, the hadrons are made up of one another. The hadrons acquire forces by exchanging hadrons, thus producing hadrons as bound states or resonances. The list of hadrons is the same in all these roles. That is the idea behind the "bootstrap" picture.

These ideas suggest that Chew's "S-matrix theory" for the hadrons and the strong interaction would have to be more general, somehow, than field theory on the mass shell and would have to allow for a hadron theory with no elementary particles. How do we distinguish elementary particles (in order to get rid of them, for example)? Here Chew and Frautschi made use of the appearance of particles as poles of the S-matrix in the complex plane of angular momentum. They proposed that elementary particles would correspond to fixed poles, not moving as energy changes, while composite particles would correspond to moving "Regge" poles on "Regge trajectories."

Meanwhile, experimental evidence was starting to accumulate that observable hadrons, including the nucleon, were in fact on Regge trajectories with angular momentum J approximately linear in the mass (for fermions) or quadratic (for bosons). But around 1962 Murph Goldberger and I, together with some collaborators, showed evidence that, in field theory with vector (spin one) bosons, an elementary particle can turn into a Regge pole rather than a fixed one when radiative corrections are included. The distinction between fixed and moving poles is not the key to what is elementary.

So, how do we implement the bootstrap process? Perhaps we can make use of the high-energy behavior of the amplitudes, regulating how many subtractions there are in the dispersion relations. But in any case the emphasis in theoretical research shifted. Bootstrap enthusiasts pursued little simplified models of the hadron system. The favorite one involved the rho meson as a compound of two pions in a p-state, with a pi-pi force coming from the exchange of a rho meson and leading to an unstable bound state which is again the rho meson, etc.

I was unhappy about this as a starting point and suggested that, instead of beating to death one or two particle states, one should use an infinite set of meson states as constituents, force creators, and bound states (initially with zero width), in a self- consistent bootstrap picture. (One should, of course, go on to include an infinite set of baryons as well.) This idea for a more sophisticated model was used by my three post-doctoral fellows Dolen, Horn, and Schmid in their well-known "duality" paper. That was how I played a minor role in the prehistory of string theory (apart from finding jobs at Caltech for some excellent string theorists). Then Veneziano published his beautiful dual model that corresponded to a string theory in 26 dimensions. This otherwise excellent starting point was marred by the appearance of a meson state of negative mass squared and also, of course, by the absence of fermions. Both difficulties were removed after Neveu and Schwarz produced their superstring model after the important contributions from Pierre Ramond. But there remained another problem, the appearance of a spin 2 particle with zero angular momentum, an embarrassment in a theory of hadrons and the strong interaction.

Then came the brilliant suggestion, for example from Joel Scherk, that one was dealing here with a model theory of all particles and all the forces. That mysterious spin 2 boson was the graviton! (Of course the theory now contained Regge trajectories with slopes reflecting a fundamental energy some nineteen orders of magnitude larger than the one involved in the hadron trajectories.) Indeed, the superstring theory yields Einsteinian gravitation in a suitable approximation and fits it into quantum mechanics without encountering, at least in low orders, any ridiculous infinite corrections. These properties suggest that maybe a future successful unified theory might resemble superstring theory in important ways. (By the way, N=8 supergravity seems to have the same kind of finite behavior. What does that signify? Do the two models have some remarkable feature in common?)

Now what about field theory in the old way of doing business compared to field theory on the mass shell? Geoff Chew, of course, contrasted field theory and the mass shell. But there is no good evidence to support the idea that, in regular Lagrangian field theory, working on the mass shell gives anything different for the S-matrix.

Now what about superstring theory? It is always treated on the mass shell. But is there a field theory for it? My student Barton Zwiebach and others have tried very hard to construct such a theory but I don't think anyone has fully succeeded. (It would, of course, be a "multilocal" theory in which extended strings interact locally with other extended strings.) Maybe for superstrings there is a difference between the mass-shell and the traditional approaches in that only the mass-shell formulation works. But is that really true? I would like to ask our string colleagues if they have resolved that issue. (By the way, since quarks and gluons don't really have mass shells, there must be some subtleties involved.)

Speaking of quarks and gluons, we may note that in quantum chromodynamics they are permanently confined, so there are no elementary hadrons that are what I called "real" (capable of emerging singly). The situation is reminiscent of the bootstrap. I thought of that when I proposed the quarks.

I understand that there are several theorists working today on field theory on the mass shell, along with some of the issues we have been discussing. I have talked briefly with one of those theorists, Dr. Nima Arkani-Hamed, who is doing fascinating work and seems to share some of my ideas. I hope we can hear about such work at this meeting. It would make for a remarkable connection across a gap of 50 (or 55) years.

Session 2

Foundations of Quantum Mechanics and Quantum Computation

Chair: *Alain Aspect*, Institut d'Optique, Paris, France
Rapporteurs: *Anthony Leggett*, University of Illinois at Urbana-Champaign, USA and *John Preskill*, California Institute of Technology, USA
Scientific secretaries: *Thomas Durt*, Institut Fresnel, Marseille, France and *Stefano Pironio*, Université Libre de Bruxelles, Belgium

Rapporteur talk by A. Leggett: The Structure of a World Described by Quantum Mechanics

Abstract

I ask the question: What can we infer about the nature and structure of the physical world (a) from experiments already done to test the predictions of quantum mechanics (b) from the assumption that all future experiments will agree with those predictions? I discuss existing and projected experiments related to the two classic paradoxes of quantum mechanics, named respectively for EPR and Schrödinger's Cat, and show in particular that one natural conclusion from both types of experiment implies the abandonment of the concept of macroscopic counterfactual definiteness.

The title of this talk was chosen with some care. It is obvious that the theoretical account of the physical world given by quantum mechanics (hereafter QM) is very bizarre. However, a theory is only as good as the experiments which support it.

Thus, in this talk I shall not attempt to discuss the "interpretation" of the quantum formalism as such (whatever that might be taken to mean). Rather, the question I shall be asking is: What can we infer about the nature and/or the structure of the physical world

(a) on the basis of <u>existing</u> experiments which were designed to test the predictions of QM, or

(b) on the assumption that <u>all future</u> experiments will continue to confirm the predictions of QM?

Over the last few decades there have been two major areas in which the relevant experiments have been conducted, related respectively to the two famous paradoxes in the foundations of QM, which are usually associated with the names of Einstein, Podolsky, and Rosen[1] and Schrödinger's Cat.[2] The interpretation of experiments in both these areas involves (or more precisely may involve, depending on one's views) the concept of "realism". So it may be as well to start by examining what exactly we mean or could mean by "realism" in physics.

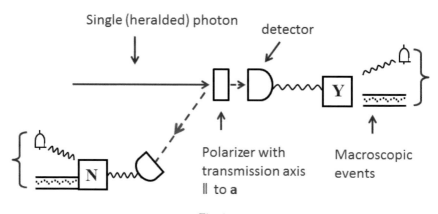

Fig. 1.

Let's consider the simple case of a two-state physical system; a typical (microscopic) example of such a system might be the polarization degree of freedom of a photon, for which a possible experimental setup is illustrated in Fig. 1. We imagine that we have a (weak) source of "heralded" photons[a] which impinge on a polarizer, let us say for definiteness set with its "transmission" axis in a direction **a** in the plane perpendicular to the photon propagation vector. If a given photon passes the

---

[a]That is, we have a means of knowing when there is a single photon in the apparatus. A typical example of such a source might be down-conversion of an incident laser beam by a nonlinear crystal: the detection of the "idler" photon informs us that there is a single "signal" photon in the apparatus.

polarizer, it will enter a detector whose activation will in turn trigger a counter (marked Y for "yes" in Fig. 1) which outputs various macroscopic events (flashing of a light, ringing of a bell, printout of a computer tape...); it is important to stress that these events may be (and in some cases are) directly observed by human agents. If on the other hand the photon in question is reflected by the polarizer,[b] it enters a different detector which triggers the counter N (for "no"), again resulting in various macroscopic and directly observable events.

To formulate a convenient language for the analysis of such experiments, one may think of the polarizer as posing a "question" to the photon, namely: Are you polarized along **a** ("yes", in which case we will assign to you the value +1 of a dichotomic variable A) or perpendicular to **a** ("no", in which case we will assign you the value $A = -1$)? It is a (highly nontrivial!) experimental fact that (assuming of course ideal conditions) each incident photon answers either "yes" or "no"; that is, that either counter Y clicks and counter N does not, or N clicks and Y does not. Thus, at least subsequent to the occurrence of one set of macroscopic events or the other, we can legitimately assign to the photon in question a value of A which is either +1 or −1, and there seems to be no ambiguity in the prescription for doing so.

Fig. 2.

But did the photon "actually possess" a definite value of A before it impinged on the polarizer? Or, better, in a situation in which there is no measurement device in place (Fig. 2), does the photon possess such a definite value? What would we mean by the claim that it does? I believe a possible (and to my mind the least unsatisfactory) answer is given by the concept of macroscopic counter factual definiteness (MCFD), which has been proposed in the literature by a number of people, perhaps most extensively by H.P. Stapp.[3] To explain this concept, let us consider the setup illustrated in Fig. 3: again we have a source of heralded photons, but now arrange to switch them, by a device which for simplicity we imagine to be actuated by some purely classical process, either into a device identical to that of Fig. 1, which "measures" the polarization parallel or perpendicular to **a**, or into a region marked "elsewhere", where we can arrange to detect it without measuring any polarization component. Consider now a particular photon which was in fact switched into the region "elsewhere", (as we verify by inspecting the output of counter E). We

---

[b]For simplicity I do not consider the possibility of absorption.

now consider the state of affairs which <u>would have</u> obtained had that photon been switched into the polarization-measuring device. Consider two possible propositions:

> P1: It is a fact that <u>either</u> counter Y would have clicked and N not (i.e. $A = +1$) <u>or</u> counter N would have clicked and Y not ($A = -1$).

Given that in all cases in which the photon is actually switched into the device one or other of these two outcomes is observed, the above proposition seems to be a truism. Contrast it now with

> P2: <u>Either</u> it is a fact that counter Y would have clicked, <u>or</u> it is a fact that counter N would have clicked, (i.e. either $A = +1$ or $A = -1$).

Are the propositions (P1) equivalent? I don't think so[c]

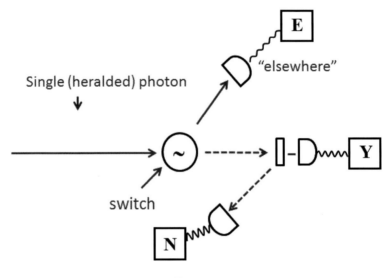

Fig. 3.

In fact, while one may argue as above that (P1) is essentially a truism, (P2) is a possible definition of the assertion of macroscopic counter factual definitiveness (MCFD), that is, the assertion that at least at the macroscopic level (the level of visible flashes, audible ringings etc.) counterfactual statements can be assigned definite truth-values. One might indeed by tempted to believe that MCFD is itself a truism; after all, we implicitly rely on it in everyday life when we make assertions such as "had I left the house five minutes earlier, I would have caught the bus" and it is also an essential ingredient of the legal system ("had the accused not pushed

---

[c]I don't think this point is a quirk of the English language: if the reader is a non-native English speaker, he/she is invited to see if it works in translation.

his victim down the stairs, she would be alive today"). Nevertheless, (P2) is not to my mind logically entailed by (P1), and as we shall see, some of the experiments designed to test QM may give reason to question it. At any rate, I believe that the concept of MCFD is a natural explication of the concept of "realism" at the microscopic level ("micro-realism"): to say that a particular photon possesses (say) a value of A equal to +1 is neither more nor less than to say that, had that photon been switched into the measuring device, then counter A would have clicked; thus, I would treat the assertion of MCFD as defining the assertion of micro-realism. An alternative point of view, apparently embraced by Stapp,[3] would be to treat the two concepts as independent, in which case one would take the view that assertion of microrealism necessarily implies assertion of MCFD but not vice versa.

Let's turn now to the class of experiments usually called "EPR-Bell". An idealized version of such an experiment is illustrated in Fig. 4: A source, typically a gas of excited atoms or, nowadays, a nonlinear crystal pumped by a laser which induces parametric down-conversion, produces pairs of photons some of which can be steered in opposite directions to distant measuring stations, which in some recent experiments have been separated by as much as 100 km; thus, the events associated with the arrival of the two photons at their respective stations are spacelike separated. Each station consists of a randomly activated switch (in some cases itself driven by a quantum process) which steers the photon in question into one of two devices, each similar to that of Fig. 1, which measure the polarization with respect to different sets of axes, and thus define different dichotomic variables A or A' at station 1, B or B' at station 2. (Thus, in the quantum-mechanical description, the operator representing $A(B)$ does not commute with that representing $A'(B')$.) While the idealized setup represented in Fig. 4 has never to my knowledge been completely realized in any existing experiment, the various so-called "loopholes" such as lack of truly spacelike separation, imperfect efficiency of the detectors etc., have, with one important exception to be addressed below, each been blocked in at least one existing experiment; and while it is not logically excluded that in a future (actually perhaps not too far in the future) experiment which blocks all of them simultaneously the result will be qualitatively different, I suspect this is not a scenario which appeals to many physicists. Thus, we may tentatively take the existing experimental results to be representative of those which would be obtained in the idealized setup of Fig. 4.

What is the point of this class of experiments? It is the following: there exists a class of general theories about the physical world ("objective local" theories, hereafter OLT's), which enable one to prove the celebrated CHSH inequality[4]

$$\langle AB \rangle + \langle A'B \rangle + \langle AB' \rangle - \langle A'B' \rangle \leqslant 2 \tag{1}$$

where the paired brackets indicate experimentally measured correlations. (Note that each of these correlations is measured on a different subensemble of the total ensemble of photon pairs, e.g. $\langle AB \rangle$ is measured on that subensemble for which photon 1 is switched into counter A and photon 2 into counter B, whereas for the measurement

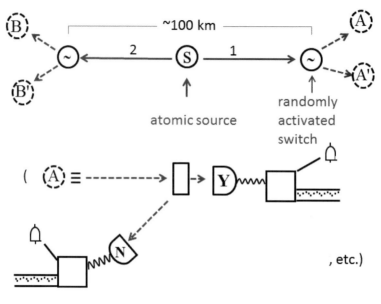

Fig. 4.

of $\langle AB' \rangle$ 1 is still switched into counter A but 2 is now switched into counter B';
etc.) The point is that the inequality (1) is <u>violated</u> by the predictions of QM, and
more importantly in the present context by the existing experimental data. Thus
the class of OLT's appears to be untenable.

What is the appropriate definition of the class of OLT's? That is, what is a set
of necessary and sufficient conditions for the inequality (1) to hold? This is a not
entirely uncontroversial issue, and in fact there is a (small) minority of workers in the
field who (in my understanding) claim that there is <u>no</u> set of "physically reasonable"
conditions which permits the demonstration of the inequality. However, I believe
that the overwhelming majority of physicists interested in these questions would at
least agree that the following set of conditions is <u>sufficient</u>.

(1) Induction
(2) Einstein locality
(3) MCFD

I now comment briefly on the meaning of these three postulates. The first is crudely
speaking, the assertion of our normal conceptions concerning the "arrow of time":
the properties of an ensemble at a given time cannot be affected by a choice of what
properties to measure on it which lies in the future. The second postulate is the
one basic to the special theory of relativity, namely that if two events are spacelike
separated (cannot be connected by any light signal), then neither can exert any
casual influence on the other. Note that within the framework of special relativity
Einstein locality implies induction; however, within a more general theory (which

might for example explicitly forbid causal chains consisting partly of "forward" and partly of "backward" links) the two postulates may be independent. Postulate MCFD has already been discussed. (In the literature, the postulate of MCFD is often replaced by that of (micro-) realism; see above for the relation between the two concepts).

Given postulates (1)-(3), the proof of the inequality (1) is so simple that it is worth sketching its main ingredients here: (a) For each individual photon pair, postulate (3) implies that each of the four quantities A, A' (for photon 1) and B, B' (for photon 2) exists and takes one of the values ± 1. Postulate (2) implies that its value is independent both of the choice of what to measure at the "distant" station and of the outcome of that measurement. (b) It then follows by trivial arithmetic that for each pair the inequality $AB + AB' + A'B' - A'B' \leqslant 2$ is satisfied, and thus that the inequality (1) for the correlations is satisfied provided all expectation values are taken on the same ensemble (i.e. on the full ensemble of photon pairs). (c) Finally, by (1), the subensembles on which the various correlations are actually measured are statistically equivalent to the full ensemble of photon pairs. Thus the CHSH inequality (1) holds for the experimentally measured correlations, QED. While, as remarked, a minority of physicists objects to the validity of this inequality, it seems to the present author that their objections are not based on the demonstration of any error in the above argument but rather on an implicit rejection of one or more of the postulates (1)-(3) (usually 1)) as "reasonable".

A different and more interesting question is whether all three of the postulates (1)-(3) are actually <u>necessary</u> for the demonstration of the CHSH inequality (1). In particular, the necessity of (3) has sometimes been questioned in the literature, most recently and trenchantly by N.Gisin.[5] His alternative proof rests heavily (as does the original CHSH argument) on the concept of probability (which I note was not used explicitly in the above argument), and seems to me to raise some important and delicate issues concerning the use of this concept in the absence of an implicit assumption of realism; but in view, inter alia, of space limitations I do not attempt to discuss these issues here.[d]

At any rate, I think the vast majority of physicists who have studied these questions would agree that if indeed a future "ideal" experiment along the lines of the idealized model of Fig. 4 confirms the results of the existing ones, then at least one of the postulates (1)-(3) has to go. Since (1) seems, at least at first sight, a vital ingredient in the very description of experimental practice, and a denial of (2) would challenge the foundations of the special theory of relativity, I believe that the majority opinion would be that it is (3) which has to give – a view nicely encapsulated in the title of a paper[6] by the late Asher Peres, "Unperformed experiments have no results."

However, before drawing this conclusion or an equally unpalatable one, let us note that there remains one "loophole" which has not to my knowledge been closed

---

[d]I hope possibly to do so elsewhere.

in any existing EPR-Bell experiment, and which provides a natural link to the second part of this talk. The loophole in question, which has been noted by several authors, is sometimes called the "collapse locality" loophole. It refers to the following consideration: In invoking Einstein locality to argue that the outcome of a measurement of A on photon 1 is not causally effected by the outcome of a measurement of (say) B on photon 2, we had to make some implicit assumption about the point at which a definite value of B is "realized." Yet if we take the formalism of QM seriously, there is no point at which we can say that any such realization has taken place! Certainly, it should not take place when the photon is presented with a polarizer and prima facie has to "decide" whether to be transmitted or reflected – in other types of experiments we can demonstrate interference between the transmitted and reflected components. Can we then say that a definite choice was made when the photon had to "decide" whether to be registered (e.g.) in the cathode of the photomultiplier constituting detector Y or that of detector N, thereby triggering a cascade of electrons in the relevant device and not in the other one? Well, there exists, at least as of now, no <u>experimental</u> evidence against such an assumption, but as pointed out by Schrödinger in his celebrated "Cat" paper,[2] it would be contrary to a consistent interpretation of the formalism of QM. And so on; it is clear that by delaying the point of "realization" to a sufficiently late stage (e.g. the stage at which the results of measurements at the two mutually distinct stations are correlated in a central coincidence counter) we can ensure that the two "events" or realization are no longer spacelike separated, so that postulate (2) can no longer be invoked. To be sure, in the case of some existing experiments, such as that of ref. 7, this would put us in the possibly awkward position of having to assert that in the first few microseconds following the arrival of the photon the relevant computer record was not in a definite state – but we have no experimental evidence, as distinct from prejudice, to the contrary! In principle, this loophole could definitely blocked by a long-baseline EPR-Bell experiment (Fig. 5) with the outcomes directly observed by human subjects, since (at least for those of us who are not adherents of the "many-worlds" scenario!) the outcome must have been "realized" by the time we consciously observe it. Given the spectacular progress of the last few years in laser ranging etc., this experiment may not be so many centuries or even decades in the future. In the meantime, most physicists will probably be content to assume that it will, like the existing experiments, give the results predicted by QM and thus violate those of the class of OLT's.

However, the considerations of the last paragraph raise a question which is interesting in its own right: can we put <u>experimental</u> bounds on the point at which definite outcomes are realized? ("can we build Schrödinger's Cat in the laboratory?") This has been the second major area of experimentation on the foundations of QM over the past few years, and I now sketch the general principle involved, that of "macroscopic quantum coherence".

Suppose we have some macroscopic system in which we can identify two states as "macroscopically distinct" by some reasonably agreed criterion. A typical example

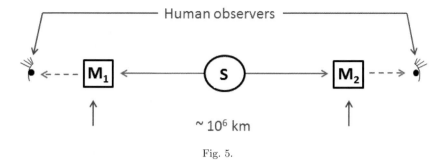

Fig. 5.

is a "flux qubit" (a thick superconducting ring interrupted by a single Josephson junction, see Fig. 6); in this case the states in question correspond to currents of the order of a few microamps which circulate clockwise in one of the two states and anticlockwise in the other. Te be sure, a microamp is not in itself a very "macroscopic" current (the current carried by an electron in a H atom in the 2p state is already $\sim$ 1mA!), but the point is that the current is carried by a large number of Cooper pairs, which are therefore in some intuitive sense all "behaving differently" in the two states in question.[e] A somewhat similar situation is realized in the Talbot-Lau interferometry of heavy molecules, with the two states now corresponding to molecular positions at the two diffracting "slits" of the interferometer, with the difference that unlike in the flux qubit case the two slits in question are not physically identical at different points in the experiment but correspond to three different "screens". While to date the molecular weight of the molecules involved in this kind of experiment has been only a few thousand,[10] there seem to be realistic prospects[11] for extending it to $\sim 10^7$, which would presumably meet the objection that the two states are not really "macroscopically" distinct.

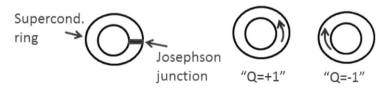

Fig. 6.

In any case, let's imagine that we have identified some physical system and two of its states which are agreed to satisfy the criterion for macroscopic distinctness, and label them with a dichotomic variable Q which takes a value $+1$ and -1 for

---

[e]The precise meaning of "behaving differently" is itself a vexed question, see refs. 8 and 9.

the two states respectively. We first check, in a preliminary set of experimental runs, that whenever measured the variable Q indeed always takes the value ±1. We further check (Fig. 7) that if we start the system in state +(Q= +1), leave it for an appropriate time Δt and then measure Q, we find the values +1 and −1 with approximately equal probability, and that the same result is found if the initial state was −(Q = −1) (and the time interval again Δt).

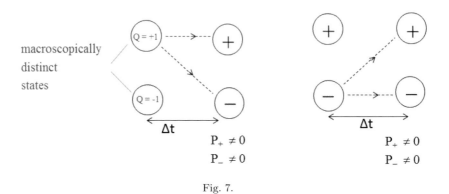

Fig. 7.

Now comes the "crunch" experiment (see Fig. (8)): We start the system at time t_i in (say) the state Q = +1, leave it alone for a time $2\Delta t$ and measure Q at time $t_i + 2\Delta t$. What do we expect to see? Consider the state of the system at the intermediate time $t_{int} \equiv t_i + \Delta t$.

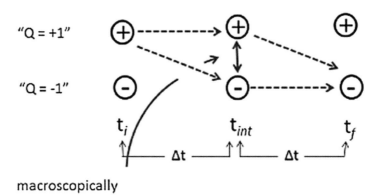

Fig. 8.

We know from our preliminary experiments that had we measured Q at that time, we would have found the results Q = ± 1 with roughly equal probability, and it is then tempting to assume that on each of the runs of the main experiment (in which we do <u>not</u> measure at t_{int}) the system "really was" in one or other of the two states. Let us call this view "macrorealism" (MR) (it will be defined more carefully below). But if this is so, then the probability of reaching the $+$ state at the final time t_f should just be the average of the probabilities of doing so from the $+$ state and from the $-$ state, which we have already seen is about 50%. Thus, on this scenario, we predict that measurement of Q at time t_f will yield $+1$ with a probability of about 50%. On the other hand, application of the QM formalism to this problem yields the result that the state at t_{int} is not a classical mixture of $+$ and $-$ (as was implicitly assumed in the above argument) but rather a <u>quantum superposition</u>, and that the amplitudes for the processes $+ \to +$ and $- \to +$ <u>interfere</u> destructively; thus, QM predicts that the probability of finding $Q = +1$ at time t_f is zero.

Several experiments along these general lines, using Ramsey-fringe interferometry, have been conducted on flux qubits; see in particular[f] ref. 12. In each case, the experimental data agree with the predictions of QM within the error bars. Thus, one can say that our current experimental information is <u>consistent</u> with the belief that the QM description continues to work at the level of flux qubits. Since the formalism of QM is prima facie inconsistent with the idea of macrorealism, this state of affairs would suggest that the latter has to be discarded, at least at the level in question. However, it is important to emphasize that the existing experiments do not rigorously establish this; for example, the conclusion of the argument of the last paragraph might be disputed by invoking a possible history-dependence in the behavior of the system in a macrorealistic theory. Thus, one would like ideally to construct an experimental test which, if it comes out according to the predictions of QM, must automatically exclude a macrorealistic world view (of course at the level of the experiment).

Such a test was indeed proposed in ref. 13, and the argument tightened in ref. 14. It turns out that if we define the class of macrorealistic (MR) theories in an appropriate way, we can demonstrate the analog of the CHSH theorem for the time-dependent behavior of our 2-state system, with the role of different polarizer settings being played by different times of measurement. Namely, any suitably defined MR theory must make the prediction[g]

$$\langle Q(t_1)Q(t_2)\rangle + \langle Q(t_2)Q(t_3)\rangle + \langle Q(t_3)Q(t_4)\rangle - \langle Q(t_1)Q(t_4)\rangle \le 2 \qquad (2)$$

where in each case the correlation in question is obtained by measurements only at the two specified times, with no intermediate measurement or interference with the system. It is easy to show that for an ideal 2-state system described by QM inequality (2) can be violated by suitable choices of the times t_i, and moreover that

---

[f]In the case of heavy molecules, although this precise experiment has not been done, related effects have been seen in a 2-slit diffraction setup.[10]

[g]In this case (unlike that of the original CHSH inequality) we can actually set (e.g.) $t_2 = t_3$ without loss of generality, thereby simplifying (2) somewhat.

small departures from ideality do not remove the violation; thus an experimental discrimination between the predictions of QM and those of the whole class of MR theories is possible.

What are the minimal defining postulates which will enable us to probe the inequality (2)? I believe the following set is sufficient:

(1) Induction
(2) Macro-objectivity ("macrorealism per se")
(3) Noninvasive measurability

I comment on these in turn: (1), as in the EPR-Bell context, is essentially the usual assumption about the "arrow of time": the properties of a physical ensemble at any given time are independent of any measurements which may or may not be made on it in the future. (2) is just the statement that a macroscopic system which has available to it two (or more) macroscopically distinct states must al all times[h] "actually be" in one or other of these states (we will explore below the precise meaning of that statement). The third postulate, which essentially plays the role in the derivation of (2) taken in the original argument for (1) by the postulate of Einstein locality, is perhaps the least transparent: that it is possible in principle to determine the value of Q at a given time t without affecting the state of the system at or subsequent to t. Needless to say, this <u>not</u> a principle respected by QM! To make it plausible, let us introduce the idea of an "ideal-negative-result" measurement (see Fig. 9): We couple the system (S) to a measuring device (M) in such a way that if S is (say) in the state "+" then it interacts physically with M and induces a (macroscopic) change in it, while if S is in the state "−" there is no physical interaction and no change induced in M.

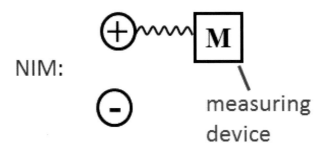

Fig. 9.

(Call this an INR(+) measurement). We then proceed as follows:[15,16] (a) we first use QM to calculate a time t at which, starting from (say) the + state at time

---

[h]A small fraction of "transit times" can be allowed for; see ref. 13.

t_i the system is guaranteed to be in the state $Q = -1$, and check by an explicit experiment (i.e. a set of runs with measurement, not necessarily noninvasive, at time t) that it is indeed so. (b) Next, we start the system from the $+$ state at time t_i and perform a set of runs in which we obtain the statistics of $Q(t')$ at various times $t' > t$ (without a measurement of any kind before t). Finally, (c) we perform a set of runs in which an INR $(+)$ measurement as described above (i.e. with S coupled to M only if the former is in the $+$ state) is conducted at t, and take the statistics of Q at various times $t' > t$ as at stage (b). If the statistics obtained at stages (b) and (c) coincide within the error bars, it would seem "reasonable" to assume that the INR $(+)$ measurement is noninvasive with respect to the state "$-$"; we can then repeat the whole procedure with the "$+$" and "$-$" states interchanged (so that we establish (or not) the noninvasiveness of an INR (-) measurement).

Once we are convinced that both INR $(+)$ and INR (-) measurements are non-invasive, we can obtain the correlations which occur in the inequality (2) entirely by a combination of runs involving these measurements in which we throw away the runs on which we got positive results (see ref. 16). All our measurements are then noninvasive, which allows the application of the inequality (2).

Because of the difficulty of obtaining a high enough measurement fidelity to make the results meaningful, the above protocol has not so far been implemented on any system for which the two states in question could reasonably be called "macro-scopically distinct".[i] An experiment which tends in this direction has recently been performed[18] in a macroscopic superconducting device ("transmon") and has given results consistent with the predictions of QM, however, it used a so-called "weak measurement" technique,[19] and in order to interpret the results as excluding realism at the relevant level the authors had to interpret the concept of "noninvasive measurability" in a way which is different from that envisaged in ref. 13 and above, and which (in the present author's opinion) represents a very much stronger constraint on the class of theories being tested. Thus, there is currently no experiment which excludes macrorealism as defined above in any physical system. Nevertheless, it seems probable that such a "complete MQC" experiment will become feasible within the next few years, and in the light of the fact that the rather closely related experiments already conducted (see e.g. ref. 12) bear out the predictions of QM. I suspect that most physicists (including the present author) would bet rather heav-ily that it will also do so and thereby definitively refute the class of macrorealistic theories defined above. Let's assume for the sake of argument that this is correct: what would we then have learned about the structure of the physical world?

To examine this question, it is necessary to ask about the same question about postulate (2) as we asked at the microscopic level: What does it <u>mean</u> to affirm that at (almost) all times the macroscopic system "actually is" in one or the other of its two macroscopic states? As in the microscopic case, it seems that the most

---

[i]An elegant experiment along these lines at the level of individual nuclear spins has been recently reported;[17] not unexpectedly, the results exclude (micro) realism in this system.

natural way of explicating this assertion is to interpret it as an affirmation of MCFD: on any given run, one of the statements "had I measured the value of Q at the time t (which I did not) I would have attained the value +1" or the corresponding statement with +1 replaced by −1 is true. If it is objected that such an interpretation makes macrorealism a trivial extension of microrealism, it may be replied that the difference is that the choice of what to measure or not is now delayed to the point where the basis which is chosen for the measurement already corresponds to states which are macroscopically distinct.

Suppose a carefully constructed complete MQC experiment (including the preliminary part of the protocol) indeed comes out according to the predictions of QM. A possible response might be to continue to assert postulate (2) above (macroobjectivity) while rejecting postulate (3) (non-invasive measurability). We would then presumably be committed to the following position: Whenever the system is in a state such that a measurement is guaranteed to give (say) Q = −1, then an INR (+) measurement is guaranteed to have no effect on the subsequent behavior of the system. However, there exist states (typically those described in QM by a quantum superposition of + and −) such that even in these cases for which a (counterfactual) measurement would have revealed that Q = −1, an INR (+) measurement still affects the subsequent behavior. Such a conclusion might seem to fit rather naturally into the Bohmian approach to QM, in which even though on any given occasion the system "actually is" in (say) the − state, there is still some (pilot-wave-like) amplitude associated with the + state, which can exert an influence on the subsequent behavior; but, of course, this does not get around the problem of explaining what becomes of this amplitude once Q is actually measured to have the value −1. (cf. ref. 20)

To summarize the conclusions of this paper

1. From the existing EPR-Bell experiments, one must either

   (a) reject at least one (and possibly more than one) of
       induction
       locality
       MCFD
   or (b) invoke the "collapse locality" loophole.

2. If a future long-baseline EPR-Bell experiment verifies the predictions of QM, option (b) will become unviable.

3. If a future "complete MQC" experiment using the protocol described above verifies the predictions of QM, we will be forced to reject (or at the relevant level) at least one of
   induction
   macrorealism
   NIM

An intriguing logical possibility is that the result of the MQC experiment might come out according to QM but that of the long-baseline EPR-Bell experiment not. That would then raise in acute form the question of whether human observers are in some sense "special" and not describable by the principles which apply to the inanimate physical world. In the very (very!) distant future it is conceivable that this question might be resolvable by a real-life implementation of the so-called "Wigner's friend" variant[21] of Schrödinger's Cat; in the meantime, a very small step in that direction is being taken in an experiment being set up at UIUC to verify (or not!) that human observers react to quantum superpositions in the same way that inanimate counters do.

A final thought: is "induction" sacred? All the considerations above here left unchallenged our normal motions concerning the "arrow of time." However, in the present author's opinion this issue is at least as sticky as any of the discussed above, and it seems conceivable that a completely new approach to this issue may be the key to a better understanding of the physical universe. Only time will tell. . .

References

1. A. Einstein, B. Podolsky and N. Rosen, *Phys. Rev.* **47**, p. 777 (1935).
2. E. Schroedinger, *Die Naturwissenschaften* **23**, p. 807 (1935).
3. H. Stapp, *Am. J. Phys* **53**, p. 306 (1985).
4. J. Clauser, M. Horne, A. Shimony and R. Holt, *Phys. Rev. Lett.* **23**, p. 880 (1969).
5. N. Gisin, *arXiv* **0901.4255**.
6. A. Peres, *Am. J. Phys* **46**, p. 745 (1978).
7. G. Weihs, T. Jennewein, C. Simon, H. Weinfurter, and A. Zeilinger, *Phys. Rev. Lett.* **81**, p. 5039 (1998).
8. J. Korsbakken, F. Wilhelm and K. Whaley, *Europhys. Lett.* **89**, p. 30003 (2010).
9. A. Leggett, *unpublished* .
10. M. Arndt, K. Hornburger and A. Zeilinger, *Physics World* **18**, p. 35 (2005).
11. O. Romero-Isart, A. Pflanzer, F. Blaser, R. Kaltenbaek, N. Kiesel, M. Aspelmeyer and J. Cirac, *Phys. Rev. Lett.* **107**, p. 020405 (2011).
12. I. Chiorescu, Y. Nakamura, C. Harmans and J. Moorij, *Science* **299**, p. 1869 (2003).
13. A. Leggett and A. Garg, *Phys. Rev. Lett.* **54**, p. 857 (1985).
14. A. Leggett, *J. Phys.: Cond. Matt.* **14**, p. R415 (2002).
15. A. Leggett, *Found. Phys.* **18**, p. 939 (1988).
16. M. Wilde and A. Mizel, *Found. Phys.* **42**, p. 256 (2012).
17. G. Knee, S. Simmons, E. Gauger, J. Morton, H. Riemann, N. Abrosimov, P. Becker, H.-J. Pohl, K. Itoh, M. Thewalkt, A. Briggs and S. Benjamin, *Nature Communications* (in press).
18. A. Palacios-Laloy, F. Mallet, F. Nguyen, P. Bertet, D. Vion and D. Esteve, *Nature Phys.* **6**, p. 442 (2010).
19. R. Ruskov, A. Korotkov and A. Mizel, *Phys. Rev. Letters* **96**, p. 200404 (2006).
20. A. Leggett, *Found. Phys.* **29**, p. 445 (1999).
21. E. Wigner, *Scientific Essays* (1970).

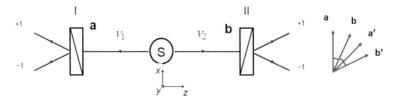

Fig. 1.   EPR Gedanken Experiment with photons. A pair of photons (ν_1, ν_2) correlated in polarization, emitted from the source S and propagating along $-z$ and z respectively, is analyzed by the polarizers I and II, followed by photon detectors in the channels $+1$ and -1 (corresponding to a polarization respectively parallel or perpendicular to the orientation $\mathbf{a}$ – respectively $\mathbf{b}$ – of the polarizer). For an EPR state of the form (1) Quantum Mechanics predicts that the results of the measurements on each side are random, but correlated according to Eq. (2). For the set of orientations indicated on the right part of the figure (the unit vectors $\mathbf{a}$, $\mathbf{b}$, $\mathbf{a'}$, $\mathbf{b'}$, indicating the orientation of the polarizers, are in a plane perpendicular to the z-axis) the correlation coefficients predicted by Quantum Mechanics violate Bell's inequalities, demonstrating a contradiction between Quantum Mechanics and a local realist description *à la Einstein*. In that point of view, the correlation is explained by common properties of the two photons of a same pair, acquired at the time of the emission, and determining the outcome of each measurement. Relativistic separation of the measurements can be imposed by changing rapidly the orientation of each polarizer while the photons are in flight, so that these changes are separated by a space-like interval.

Prepared comments

A. Aspect: From EPR to Bell, and experimental tests

After many tries to find a self-contradiction in the quantum mechanical description of *a single particle*, Albert Einstein took another angle of attack in 1935, when he discovered, with his collaborators Boris Podolski and Nathan Rosen, the amazing properties of *a pair of entangled particles*. Figure 1 describes an ideal Einstein-Podosky-Rosen (EPR) GedankenExperiment, in which a pair of photons (ν_1, ν_2) is analyzed in polarization. A pair of photons whose polarization state is

$$|\Psi(\nu_1, \nu_2)\rangle = \frac{1}{\sqrt{2}}(|x, x\rangle + |y, y\rangle) \tag{1}$$

is entangled, i.e. it is impossible to assign a given value to the polarization of each individual photon (the state cannot be factorized). Elementary quantum mechanical calculations then show that while each polarization measurement yields random results (either $+1$ or -1 with equal probabilities), a correlation is predicted between these random results. The correlation coefficient depends on the relative orientations of the two polarizers according to the formula

$$E_{\mathrm{QM}} = \cos 2(\mathbf{a}, \mathbf{b}). \tag{2}$$

For parallel directions of analysis ($[\mathbf{a}, \mathbf{b}] = 0$), the correlation is total: when the result $+1$ is found at polarizer I, $+1$ is also found at polarizer II, and similarly for the -1 outcomes. For Einstein, whose world view excludes any possibility of a direct connection between the space-like separated measurements, this means that both photons had the same property determining the polarization since their emission at the source. This common property, which allows us to understand the correlation, changes randomly from one pair to the next one. Such a model as an explanation of the EPR correlations amounts to completing the QM formalism, since there is a supplementary parameter which assumes different values for pairs all described by the same quantum state of equation (1). This is why Einstein gave a negative answer to the question that constitutes the title of the EPR paper.[a] Niels Bohr immediately opposed that conclusion[b] and the debate never ceased until the death of these giants. What was at stake was Einstein's local realist world view, in which each object localized in a finite volume of space time has a behavior totally determined by a set of parameters belonging to that space-time region, and cannot be affected by events space-like separated. Bohr argued that such a point of view was in contradiction with Quantum Mechanics. The debate, however, remained philosophical (epistemological), since the two debaters disagreed about the interpretation of the results of the calculation, but both agreed about the results of the quantum calculations.

The situation changed in 1964 when John Bell discovered that Einstein's point view eventually led to a conflict with some quantum mechanical predictions. More precisely, assuming that the correlations can be explained by assigning to the two photons of the same pair a common property that will determine the results of the measurements, Bell could demonstrate that a certain combination of several correlation coefficients cannot exceed a certain value. It turns out that this limit is violated by the quantum mechanical predictions of equation (2) for e.g. the set of orientations shown in Fig. 1. The debate then changes from an epistemological discussion into a question, about the behavior of nature, that can be settled by experiment. A series of experiments closer and closer to the ideal scheme[c] have shown a clear violation of Bell's inequalities, and although there is still room for improved experiments (to close simultaneously all loopholes which have been addressed separately), it is reasonable to conclude that one must renounce the local realist description of the world advocated by Einstein. One may then ask whether it is possible to decide which of the two concepts, locality or physical reality, should be abandoned? Here again we enter the domain of personal opinion. Mine is that in the Einstein's world view these two notions are not independent: it seems hard to figure out what would be

---

[a] A. Einstein, B. Podolsky, and N. Rosen, Can quantum-mechanical description of physical reality be considered complete? *Phys. Rev.* **47** (1935) 0777.
[b] N. Bohr, Can quantum-mechanical description of physical reality be considered complete? *Phys. Rev.* **48** (1935) 696.
[c] A. Aspect, Bell's inequality test: More ideal than ever, *Nature* **398** (1999) 189.

the physical reality *à la Einstein* of a system which could be affected by something happening out of its past light-cone. Local realism as a whole would have given a consistent image of the world that I find palatable, but renouncing one of the two concepts is not less weird than Quantum Mechanics itself. We have to live with it, and acknowledge the fact that when such a weird feature of nature is discovered, it is not a bad idea to try to use it for something new, as is done in quantum information research.[d]

G. 't Hooft: Emergent Quantum Mechanics and Bell's inequalities

To understand quantum mechanics, it is of importance to distinguish two questions: one, the occurrence and physical relevance of quantum-superimposed, and entangled, *states*, and two: the question whether such states can *evolve* from ontological, non entangled states. The latter is a question of dynamics, and is often overlooked. Bell's inequalities, instead, refer to the first question. The reason why these inequalities are often violated may well be that states have always been entangled, from day one. This then is a question of the statistical correlations of the initial conditions on the universe, and if so, no transitions from ontological states to entangled ones are needed.

Keeping this in mind, quantum entangled states can be introduced easily in a classical theory. All one has to do is map the classical states it can be in, one-to-one, to elements of a basis in Hilbert space. For instance, if the classical theory has states evolving as $(1) \to (2) \to (3) \to (1)$, then the evolution operator is

$$U = \begin{pmatrix} 0 & 0 & 1 \\ 1 & 0 & 0 \\ 0 & 1 & 0 \end{pmatrix} = e^{-iH\delta t},$$

where δt is the duration of the time steps, and H is a quantum hamiltonian obtained simply by taking the logarithm of U. Note, that the eigenvalues of H are equidistant, which makes that this system can serve as a primitive model for a Zeeman atom. Indeed, by taking the discrete Fourier transforms of its three eigenstates, one can verify that the Zeeman atom evolves deterministically in regular time steps.

In general, if a *quantum mechanical* model happens to allow for the construction of a non-trivial set of operators $\mathcal{O}_i(t)$ at different times t, such that they all commute: $[\mathcal{O}_i(t), \ \mathcal{O}_j(t)] = 0, \ \ \forall \ t, \ t', \ i, j$, then we say that this is a deterministic and ontological model. There are various examples of such models, which look quantum mechanical but are deterministic, for

---

[d]J. Preskill, this volume.

instance: all harmonic oscillators, and massless, non-interacting, second-quantized fermions.

Time may be discrete or continuous. In the latter case, it is easy to find the quantum hamiltonian that generates an evolution law of the form

$$\frac{d}{dt}q_i(t) = f_i(\vec{q}) , \quad H = \tfrac{1}{2}\sum_i \left(p_i f_i(\vec{q}) + f_i(\vec{q})p_i \right) .$$

Operators such as $p = -i\frac{\partial}{\partial q}$ and $\sigma_x = \begin{pmatrix} 0 & 1 \\ 1 & 0 \end{pmatrix}$ etc. , are useful, even though they look quantum mechanical while our model nevertheless may be deterministic.

We now claim that, along the lines briefly sketched above, deterministic models may exist that describe the quantum mechanical world as we know it. Quantum entangled states occur because the initial conditions were quantum entangled; these may be described as a density matrix, which expresses the fact that we do not know the initial state exactly. Even if this initial state looks entangled to us, this could be an optical illusion: it is entangled because the templates that we use to describe it, such as photons, protons, and even quantum fields, are not the ultimate ontological ones. *They* are entangled states! In terms of the ontological observables, the "wave function of the universe" contains a one and many zeros. Indeed, this beautifully explains why this wave function collapses in terms of macroscopic observables (which presumably *are* ontological as well), and also the Born rule, which is the law that connects the norm of a wave function squared to the probabilities.[e]

We suspect that ontological models of this sort may be the only completely consistent schemes for reconciling quantum mechanics with general relativity and cosmology, but even if no such ontological model can be identified in the near future, the picture we sketched here can serve very well to help us interpret the theory of quantum mechanics, without the need of introducing "parallel universes", or any modifications of the Schrödinger equation to accommodate for the collapse of the wave function, since such collapses happen automatically here.

J. Hartle: The Impact of Cosmology on Quantum Mechanics

When quantum mechanics was being discovered in the '20s another great revolution in physics was just starting. It began with the discovery by Lemaître, Hubble and others that the universe is expanding. For a long time quantum mechanics and cosmology developed independently of one another. Quantum mechanics was concerned with the results of observa-

---

[e]G. 't Hooft, arXiv:1112.1811[quant-ph], to be published.

tions in the laboratory, and cosmology with observations of the structure of the universe on very large scales. Yet the very discovery of the expansion would eventually draw the two subjects together. It means that there was a time — the big bang — where quantum mechanics was important for cosmology and for our observations today.

One prominent example is the the application of quantum mechanics to understand the present large scale structure of the universe. Initial quantum fluctuations away from homogeneity and isotropy grew by gravitational attraction to be the galaxies, stars and CMB irregularities that we see today.

There isn't time to review what quantum mechanics has done for cosmology. Rather I want to raise a few questions for discussion about what cosmology implies for quantum mechanics.

The usual Copenhagen formulations of quantum mechanics are inadequate for cosmology. Characteristically, these formulations assumed a division of the world into "observer" and "observed". They assumed that outcomes of "measurements" are the only objectives of quantum mechanical prediction. They posited the existence a separate classical world. However, in a theory of the whole thing there can be no fundamental division into observer and observed. Measurements and observers cannot be fundamental notions in a theory that seeks to describe the early universe when neither existed. In a basic formulation of quantum mechanics there is no reason in general for there to be any variables that exhibit classical behavior in all circumstances.

A formulation of quantum mechanics general enough for cosmology was started by Everett and developed by many. That effort has given us a framework — decoherent (or consistent) histories quantum theory — that is logically consistent, in agreement with experiment as far as is known, applicable to histories, applicable to cosmology, consistent with the rest of modern physics including special relativity and quantum field theory, and generalizable to include semiclassical quantum gravity. Under suitable assumptions it implies the Copenhagen quantum theory of laboratory experiment in measurement situations. It is the only presently available formulation of quantum theory with all these properties. But it may not be the only one.

With this in mind the following questions are of interest:

- Do you expect quantum mechanics (say as represented by decoherent histories extended to include gravity) to successfully apply to the whole universe, or will quantum mechanics have to be further modified? If so, what kinds of modifications do you expect, and what are the experiments whose outcomes would drive them?

- Quantum states of the universe like the no-boundary wave function are superpositions of different possible classical histories. Is there any experimental way of testing this? Is there any compelling theoretical reason to object to it?

- Many approaches to a final theory have separate theories of the dynamics and the state of the universe. Which features of the universe do you expect to be traceable to the quantum state and which to the dynamics? State and dynamics are connected in theories like the no-boundary wave function. Do you expect a final theory to give a unified picture of both the universe's quantum state and dynamics? Do you expect it to be a quantum theory at all?

W. Phillips: The Quantum Measurement Problem, Interpretations of Quantum Mechanics, and All That: An opinion from an acolyte of the "Church of Shut Up and Calculate"

I have already revealed my biases about the interpretation of quantum mechanics by saying that I subscribe to the "shut up and calculate" point of view. Let us examine what the question is. The left hand side of the figure illustrates schematically an experiment we do routinely in the lab. We have an atom or a group of atoms, essentially at rest, represented by the blue circles. We shine a pulse of counter-propagating laser beams, represented by the wavy lines, onto the atoms. One of the things that can happen is that the atom will absorb a photon from one of the laser beams (the blue beam) and emit a photon into the other (the red beam) so that the atom is moving, having acquired the momentum of the two photons. On the other hand it is possible that nothing happens. Higher order processes in which multiple photons are absorbed and emitted can happen as well, but we ignore them here. When we do this in the lab, the typical result is that some of the atoms absorb and emit photons and some don't. So some of the atoms are moving and some are not. Of course that is not quite a proper way of describing what happens. A proper description is that every atom is put into a superposition state of having absorbed and emitted the photons and not having absorbed and emitted the photons. We do this experiment with about a million atoms at a time and we take a picture of the group of atoms at a certain time-of-flight after the laser pulse. The position of the atoms in the photograph indicates how far they have traveled during the time-of-flight, and that tells us their momentum. The photograph on the right of the figure is an example of such a measurement.

Before we make a measurement (before we take the picture) the many-body state is a very simple state that is N (N being about a million) times the single atom state, and that single atom state is in a coherent superposition of being in two different places, as shown in the equation above the photograph. Every atom has to "decide" where it is going to be when you measure it. Or, God throws the dice and decides for every atom where it is going to be. A very simplified statement of the quantum measurement problem is: *How in the world does that happen?* How do you go from a situation where an atom is in two macroscopically different places at the same time and then all of a sudden it is only in one place? How does the wave function collapse so as to have that happen? Of course if you don't believe the wave function has any reality then that is not a problem. But a lot of people like to think of the wave function as having a reality, and for them, there is a problem. So how do people answer this? I am going to give an over-simplification of only some of the answers: One of the answers is the *many worlds* approach: nothing really collapses. It is just that there are other versions of the universe, other versions of reality, where other things are happening. I don't like the many worlds approach because it seems like a tremendously complicated answer to a very simple question. Why do I have to accept a gazillion different universes just to answer the question of how I produce this photograph in my laboratory?

Another possibility is to consider that a possible description of the atom is as a density matrix. The density matrix has diagonal elements that describe the probability that it will be in one state or another and it has off-diagonal elements that describe the coherence between the different states. If it were not for that coherence this would not be a quantum problem; it would be a classical problem. Now I allow my quantum system to interact with the environment, with a measurement apparatus, and I trace over the environment. Now I have gotten rid of the off-diagonal elements of the density matrix and I have a classical problem. The problem of whether the atom is in one place or another is a classical problem now, because there are not any off-diagonal elements. It is just like the classical problem of flipping a coin. Well, I don't like this *decoherence* explanation either because it just sweeps the problem under the rug. The difficulty is all put into the entanglement between the atom and the apparatus, and because of my ignorance, I trace over the environment.

So my advice is to *shut up and calculate*. Quantum mechanics gives the right answer and all these questions about interpretation are just so much *gum flapping*. It is just idle talk with no real significance. Does that mean I don't like to think about the wave function when I do an experiment? No! I like to think about the wave function because it gives me a framework in which to think about experiments, and, more importantly, in which to think about new experiments. In fact one of my favorite frameworks for thinking is the so-called *quantum Monte Carlo wave function* technique. I like this because it is doing quantum mechanics the way God is supposed to do it. That is, at every point in the calculation you evolve the wave function deterministically according to the Schroedinger equation. Then, playing God, you throw the dice to decide whether some kind of quantum

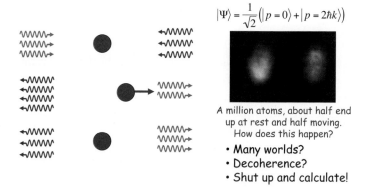

$$|\Psi\rangle = \frac{1}{\sqrt{2}}\left(|p=0\rangle + |p=2\hbar k\rangle\right)$$

A million atoms, about half end
up at rest and half moving.
How does this happen?

• Many worlds?
• Decoherence?
• Shut up and calculate!

Fig. 1.   The Quantum Measurement Problem.

jump is going to happen. Whether it jumps or not, you continue from
there to evolve the wavefunction deterministically until the next jump. If
you do this for one particular trial (trajectory) you will get one example
of what might happen in one case. If you do it for many trajectories and
average over them, you will get the right answer, just as if you had done the
density matrix calculation. It is completely equivalent to standard quantum
mechanics; in fact, it IS standard quantum mechanics. But it is a wonderful
way of thinking about things. It avoids many kinds of errors and stupid
mistakes, mistakes that have been made and that would have been avoided
had people thought in this quantum Monte Carlo way.

Discussion

A. Aspect Thank you very much Bill (Phillips), it is clear that you work in an Institute of Standards because your timing was perfect. All right, now comes the part devoted to discussion...

A. Zeilinger I have a comment and/or a question to Tony Leggett: you showed that there are three possibilities related to Bell's arguments, or at least to CHSH's arguments, and you said that at least one among those three possibilities must be false. I would like to remark that it has been pointed out that actually there are two other possibilities, which are probably even more severe that could logically be considered.

The first one is that the world might be completely deterministic, that is, there is no choice on the side of the experimentalist. This is somewhat analogous to the counterfactual definiteness question but not completely the same.

The other point that certain people have argued is that in order to derive Bell's inequalities you need Aristotelian logic. This could also break down. This is a very far reaching idea but it has been said.

Concerning the long distance experiment it will be very hard to do the experiment on such long distances because one would have all kinds of problems regarding collection efficiency which opens up big other loopholes. I would like to have your opinion about two other possibilities: firstly, one would have an experiment on earth, with the two spacelike separated regions not far away from each other but with signals coming from far away which are independent. These signals could be coming from human observers (from humans), and come independently in two separated locations such that they are completely independent in space time, or one could use light from distant sources like fluctuations in the light of quasars and so on...

A. Leggett Yes, I do agree that a completely deterministic universe would invalidate the argument. I was actually implicitly counting that among the loopholes which, I would claim, have been blocked, simply because of our direct consciousness, that we decide which way to turn a switch, etc.

For what concerns your second suggestion, this is a very interesting one but I do not think that, if the two observing stations are not widely separated from the source, it would resolve the possibility of a late collapse; that is, I am not sure but I think that you could still use the collapse locality loophole to get around that...

D. Kleppner Bill Phillips discussed the question of the reality of the wave function. There are two views of that. The first one is to consider that the wave function represents a property of the system, a well established property but that belongs purely to the system; the second one is to consider that it represents our understanding of the system. Both interpretations lead to the same prediction but for some of us that leads to more mental comfort and for some of us that leads to distress. I would like to formulate the following question:

How do people view the reality of the wave function? What is it? Is it a property of the system or is it just a statement about what we know about the system?

W. Zurek I think that this is a very interesting point that was just brought up, and it calls for reflexion. In a way, there are two different sorts of states, already in classical physics. If one does classical statistical physics one does it in phase space and there are probability distributions. They are also states, and of them one doesn't require reality. One only requires reality at the bottom of a phase space point. My view of what we are trying to get at when we are talking about quantum mechanics is the degree to which quantum states reflect one or the other of these features of classical states. In classical physics there was a nice separation: this phase space point existed irregardless of what we knew about it; in quantum physics that sort of "church and state separation" is not there, is broken...

Before I leave the floor I would like to point an inconsistency in Bill (Phillips)'s presentation:

- on one hand he doesn't like decoherence because it sweeps the problem under the rug,
- on the other hand he admits there is a problem when he says "shut up and calculate, don't think about it"; so I think that there is something there.

A. Aspect At this point, knowing well Bill Phillips, I would like to add that actually he often claims that indeed he belongs to the church of "shut up and calculate", but only on Sundays. This reminds me of John Bell saying once: "during the week I am a quantum engineer"; he meant: I belong to the school of shut up and calculate... But he added: "but on Sundays I have principles".

D. Gross Just a comment, what is the opposite of realism? unrealism? These discussions are slightly loaded because it seems that quantum mechanics which we all love and understand and not only calculate is put against realism. I must say that the attempts that I heard including Gerard's ('t Hooft) clever attempts to construct a realistic theory sound to me awfully unrealistic. So, I just don't see what is wrong with quantum mechanics. Of course we shall have to give up some aspects of classical reality, of realism but so what? After all we had to give up the notion of simultaneity of events: for some events, the notion of simultaneity makes no sense, and I ask: "And so what?" That is the way that the world is. Similarly, why do we hold on to classical realism if that is what you mean by realism? Why do we not simply go with quantum realism which seems perfectly realistic to me.

A. Aspect Is this a question?

D. Gross It is really a question to those who seem to be bothered by...

A. Aspect I am one of them, so I will react. Indeed, it is hard for me to think of a world where the word physical reality has no meaning. I became a physicist because I wanted to understand how physical reality works. What I learned from the violation of Bell's inequality is that the very comfortable vision *à la* Einstein of a local reality, that is to say, a physical reality embedded in well-defined portions of space time, is a notion that we have to abandon. I think that this is very important because it shows that entanglement is something yet more weird than what we thought, about quantum mechanics. It is different in nature from wave particle duality for a single particle. It has eventually lead to the idea of quantum computation. Now, what is my position relatively to the question "where do I need realism and what kind of realism am I ready to accept?" At the moment, personally, I am ready to abandon locality and to adopt a model which is both realistic and non-local. That model consists of considering that the wave function describes physical reality. Then of course I have to accept non-locality but, at least, I have an image... When I think of a physical situation, I have this image in my head, and then of course I do the calculation as everybody else, but this image allows me to have an intuition and, in the good days, to imagine a new situation that may turn out to be interesting.

G. 't Hooft I need to react on what David (Gross) said; he claimed to be very happy with quantum mechanics as it is. But I have some objections: you have to believe in the realistic nature of the wave function which none of us can measure, so this would be a big step to take. The second big step is that if you consider the entire universe as Jim (Hartle) has done, then to explain his talk one would need to consider the entire Hilbert space of all universes, which is somewhat disturbing, particularly because the interpretation of the probabilities given by any state in that Hilbert space cannot be meaningful for an early universe, where you cannot do an unlimited amount of measurements. The real meaning of the wave function of the universe is therefore problematic. Another reason why I say this is that I think that it is possible to write down a realistic theory if you assume that quantum states, however entangled they are, just represent the things we know. The wave function may just represent the things that we know about Nature. In particular this should apply to the vacuum, the vacuum being a superposition of all possible states of nature; we know that there are vacuum fluctuations so the vacuum cannot be a single real entity, it is something much more complicated than that; there is no reason why it should not be. Think of probabilities that are invading everywhere in our present description of the world; our understanding of the world, today, is imperfect; of course nobody should be surprised by that: we physicists are limited by brains that are totally insufficient to really comprehend the universe in all its details. We do not know the initial state of the universe exactly, we do not know many other things. Why then would the wave functions we use to describe reality be anything else than our best guess about the state nature is in? This does not mean that reality has to evolve in a fundamentally quantum mechanical sense where pure states evolve into quantum superpositions; if we use the right basis we can perhaps avoid that. This is my proposal: to search for the right basis in which all operators that are diagonal at one time remain diagonal at all times, and I claim that this should be possible – there is no reason why it should not be – it does not contradict the experiments today which violate Bell's inequalities; Bell's inequalities may perhaps not apply just because the vacuum state itself is a quantum entangled state. This is an intriguing possibility that it is worth pursuing for instance to solve problems such as the problems that David (Gross) has been working on: trying to quantize gravity. Indeed, in the present day situation, with our understanding of quantum mechanics, it seems to be extremely difficult to get a theory where gravity agrees with quantum mechanics. String theory is on its way, but it so happens that string theory is one of the few theories where my proposal has a good chance to work, which is what I would like to emphasize.

S. Das Sarma While accepting the fact that, maybe, applying the quantum theory to the whole universe and to cosmology is a technical problem, I do not see any reason to invoke the concept of reality. So, I kind of agree with David Gross because we know from the history of Science that the concept of reality has changed repeatedly. After all, for two thousand years, everybody believed until Newton that you need to push things for letting them move... Newton dispelled that notion and now we take it for granted that classically you do not need a force to keep objects moving. Similarly, we learn from special relativity that simultaneity disappears. Considered so, reality is a tricky thing and it seems to me that trying to preserve realism does not make much sense. We can still admit that combining quantum mechanics with gravity is a problem but reality is probably not the issue...

Actually, I have a question to Tony (Leggett). I was a bit disappointed by your talk because I thought that you believed that at some macroscopic scale quantum mechanics may actually fail. Now, that is a quite more concrete statement that I would like to think about, rather than thinking about reality. In fact, the last time that we discussed about quantum mechanics you even gave me a number like ten to the power sixteen and you think that there is a fundamental concept in Nature that we have not discovered yet. You did not mention any of these provocative statements in your talk and I would like to hear your opinion about this. I must say that I completely disagree with you on that point and actually that you are wrong but this is the kind of "error" that is exciting.

A. Leggett The reason why I did not mention this in my talk is quite simple: we you construct an approximate single particle or many particles or whatever it is wave function. The only thing that has a real wave function is the universe. Possibly that is even not a wave function, if you have a multiverse to which the universe belongs, where the different parts interact rather than being independent. In that case you have a density matrix for the universe and you have to work from that. But let us for the moment ignore that complication and just talk about the wave function of the universe. Based on the wave function of the universe you can in some cases construct special wave functions. When everything else has been doing something, the part that you are considering has a wave function, but it is a construct, it has been pointed long ago in the ninety fifties. Related to that is the question why to bother then to talk about many worlds, unless those many worlds interact with one another. As I just said one minute ago there is no point in worrying about them at all: as long as they do not interact they have no effect on the theory. So, what we can say is much simpler then. It is to talk about histories of ONE universe, alternative histories of one universe... And that is the simplest interpretation and if you include eliminating the interference terms by decoherence then you have a perfectly sensible description of what is going on. This decoherent-histories approach seems to work. There does not seem to be any objection against it, and I do not really see why it is necessary to introduce principles of realism and so on and so forth that are constraining. Why instead of pushing quantum mechanics around do we not learn from quantum mechanics? Quantum mechanics is there to teach us and I do not understand why it needs to be brutalized... The last thing that I want to say is that in the two-photon experiment, a lot of strange things have been said and written about it that illustrate these difficulties that I try to emphasize. People say that locality is violated somehow and, as was expressed very beautifully by my late colleague Wodkiewicz at the university of New Mexico, these questions about non-locality, this issue on non-locality, comes from trying to impose a classical situation on a quantum mechanical problem. If what is happening was classical, then you would have, because of Bell's theorem and so on, then you would have to drop either locality or positive probability or both, that is perfectly true. But it has to be introduced by the phrase... "if the problem were classical...", if it is not it is not! Then you do not have to drop anything. Then, finally there is this weird thing that Jim Hartle and I are coming up with which is to say that if you really want to go to some different philosophical position and still have all the benefits of quantum mechanics you can do it perhaps by saying that there is a single real history but you cannot find it. It is buried in an infinite number of different, say, descriptions that are given by the system. And you can do it in that way just as, in statistical mechanics you use a statistical distribution with probabilities, the so-called ensemble. Instead of using that ensemble to represent the actual universe, then you can get an alternative interpretation that satisfies all these things. It does not improve anything I can tell, but it may make some people happier. So, (a) we do not care about the people being happy, that is the first thing, and (b) the second thing is, probably we can actually make them happy if we try. All this, again, without disturbing quantum mechanics which seems to be perfectly correct.

A. Aspect Thank you very much Prof. Gell-Mann. I would like to tell you respectfully that I agree with most of what you said, except for one thing: you said "we do not care that Einstein said that, it was useless...". I fully disagree on this point, it was very useful. It was because of the question that Einstein asked that Bell came... it is because Bell came that you can say "if it was classical we would have this but it is not classical"... Before Bell's answer to Einstein's question, the difference between the classical point of view and the quantum world was not certain. It is only when this difference was doubtless that some people came with the idea of quantum information.

M. Gell-Mann I did not use the word useless, I admire Einstein of course enormously and his contribution was the contribution of an extremely brilliant person. That he was faced with this and that he came to the wrong conclusion is just a part of the story.

original packet, whose phases (and much more structure[l]) are determined by the gauss sums of number theory . This intricate 'quantum carpet'[m] has its counterpart in optics: the Talbot effect,[n] observed in 1836 (near Bristol, as it happens) but understood only recently.[o]

As well as the physical phenomena associated with nonanalyticity, explorations of the classical limit have led to advances in mathematics: deeper understanding of divergent series,[p] especially in situations previously regarded as not summable and in ways that lead to super-accurate numerics, and tantalizing progress towards the Riemann hypothesis of number theory[q] which we now understand as intimately connected with quantum chaos.[r]

Wojciech Zurek: Quantum Origin of Quantum Jumps

Superposition principle applies to isolated quantum systems, but it is famously violated in measurements. Quantum system can exist in any state, but measurements force it to choose between a set of orthogonal outcomes. As Dirac put it in his textbook "... a measurement always causes the system to jump into an eigenstate of the dynamical variable that is being measured... " I show that this restriction (usually imposed by the *collapse postulate*) can be *derived* when information transfer is modeled as a *unitary* quantum process.

In addition to the quantum principle of superposition and the Schrödinger-like unitary evolutions there is one more assumption we shall make: We assume (as do textbooks, e.g. Dirac) that a measurement repeated immediately yields the same outcome. Thus, we assume there are states $|\heartsuit\rangle$ and $|\spadesuit\rangle$ of the system that are left untouched:

$$|\heartsuit\rangle|A_0\rangle \Longrightarrow |\heartsuit\rangle|A_\heartsuit\rangle, \quad |\spadesuit\rangle|A_0\rangle \Longrightarrow |\spadesuit\rangle|A_\spadesuit\rangle$$

Quantum evolutions are unitary: They preserve scalar products. So, the

---

[l]Berry, M. V.,1996, Quantum fractals in boxes, *J. Phys. A* **26**, 6617-6629.

[m]Berry, M. V., Marzoli, I. & Schleich, W. P.,2001, Quantum Carpets, carpets of light, *Physics World*, 39-44.

[n]Talbot, H. F.,1836, Facts relating to optical science. No IV, *Phil. Mag.* **9**, 401-407.

[o]Berry, M. V. & Klein, S.,1996, Integer, fractional and fractal Talbot effects, *J. Mod. Opt.* **43**, 2139-2164.

[p]Berry, M. V.,1989, Uniform asymptotic smoothing of Stokes's discontinuities, *Proc. Roy. Soc. Lond.* **A422**, 7-21; Berry, M. V.,1989, Stokes' phenomenon; smoothing a Victorian discontinuity, *Publ. Math.of the Institut des Hautes Études scientifique* **68**, 211-221; Berry, M. V. & Howls, C. J.,1991, Hyperasymptotics for integrals with saddles, *Proc. Roy. Soc. Lond. A* **434**, 657-675; Berry, M. V.,1992, *Asymptotics, superasymptotics, hyperasymptotics in Asymptotics beyond all orders* eds. Segur, H. & Tanveer, S. (Plenum, New York), pp. 1-14.

[q]Titchmarsh, E. C.,1986, *The Theory of the Riemann Zeta-function* (Clarendon Press, Oxford); Edwards, H. M.,1974, *Riemann's Zeta Function* (Academic Press, New York and London).

[r]Berry, M. V.,1986, Riemann's zeta function: a model for quantum chaos?, *Quantum Chaos and Statistical Nuclear Physics*, eds. Seligman, T. H. & Nishioka, H., Vol. 263, pp. 1-17.; Berry, M. V. & Keating, J. P.,1999, The Riemann zeros and eigenvalue asymptotics, *SIAM Review* **41**, 236-266.

scalar product of the states of the system+apparatus before and after the information transfer must be the same, which leads to:

$$\langle \heartsuit | \spadesuit \rangle = \langle \heartsuit | \spadesuit \rangle \langle A_\heartsuit | A_\spadesuit \rangle$$

Above, we have recognized that $\langle A_0 | A_0 \rangle = 1$. A natural temptation is to "simplify" and divide both sides by $\langle \heartsuit | \spadesuit \rangle$. This yields $\langle A_\heartsuit | A_\spadesuit \rangle = 1$. This is a disaster – we have just proved that the measurement failed, as $\langle A_\heartsuit | A_\spadesuit \rangle = 1$ implies $| A_\heartsuit \rangle = | A_\spadesuit \rangle$ – the two states of the apparatus are *identical*, so it gained no information about the states it was supposed to distinguish. Only one when $\langle \heartsuit | \spadesuit \rangle = 0$ one cannot "simplify".

Our simple equation shows that measurement can transfer data that lead to prediction of the future outcome with certainty only when the states corresponding to the outcomes are orthogonal – $\langle \heartsuit | \spadesuit \rangle = 0$. We have just proved that only orthogonal states of the system can be found out without getting perturbed in the process: Now $\langle A_\heartsuit | A_\spadesuit \rangle$ can take on any value (including 0 that corresponds to a perfect measurement).

This derivation is based on information flow in the most primitive sense – the system and the apparatus become correlated – but it does not appeal to any high-level information concepts or even rely on probabilities: The only values of scalar product we have used are "0" and "1". Both reflect certainty. So, the uncontroversial quantum postulates – (i) states that live in the Hilbert space and (ii) unitarity – result, along with (iii) repeatability, in a discrete set of orthogonal outcomes.

This discreteness is why "quantum jumps" happen, and why an unknown quantum state cannot be "found out" – conclusion usually justified by the controversial "collapse" axiom. We have recovered symptoms of "collapse": When we choose an apparatus that yields repeatable measurements, we simultaneously pick out a menu of possible outcomes. This discussion can be made more precise and more general.[s]

Information flow is key: Whether it flows to an apparatus, observer, or an environment is immaterial. In this last setting the above derivation offers another view of the emergence of pointer states. We shall have this setting (where the environment acts as an apparatus) in mind below, as we show how probabilities emerge in quantum physics from quantum correlations using symmetries of entangled states.

Consider a perfectly entangled state. One can use its symmetries (see figure) to prove that outcomes of measurements on the two subsystems must be equiprobable. The crux of the proof is straightforward: The correlations between the possible outcomes on the two ends (system and environment

[s]W. H. Zurek, Quantum origin of quantum jumps: Breaking of unitary symmetry induced by information transfer and the transition from quantum to classical, *Phys. Rev. A* **76**, 052110 (2007); Quantum Darwinism, *Nature Physics*, vol. 5, pp. 181-188 (2009); Actionable information, repeatability, quantum jumps, and the wavepacket collapse, arXiv:1212.3245 [quant-ph].

– $\mathcal{S}$ and $\mathcal{E}$) can be manipulated locally. Thus, one can swap $|\heartsuit\rangle$ and $|\spadesuit\rangle$ in the states on the left in the figure by a unitary acting only on the system $\mathcal{S}$. Such a swap exchanges probabilities of the two possible results, $\heartsuit$ and $\spadesuit$. This is obvious, as $\mathcal{E}$ was untouched. Therefore, the "new" probabilities of $\heartsuit$ and $\spadesuit$ (that before matched probabilities of $\diamondsuit$ and $\clubsuit$, respectively) now match the (unchanged) probabilities of $\clubsuit$ and $\diamondsuit$ instead.

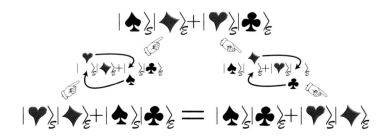

However, the initial state of the whole composite $\mathcal{SE}$ can be restored by "swapping" states in $\mathcal{E}$. This means that the probabilities of $\heartsuit$ and $\spadesuit$ are at the same time exchanged (by the swap on $\mathcal{S}$) and unchanged (because one can restore the whole entangled state without touching $\mathcal{S}$, as is also seen in the figure). This "exchanged, yet unchanged" requirement can be satisfied only when the probabilities are equal, $p_\heartsuit = p_\spadesuit$. When certainty corresponds to probability of 1, then $p_\heartsuit = p_\spadesuit = \frac{1}{2}$.

The symmetry of entangled quantum states we have used above is known as *entanglement-assisted invariance* or *envariance*, for short. In the quantum world unpredictability can be a consequence of knowing about the wrong thing. Completely known state of a composite classical system necessarily consists of parts that are also perfectly known. (Such pure classical state is a "Cartesian product" of states of its parts.) By contrast, quantum entangled states can be completely known but – as we have just seen – information about the whole can be completely incompatible with the information about their parts.

This derivation of equiprobability in entangled states leads directly to Born's rule.[t] All one needs is a bit of simple algebra, and $p_k = |\psi_k|^2$ follows. Indeed, one can also show its inverse, i.e. prove that the amplitude of an outcome state corresponding to a certain frequency of outcomes in the state overall vector must be proportional to the *square root* of that frequency. The two simple results discussed above show how the phenomenology of

---

[t]W. H. Zurek, Environment - Assisted invariance, causality, and probabilities in quantum physics, *Phys. Rev. Lett.* **90**, 120404 (2003); Probabilities from entanglement, Born's Rule from envariance, *Phys. Rev. A* **71**, 052105 (2005); Entanglement symmetry, amplitudes, and probabilities: Inverting Born's Rule, *Phys. Rev. Lett.* **106**, 250402 (2011).

measurements – including symptoms of collapse and Born's rule – can be deduced from the basic "quantum" postulates of quantum theory.

David Wineland: Foundations of Quantum Mechanics and Quantum Computation

The quantum/classical barrier continues to intrigue many physicists. Experimentalists are obliged to take a pragmatic view about measurements in order to get on with the day ("shut up and calculate"), but there remains in the background the thought that there is something fundamental missing in our understanding of measurements. As experimentalists gain control of increasingly complex quantum systems, these issues will become more apparent. In many experiments now, detection of a quantum system is realized through a von Neuman chain.[u] In simple form, suppose the system to be measured has two states $|\downarrow\rangle$ and $|\uparrow\rangle$ that are in a coherent superposition $\alpha|\downarrow\rangle + \beta|\uparrow\rangle$. We assume the system can be coupled to a quantum meter M such that the meter's states become entangled with the system's states

$$(\alpha|\downarrow\rangle + \beta|\uparrow\rangle) \otimes |M\rangle \to \alpha|\downarrow\rangle|M_\downarrow\rangle + \beta|\uparrow\rangle|M_\uparrow\rangle. \tag{1}$$

The meter is then coupled to the (macroscopic) environment E, whose states through an appropriate coupling can be entangled with those of the meter

$$(\alpha|\downarrow\rangle|M_\downarrow\rangle + \beta|\uparrow\rangle|M_\uparrow\rangle) \otimes |E\rangle \to \alpha|\downarrow\rangle|M_\downarrow\rangle|E_{M_\downarrow}\rangle + \beta|\uparrow\rangle|M_\uparrow\rangle|E_{M_\uparrow}\rangle, \tag{2}$$

where we assume $\langle E_{M_\uparrow}|E_{M_\downarrow}\rangle = 0$. The argument is then made that the environment is so complicated that we must trace over its uncontrolled degrees of freedom which effectively collapses the entangled superposition of Eq. (2) to a mixed state and establishes the quantum/classical boundary. In the example[v] presented at the meeting, the system corresponded to two internal state of an $^{27}\text{Al}^+$ ion, whose states are coupled to two internal states of a $^{25}\text{Mg}^+$ ion, which serves as the meter. The environment states are distinguished by those in which the $^{25}\text{Mg}^+$ ion either scatters or does not scatter many photons (as registered by a nearby photomultiplier tube), correlated with the internal state of the $^{25}\text{Mg}^+$ ion. A number of other experiments in atomic physics and quantum optics conform to this basic scheme. In many of these experiments the entangling operation of Eq. (1) is coherent and can be reversed. Moreover, if the relevant environment is not too complicated, one might argue that inability to reverse the entanglement of Eq. (2) is simply a technical, not fundamental limitation. For example,

---

[u]W. Zurek, *Physics Today* **44**(10), 36 (1991).
[v]P. O. Schmidt *et al.*, *Science* **309**, 749 (2005).

one might store the emitted photons in a cavity such they can be coherently put back into the atoms.[w]

As technology improves, one expects that the system and meter will become progressively larger. For example, atomic physics experiments can already make states (for small N) of the form

$$\Psi = \tfrac{1}{\sqrt{2}}\Big[|\downarrow\rangle_1 |\downarrow\rangle_2 \cdots |\downarrow\rangle_N + |\uparrow\rangle_1 |\uparrow\rangle_2 \cdots |\uparrow\rangle_N \Big]. \tag{3}$$

If we separate atom 1 from the others, and if N can be made very large, the quantum meter composed of atoms 2 through N becomes macroscopic. We would realize a situation like Schrödinger's cat where the state of a microscopic system (atom 1) is entangled with a macroscopic system, whose properties (e.g. orientation of its magnetic dipole) can be viewed classically. This effectively pushes the quantum/classical boundary farther out, and begs the question of how big N must be before this trend breaks down and the coherence in the superposition of Eq. (3) is lost. Many experimentalists would argue that N is limited only by technical problems, whose solutions are difficult but straightforward to solve. However, the trend might break down because of some as-of-yet-unobserved collapse mechanism that acts with small probability on small, simple systems, but with high probability on large, complex systems.[x] Would this depend on the system being macroscopic,[y] and if so, what defines macroscopic? Would it depend on the number elementary constituents in the system, the number of degrees of freedom that are entangled, or the overall spatial exent of the system? Or might the break-down depend on some other criterion; for example, perhaps only apply to systems whose states can be distinguished by humans.[z] It appears that at the moment, these questions cannot be easily answered. Therefore, for experimentalists, it is interesting to expand the conditions under which superposition and entanglement can be observed. Perhaps someday we will observe and confirm a fundamental breakdown in our ability to engineer quantum systems, which might signal the quantum/classical transition.

---

[w]S. Haroche and J.-M. Raimond, *Exploring the Quantum* (Oxford University Press, Oxford, U.K.), 2006.

[x]See, for example A. Bassi and G. Ghirardi, *Phys. Rep.* **379**, 257 (2003).

[y]A. J. Leggett, *J. Phys.: Condens. Matter* **14**, R415 (2002).

[z]One might argue that this is already ruled out in atomic physics where for small numbers of atoms, entanglement can be made manifest but, for example, the fluorescing and non-fluorescing states of even a single ion can be distinguished by the human eye.

Discussion

M. Gell-Mann I just want to make a tiny remark about Schrödinger's cat. The idea, as was just mentioned, is that a quantum change can be coupled to a massive, classical change, like killing a cat or not killing it, or destroying a city with a thermonuclear weapon or not doing so. Any such big event could be controlled by a quantum fluctuation, this is perfectly true. The other part of the argument, though, where the cat is in this one-over-the-square-root-of-two alive cat plus one-over-the-square-root-of-two dead cat wavefunction is absurd. The cat, if it is alive, has to eat, to excrete, and so on; if it is dead, it rots. It is thus interacting heavily with the environment and there is no such thing as a wavefunction for the cat. There is of course a wavefunction for the universe including the cat. In Wineland's experiment I do not know whether this effect is very important. I think that with these magnesium atoms, there is not a massive interaction with the outside world, which makes an enormous difference as to whether there is a suitable local partial wavefunction for the analogue of the cat.

A. Aspect Thank you for this comment. Dave (Wineland), do you want to add something?

D. Wineland I certainly agree with that comment. We have to be careful about isolating the system. Right now, we do not do a great job of isolating these magnesium and aluminum ions. But I viewed it more as a practical problem. We should be able to do it. I don't think that there is anything fundamental there, it is just a technological limitation at present. I do not have any great answers, except to say that we should be able to do it as experimentalists.

D. Kleppner I do not want to beat the Schrödinger cat to death, but aside from the absurdity of writing down a wavefunction for a cat, what we know about the cat is that the cat is as likely to be alive as it is to be dead. This has nothing to say about the cat itself, it is just about our knowledge. I think that from that point of view there is really no paradox at all. This comes back to this question of whether the wavefunction is a statement of our knowledge or a statement about the system itself.

W. Phillips There are so many things to say. If we were to make a macroscopic superposition, that is, a Schrödinger cat, to verify that we had such a thing, it would not just be a question of making something that had an equal probability of being alive and dead. You would have to show the coherence. This would imply that there exists some process that could be applied to the cat such that after that process, we would know for sure that the cat was either alive or dead and that we get the expected result when we measure it. Thus the process of verifying a Schrödinger cat involves both of those things. This is not so easy because of the connection to the environment, but it is one of the things that we are trying to do as experimentalists. On an other subject, I really wanted to ask a question to Wojcieh (Zurek) about the calculations that he described to try to reduce the number of axioms in our postulates in quantum mechanics. Particularly, you talked about a quantum system that was measured by a quantum measuring apparatus. I do not understand what that means because how does the quantum measurement apparatus know what the measurement is, unless someone else or some other process looks at it. For example in the case of Dave (Wineland)'s experiment, his magnesium ion was his quantum measuring apparatus but the thing that really made the measurement was the scattering of the photon. Could you elucidate this point?

W. Zurek Thank you, it is a very good question. It is not a matter of who knows what. What you know is that if you measure the system a second time, it is guaranteed to yield the same result. This is going in particular to constrain interactions in the end; it would mean de facto that if you are measuring a non-demolition observable then it would commute with the appropriate Hamiltonian, and so on. But the conditions come through repeatability. I think that the repeatability is in a sense really relevant — not at the microscopic level because you know how expensive it is to do non-demolition measurements at the levels of photons, atoms, etc — but at the macroscopic level where there is a transition to classicality, because we expect things to be there when we look at them a second time. So what you want to do is translate, and I do not have time to talk about it in detail here, this derivation in the formalism of density matrices, subspaces of Hilbert spaces and so on. This is possible. What you obtain then is a derivation of the fact that you are going to have a distinguished set of results. Things are going to click and if they click, the clicks are repeatable and are distinct, orthogonal to clicks that signify different results.

W. Phillips But somehow a click sounds like something irreversible has happened. If it is a purely quantum system, then I would think that it should be reversible.

W. Zurek All of what I said was reversible. In my derivation I am using explicitly unitarity. I split the collapse axiom in two pieces as I tried to do on the transparencies. The first piece says that you are only able to acquire a set of states which are pre-determined by your measuring apparatus...

W. Phillips ... the pointer states ...

W. Zurek ... and I can prove that these states have to be orthogonal. If the information transfer is going to be such that it guarantees or allows repeatability, they have to be orthogonal. What I have not addressed is the part which Everett addresses for me. I am not completely comfortable with the language of many worlds, but I am quite comfortable with the language, which is also consistent with Everett, of the relative-state interpretation. Basically, this boils down to the idea that if I am in a particular state, the rest of the universe is in a particular state which is consistent with my own records: if I were to imagine myself as one of these apparatuses and got an outcome, say, "A17", then I would be guaranteed that the rest of the universe is in the corresponding state. Now, whether at this point, I am allowed to ask questions about reality of the wavefunction of the universe, etc., I do not know. This goes back to what we discussed before. Is the wavefunction a real state in the sense in which a point in a phase space is a real state? Or is it something in between a real state and information about it, as a density distribution in phase space is?

S. Sachdev I want to address this question of an infinite chain of measurers measuring the measurement apparatus until we reach the entire universe. Is it not more logically clearer to invoke the idea of symmetry-breaking and think of a relatively small apparatus essentially measuring itself? An isolated system because of tiny variations in the initial conditions will collapse effectively even though it obeys Schrödinger equation into one of two possibilities. Basically, that is something like a phase transition and this is all you need. You can terminate your considerations at this level and do not have to keep measuring any further.

A. Aspect This question of symmetry-breaking is a matter of church. In the church of condensed matter physics, they like symmetry-breaking. In the church of quantum optics, it is not so popular.

C. Bunster I have an embarrassingly naive questions for the experts. Do you think that quantum mechanics was working before man appeared on earth?

A. Aspect, For somebody who is a realist like me yes, no doubt. Who wants to answer?

D. Gross For an unrealist yes.

W. Zurek For an undecided, yes.

W. Phillips Absolutely, yes... for whatever I am!

S. Das Sarma I very carefully and respectfully enter your church and ask a question to the experts who worry about the foundations of quantum mechanics — personally, I do not. Bell inequalities were alluded to several times. All the reasonably compelling experiments testing Bell inequalities that I have come across use photons, like Alain Aspect's experiment and the follow-up to that experiment...

A. Aspect Dave Wineland did an experiment with ions.

S. Das Sarma Exactly, so my question is what is the status of these experiments with massive objects? I am familiar with Wineland's experiment, but it is nowhere as compelling as experiments done with photons. However, it would be much better to do Bell inequality tests with massive objects, because a photon is classically a wave and waves are non-local objects classically. These experiments with photons are still very interesting, but they are interesting only if you know all of quantum mechanics. So my question is are there problems in doing Bell inequality tests with massive objects, like uranium atoms or something like that?

A. Aspect Dave Wineland did an experiment with ions, but it is still with internal degrees of freedom. But you can easily design an experiment — I say design, because to do it is another story — where you test a Bell inequality with massive particles entangled with respect to mechanical degrees of freedom, like momentum. I even argue in one of my proposals that doing these experiments with light atoms, like Helium, and doing it with a heavy atom, like Rubidium, would be interesting.

S. Das Sarma But do I understand you right that it has not been done yet?

A. Aspect Not that I know.

A. Zeilinger On this question, I would like to remark first that Bell inequalities have nothing to do with the nature of the underlying system. Bell inequalities can be derived independently of quantum mechanics. It is just a statement about possible measurement results. Now, if I remember correctly there was an early important proton-scattering experiment by Lamehi-Rachti and Mittig, but it is still being discussed how conclusive it was. In term of experimental progress, however, your question is timely because there are a number of interesting experimental proposals to test Bell inequalities with massive objects. There are for instance very realistic ideas on how to put two micro-mechanical levers into an entangled state. This is in my eyes just a question of technological development. There are also ideas to close some of the loopholes. In one experiment, the idea is to put two atoms which are located at a large distance from each other into an entangled state. Weinfurter for example is thinking about this...

A. Aspect ... Chris Monroe also in Maryland...

A. Zeilinger ... Chris Monroe is also thinking about that. These are the two proposals which come to my mind, but they are probably not the only thing. I expect that within the next couple of years, we will have a few experiments like that.

A. Aspect Because you are very worried about this question, Sankar (Das Sarma), there will be an answer Sunday. Now it is time to move to the last part and if he is ready we are going to listen to the next rapporteur John Preskill.

Rapporteur talk by J. Preskill: Quantum Entanglement and Quantum Computing

Abstract

Quantum information science explores the frontier of highly complex quantum states, the "entanglement frontier." This study is motivated by the observation (widely believed but unproven) that classical systems cannot simulate highly entangled quantum systems efficiently, and we hope to hasten the day when well controlled quantum systems can perform tasks surpassing what can be done in the classical world. One way to achieve such "quantum supremacy" would be to run an algorithm on a quantum computer which solves a problem with a super-polynomial speedup relative to classical computers, but there may be other ways that can be achieved sooner, such as simulating exotic quantum states of strongly correlated matter. To operate a large scale quantum computer reliably we will need to overcome the debilitating effects of decoherence, which might be done using "standard" quantum hardware protected by quantum error-correcting codes, or by exploiting the nonabelian quantum statistics of anyons realized in solid state systems, or by combining both methods. Only by challenging the entanglement frontier will we learn whether Nature provides extravagant resources far beyond what the classical world would allow.

1. Introduction: Toward Quantum Supremacy

My assignment is to report on the current status of *quantum information science*, but I will not attempt to give a comprehensive survey of this rapidly growing field. In particular, I will not discuss recent experimental advances, which will be covered by other rapporteurs.

To convey the spirit driving the subject, I will focus on one Big Question:

Can we control complex quantum systems and if we can, so what?

Quantum information science explores, not the frontier of short distances as in particle physics, or of long distances as in cosmology, but rather the frontier of highly complex quantum states, the *entanglement frontier*. I will address whether we can probe deeply into this frontier and what we might find or accomplish by doing so. This Big Question does not encompass everything of interest in quantum information science, but it gets to the heart of what makes the field compelling.

The quantum informationists are rebelling against a fundamental dualism we learned in school:

The macroscopic world is classical.
The microscopic world is quantum.

We fervently wish for controlled quantum systems that are large yet exhibit profoundly quantum behavior. The reason we find this quest irresistible can be stated succinctly:

Classical systems cannot in general simulate quantum systems efficiently.

We cannot yet prove this claim, either mathematically or experimentally, but we have reason to believe it is true; arguably, it is one of the most interesting distinctions ever made between quantum and classical. It means that well controlled large quantum systems may "surpass understanding," behaving in ways we find surprising and delightful.

We therefore hope to hasten the onset of the era of *quantum supremacy*, when we will be able to perform tasks with controlled quantum systems going beyond what can be achieved with ordinary digital computers. To realize that dream, we must overcome the formidable enemy of *decoherence*, which makes typical large quantum systems behave classically. So another question looms over the subject:

*Is controlling large-scale quantum systems merely **really, really hard**, or is it **ridiculously hard**?*

In the former case we might succeed in building large-scale quantum computers after a few decades of very hard work. In the latter case we might not succeed for centuries, if ever.

This question is partly about engineering but it is about physics as well (and indeed the boundary between the two is not clearly defined). If quantum supremacy turns out to be unattainable, it may be due to physical laws yet to be discovered. In any case, the quest for large-scale quantum computing will push physics into a new regime never explored before. Who knows what we'll find?

2. Quantum Entanglement and the Vastness of Hilbert Space

At the core of quantum information science is entanglement, the characteristic correlations among the parts of a quantum system, which have no classical analog. We may imagine a quantum system with many parts, like a 100-page quantum book. If the book were classical, we could read 10 of the pages and learn about 10% of the content of the book. But for a typical 100-page quantum book, if we read 10 pages we learn almost nothing about the content of the book; the information is not printed on the individual pages — rather nearly all the information in the book is encoded in the correlations among the pages. (See Fig. 1.) These correlations are very complex, so that recording a complete classical description of the quantum state would require a classical book of astronomical size.

Does Nature really indulge in such extravagant resources, and how can we verify it?

The issue is subtle. Yes, the Hilbert space of a large quantum system is vast, because the classical description of a typical pure quantum state is enormously long.

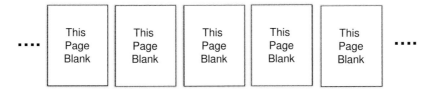

Fig. 1.   For a typical quantum state with many parts, a measurement acting on just one part collects a negligible amount of information about the state.

But we don't really care about typical quantum states, because preparing them is completely infeasible.[1] The only quantum states that are physically relevant are those that can be prepared with reasonable (quantum) resources, which are confined to an exponentially small portion of the full Hilbert space (Fig. 2(a)). Only these can arise in Nature, and only these will ever be within the reach of the quantum engineers as technology advances.

Mathematically, we may model the feasible quantum states this way: Imagine we have n qubits (two-level quantum systems) which are initially in an uncorrelated product state. Then we perform a quantum circuit, a sequence of unitary operations ("quantum gates") acting on pairs of qubits, where the total number of quantum gates is "reasonable," let us say growing no faster than polynomially with n. Equivalently, we may say that a state is feasible if it can be constructed, starting with a product state, by evolving with a local Hamiltonian for a reasonable time. Likewise, we say a measurement is feasible if it can be constructed as a quantum circuit of size polynomial in n, followed by single-qubit measurements.

These quantumly feasible states and measurements are plausibly allowed by Nature. Though far from "typical," they may nevertheless be hard to simulate classically. That is why quantum computing is exciting and potentially powerful.

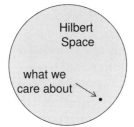

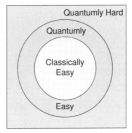

Fig. 2.   (a) Hilbert space is vast, but the quantum states that can be prepared with reasonable resources occupy only a small part of it. (b) We believe that quantum computers can solve some problems that are hard for classical computers, but even quantum computers have limitations.

3. Separating Classical from Quantum

The best evidence for such a separation between quantum and classical complexity comes from quantum algorithms that perform tasks going beyond what we know how to do with classical digital computers (Fig. 2(b)). The most famous examples are Shor's algorithms for finding the prime factors of integers and evaluating discrete logarithms,[2] which are based on using a fast quantum Fourier transform to probe the period of a function.

There are other such "superpolynomial" speedups known, in which the time required to solve a problem scales polynomially with the input size when a quantum computer is used, but faster than polynomially when a classical computer is used. For example, by efficiently simulating topological quantum field theory using a quantum computer, we can evaluate approximately certain topological invariants of links and 3-manifolds (e.g. the Jones polynomial[3,4] or Turaev-Viro invariant[5]). In fact, approximate evaluation of such topological invariants is a *BQP-hard* problem, meaning that any problem that a quantum computer can solve efficiently can be reduced to an instance of the problem of additively approximating the Jones polynomial of a link.

A superpolynomial speedup is also achieved by a quantum algorithm for computing properties of solutions to systems of linear equations.[6] For example, if A is an $N \times N$ Hermitian matrix, and x solves $Ax = b$ where x and b are N-component vectors, then a quantum algorithm can estimate $x^\dagger M x$ in a time scaling like a power of $\log N$, provided $|b\rangle$ is an efficiently preparable quantum state, A is sparse, and M is an efficiently measurable operator. This problem, too, is BQP-hard.

Someday, we hope to probe quantum physics in a previously unexplored regime by running fast quantum algorithms on quantum computers. For this purpose, it is convenient that the problems with superpolynomial speedups include some problems (like factoring) in the class NP, where the solution can be checked efficiently with a classical computer. Running the factoring algorithm, and checking it classically, we will be able to test whether Nature admits quantum processes going beyond what can be classically simulated. (However, this test is not airtight, because we have no proof that factoring is really classically hard.)

While quantum algorithms achieving superpolynomial speedups relative to classical algorithms are relatively rare, those achieving less spectacular polynomial speedups are more common. For example, a quantum computer can perform exhaustive search for a solution to a constraint satisfaction problem in a time scaling like the square root of the classical time,[7] essentially because in quantum theory a probability is the square of an amplitude. By simulating a quantum walk on a graph, a quantum computer can also speed up the evaluation of a Boolean formula,[8] and hence determine, for example, whether a two-player game has a winning strategy. But again the speedup is merely polynomial.

It seems that superpolynomial speedups are possible only for problems with special structure well matched to the power of a quantum computer. We do not expect superpolynomial speedups for the worst-case instances of problems in the NP class, such as 3-SAT or the Traveling Salesman Problem. For such problems with no obvious structure, we might not be able to do better than quadratically speeding up exhaustive search for a solution.[9]

But problems outside the class NP are also potentially of interest. Indeed, the "natural" application for a quantum computer is simulating evolution governed by a local Hamiltonian, preceded by the preparation of a reasonable state and followed by measurement of a reasonable observable.[10] In such cases the findings of the quantum computer might not be easy to check with a classical computer; instead one quantum computer could be checked by another, or by doing an experiment (which is almost the same thing).

As we strive toward the goal of quantum supremacy, it will be useful to gain a deeper understanding of two questions: (1) What quantum tasks are feasible? (2) What quantum tasks are hard to simulate classically? Conceivably, it will turn out that the extravagant exponential resources seemingly required for the classical description and simulation of generic quantum states are illusory; perhaps the quantum states realized in Nature really do admit succinct classical descriptions, either because the laws of physics governing complex quantum systems are different than we currently expect, or because there are clever ways to simulate the quantum world classically that have somehow eluded us so far.

4. Easiness and Hardness

Though we have sound reasons for believing that general quantum computations are hard to simulate classically, in some special cases the simulation is known to be easy. Such examples provide guidance as we seek a path toward quantum supremacy.

Suppose for example, that n qubits are arranged in a line, and consider a quantum circuit such that, for any way of cutting the line into two segments, the number of gates that cross the cut is modest, only logarithmic in n. Then, if the initial state is a pure product state, the quantum state has a succinct classical description at all times, and the classical simulation of the quantum computation can be done efficiently.[11,12] The quantum computation does not achieve a super-classical task, because the quantum state becomes only slightly entangled.

Correspondingly, if you receive multiple copies of an n-qubit state that is only slightly entangled, you would be able to identify the state with a feasible number of measurements. In general, quantum state tomography is hard — Hilbert space is so large that a number of measurements exponential in n would be required to determine a typical n-qubit state. But for a slightly entangled state, a number of measurements linear in n suffices.[13] We can perform tomography on segments of constant size, then do an efficient classical computation to determine how the pieces are stitched together.

Gaussian quantum dynamics is also easy to simulate.[14] Consider an interferometer assembled from linear optical elements, which can be described by a Hamiltonian quadratic in bosonic creation and annihilation operators. Suppose that a Gaussian initial state (a coherent state, for example) enters the input ports, and that we measure quadrature amplitudes at the output ports. Then the state has a succinct description at all times and can be simulated classically. But if we introduce some optical nonlinearity, or single photon sources together with adaptive photon counting measurements, then this system has the full power of a universal quantum computer, and presumably it cannot be simulated classically.[15,16] Here "adaptive" means that subsequent operations can be conditioned on the outcomes of earlier measurements.

Free fermions are likewise easy to simulate classically, and, in contrast to free bosons, adaptive measurements of the fermion mode numbers do not add computational power.[17,18] But if we add four-fermion operators to the Hamiltonian, or if we allow nondestructive measurements of four-fermion operators, then universal quantum computation is achievable.[19]

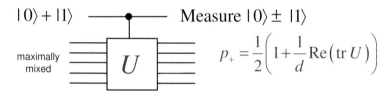

Fig. 3.   The trace of a large matrix can be computed in the "one-clean-qubit" model of quantum computation, for which the input is one pure qubit and many maximally mixed qubits.

Fig. 4.   Two quantum systems that may be hard to simulate classically. (a) A quantum circuit with commuting gates. (b) Nonadaptive linear optics with photon sources and photon detectors.

Some computational models, though apparently weaker than the full blown quantum circuit model, nevertheless seem to have surprising power. One intriguing case is the "one-clean-qubit model", in which the input to the computation is one qubit in a pure state and many qubits in a maximally mixed state;[20] see Fig. 3. The study of this model was motivated initially by the nuclear-magnetic-resonance approach to quantum computing, where the initial state may be highly mixed.[21,22] The one-clean-qubit quantum computer can evaluate the trace of an exponentially large unitary operator if the operator can be realized by an efficient quantum circuit. This capability can be exploited to approximate the Jones polynomial of the trace closure of a braid[23] or the Turaev-Viro invariant of a three-dimensional mapping torus,[24] problems for which no efficient classical algorithms are known; in fact these problems are complete for the one-clean-qubit class.

Another provocative example is the "instantaneous quantum computing" model.[25] Here all the gates executed by our quantum computer are mutually commuting, simultaneously diagonal in the standard Z basis. In addition we can prepare single qubits in eigenstates of the conjugate operator X, and measure qubits in the X basis; see Fig. 4(a). (Because all the gates commute, in principle they can be executed simultaneously.) It is not obvious how to simulate this simple quantum circuit classically, and there is evidence from complexity theory that the simulation is actually hard.[25] Even though the model does not seem to have the full power of universal quantum (or even classical) computing, nevertheless it may in a sense perform a super-classical task.

Yet another tantalizing case is linear optics accompanied by photon sources and photon detectors, but now without adaptive measurements; see Fig. 4(b). Suppose we have m optical modes, where initially $n < m$ are occupied by single photons and the rest are empty. A linear optics array mixes the m modes, and then a measurement is performed to see which of the output modes are occupied. Though this system is not a universal quantum computer, we do not know how to simulate it classically, and there is evidence from complexity theory that the simulation is hard.[26]

Such examples illustrate that there may be easier ways to achieve quantum supremacy than by operating a general purpose quantum computer. Admittedly, though, this linear optics experiment is still not at all easy — to reach the regime where digital simulation is currently infeasible one should detect a coincidence of about 30 photons, whose paths through the interferometer can interfere. Furthermore, it is not clear how the hardness of simulating this system classically would be affected by including realistic noise sources, such as photon loss.

5. Local Hamiltonians

An important task that a quantum computer can perform efficiently is simulating the dynamics of a quantum system governed by a local Hamiltonian H.[27] By "local" I do not necessarily mean *geometrically* local in some spatial dimension; instead, I mean that the Hilbert space has a decomposition into qubits (or other small systems), and H can be expressed as a sum of terms, each of which acts on a constant number of qubits (independent of the system size). More generally, the simulation is feasible if H is a sparse matrix.[28]

This capability can be exploited to measure the energy of the system, as in Fig. 5. The quantum circuit shown evolves an initial state $|\psi\rangle$ for a time t stored in an auxiliary register, then performs a quantum Fourier transform and reads out the register to sample from the frequency spectrum of the operator e^{-iHt}, a procedure called *phase estimation*.[29] The accuracy of the measured eigenvalue, in accord with the energy-time uncertainty relation, is inversely proportional to the maximal evolution time; hence, for an n-qubit system, accuracy scaling like an inverse polynomial in n can be achieved by a quantum circuit with size polynomial in n.

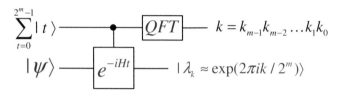

Fig. 5. The energy of a system governed by a local Hamiltonian can be measured efficiently by a quantum computer, using a procedure called "phase estimation."

If the initial state $|\psi\rangle$ has an overlap with the ground state of H which is not smaller than inverse polynomial in n, it follows that we can measure the ground-state energy to inverse polynomial accuracy in polynomial time using a quantum computer. This algorithm has noteworthy applications; for example, a quantum computer can compute the ground-state energy of a large molecule.[30]

But there is a catch — preparing an initial state that overlaps substantially with the ground state could be very hard in some cases. This is already true classically; finding the ground state of a classical spin glass is NP-hard, as hard as any problem whose solution can be checked efficiently by a classical computer. Finding the ground state for a quantum system with a local Hamiltonian seems to be even harder; it is QMA-hard,[31] as hard as any problem whose solution can be checked efficiently by a quantum computer, and we expect that QMA is a larger class than NP. Surprisingly, computing the ground-state energy seems to be a hard problem for a quantum computer even for the case of a geometrically local translationally-invariant quantum system in one dimension.[32]

A general procedure for preparing ground states is adiabatic evolution. We can prepare a state having sizable overlap with the ground state of H by starting with the easily prepared ground state of a simpler Hamiltonian $H(0)$, then slowly deforming the Hamiltonian along a path $H(s)$ connecting $H(0)$ to $H(1) = H$. This procedure succeeds in polynomial time provided the energy gap $\Delta(s)$ between the ground and first excited states of $H(s)$ is no smaller than inverse polynomial in n for all $s \in [0, 1]$ along the path. For problem instances that are quantumly hard, then, the gap becomes superpolynomially small somewhere along the path.[33]

Though the general problem is quantumly hard, we may surmise that there are many local quantum systems for which computing the ground-state energy is quantumly easy yet classically hard. Furthermore, a quantum computer may be able to simulate the evolution of excited states in cases where the simulation is classically hard, such as chemical reactions[34] or the scattering of particles described by quantum field theory.[35] Even in the case of quantum gravity, evolution may be governed by a local Hamiltonian, and therefore admit efficient simulation by a quantum computer.

6. Quantum Error Correction

Classical digital computers exist, and have had a transformative impact on our lives. Large-scale quantum computers do not yet exist. Why not?

Building reliable quantum hardware is challenging because of the difficulty of controlling quantum systems accurately. Small errors in quantum gates accumulate in a large circuit, eventually leading to large errors that foil the computation. Furthermore, qubits in a quantum computer inevitably interact with their surroundings; decoherence arising from unwanted correlations with the environment is harmless in a classical computer (and can even be helpful, by introducing friction which impedes accidental bit flips), but decoherence in a quantum computer can irreparably damage the delicate superposition states processed by the machine.

Quantum information might be better protected against noise by using a quantum error-correcting code, in which "logical" information is encoded redundantly in a block of many physical qubits.[36,37] Quantum error correction is in some ways much like classical error correction, but more difficult, because while a classical code need only protect against bit flips, a quantum code must protect against both bits flips and phase errors.

Suppose for example, that we want to encode a single logical qubit, with orthonormal basis states denoted $|0\rangle$ and $|1\rangle$, which is protected against all the errors spanned by a set $\{E_a\}$. For the distinguishability of the basis states to be maintained even when errors occur, we require

$$E_a|0\rangle \perp E_b|1\rangle, \tag{1}$$

where E_a, E_b are any two elements of the error basis. This condition by itself would suffice for reliable storage of a classical bit.

But for storage of a qubit we also require protection against phase errors, which occur when information about whether the state is $|0\rangle$ or $|1\rangle$ leaks to the environment; equivalently, distinguishability should be maintained for the dual basis states $|0\rangle \pm |1\rangle$:

$$E_a\left(|0\rangle + |1\rangle\right) \perp E_b\left(|0\rangle - |1\rangle\right), \tag{2}$$

where E_a, E_b are any two errors. In fact, the two distinguishability conditions Eq. (1) and (2) suffice to ensure the existence of a recovery map that corrects any error spanned by $\{E_a\}$ acting on any linear combination of $|0\rangle$ and $|1\rangle$.[38]

Together, Eq. (1) and (2) imply

$$\langle 0|E_a^\dagger E_b|0\rangle = \langle 1|E_a^\dagger E_b|1\rangle; \tag{3}$$

no measurement of any operator in the set $\{E_a^\dagger E_b\}$ can distinguish the two basis states of the logical qubit. Typically, because we expect noise acting collectively on many qubits at once to be highly suppressed, we are satisfied to correct *low-weight* errors, those that act nontrivially on a sufficiently small fraction of all the qubits in the code block. Then Eq. (3) says that all the states of the logical qubit look the same when we examine a small subsystem of the code block. These states are highly entangled, like the hundred-page book that reveals no information when we read the individual pages.

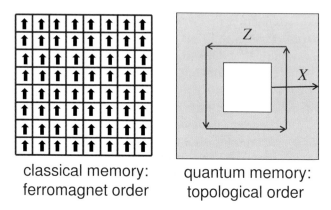

classical memory:
ferromagnet order

quantum memory:
topological order

Fig. 6.   A prototypical classical memory is a ferromagnet, and a prototypical quantum memory is a topologically ordered medium.

It is useful to formulate the distinction between classical and quantum error correction in more physical terms (see Fig. 6). The prototype for a protected classical memory is a ferromagnet, where a single bit is encoded according to whether most of the spins are up or down. The encoded bit can be read out by performing local measurements on all spins, and then executing a majority vote to protect against errors that flip a minority of the spins. Errors in the memory create domain walls where neighboring spins misalign, and a logical error occurs when a domain wall sweeps across the sample, inducing a global operation acting on many spins. The memory is robust at a sufficiently small nonzero temperature because the energy cost of a large droplet of flipped spins is large. This memory is a particularly simple physically motivated example of a classical error-correcting code; there are more sophisticated examples.

The prototype for a protected quantum memory is a medium in two dimensions with Z_2 topological order.[39] We may consider a planar sample with a large hole in the middle. In contrast to the ferromagnet, errors in the medium create point-like quasiparticles ("anyons") rather than domains walls. There are two types of anyons (which we may regard as "electric" and "magnetic" excitations), having Z_2 Aharonov-Bohm interactions with one another. The space of quantum states with no particles present is two-dimensional — this space is the encoded qubit. Logical errors can be induced by the transport of particles; a logical X acts on the encoded qubit if an electric particle travels between the inner and outer boundaries of the sample, and a logical Z error acts if a magnetic particle travels around the hole. Correspondingly, we read out the logical qubit in the X basis by measuring a nonlocal string-like operator which connects the inner and outer boundaries, simulating the propagation of an electric particle, while we read it out in the Z basis by measuring a string operator that encloses the hole in the sample, simulating the propagation of a magnetic particle.

The system is protected by a nonzero energy gap, the energy cost of creating a pair of particles. Hence the storage time is long if the temperature is small compared to the gap, but unlike the case of a two-dimensional ferromagnet the storage time does not improve as the system size increases. However, if we monitor the particles as they diffuse through the sample, then a logical error occurs only if particles propagate across the sample without being noticed, an event which *does* become increasingly unlikely as the system size grows.[40]

A topologically ordered medium on a topologically nontrivial surface is a special type of quantum error-correcting code, one that can be realized as the ground state of a system with a geometrically local Hamiltonian; in this respect its status is similar to that of the ferromagnet in classical coding theory. The locality of the Hamiltonian has advantages. For one, we might be able to realize a relatively robust quantum memory described by a Hamiltonian in the universality class of the code. From a more abstract viewpoint, we can collect information about the errors in the code block by making localized measurements, e.g. by identifying domain walls in the ferromagnet or quasiparticle excitations (anyons) in the topologically ordered medium.

7. Scalable Quantum Computing

The theory of quantum error correction establishes that quantum computing is "scalable" in principle. This means that, if the noise strength is below a critical value (the "accuracy threshold"), then we can simulate an ideal quantum circuit accurately using a circuit of noisy gates, with a reasonable overhead cost in additional gates and additional qubits.[41–45] The numerical value of the threshold, and the overhead cost, depend on the fault-tolerant scheme used and on how we model the noise.

Engineering considerations favor a two-dimensional layout with short-range interactions among the qubits, for which the computation can be protected against noise by using a topological code like the one described in Sec. 6. A topological medium can be simulated using any convenient type of quantum hardware, with the physical qubits carried by, for example, trapped ions, electron spins in quantum dots, or superconducting circuits. To encode many logical qubits, the simulated medium has many holes, and logical errors are suppressed by ensuring that the holes are sufficiently large and distantly separated from one another. A complete set of universal quantum gates can be executed on the encoded qubits; hence arbitrary quantum circuits can be simulated efficiently and reliably.[40,46]

There are many challenges to making large-scale fault-tolerant quantum computing practical, including serious systems engineering issues. There are also issues of principle to consider, such as, what is required for a fault-tolerant scheme to be scalable, and what conditions must be satisfied by the noise model? One essential requirement is some form of cooling, to extract the entropy introduced by noise.[47] Parallel operations are also necessary, so noise can be controlled in different parts of the computer simultaneously.

It is natural to describe noise using a Hamiltonian that includes a coupling between the system and its unobserved environment, and proofs of scalability require the noise to be suitably local. For example, we may write the noise Hamiltonian as a sum of terms, each acting on just a few of the physical qubits in the quantum computer, but possibly acting on the environment in a complicated way. Then the proof of scalability applies if each such term in the noise Hamiltonian has a sufficiently small norm.[44,48,49] If the noise Hamiltonian includes terms that act on $k \gg 1$ qubits in the quantum computer (and in some complicated way on the environment), the proof of scalability works if these terms decay exponentially with k, and also decay rapidly enough as the qubits separate in space. A drawback of such scalability criteria is that the condition on the noise is not expressed in terms of directly measurable properties; an advantage is that the state and dynamics of the environment need not be specified.

Alternatively, we may suppose that the environment is described by a Gaussian free field, so the noise can be completely characterized by its two-point correlation function. Then the proof of scalability goes through if the noise is sufficiently weak, with correlations decaying sufficiently rapidly in both time and space.[50] This criterion has the advantage that it is expressed in terms of measurable quantities, but it applies only for if the initial state and the dynamics of the environment obey suitable restrictions.

Thus quantum error correction works in principle for noise that is sufficiently weak and not too strongly correlated, but may fail if the noise acts collectively on many qubits at once. As quantum hardware continues to advance, it will be important to see whether the noise in actual devices has adequately weak correlations, keeping in mind that there are possible ways to suppress correlations, for example by using dynamical decoupling sequences.[51]

8. Topological Quantum Computing

To a theorist, a particularly appealing and elegant way to achieve fault-tolerant quantum computing is by using the exotic statistics of nonabelian anyons.[39,52,53] Quantum information, stored in the exponentially large fusion Hilbert space of n anyons, is well protected if the temperature is low compared to the energy gap (to prevent unwanted thermal production of anyon pairs) and if the anyons are kept far apart from one another (to prevent unwanted nontopological interactions due to quantum tunneling). Robust information processing can be achieved by exchanging the particles, exploiting their exotic quantum statistics, and information can be read out by measuring charges of anyon pairs (for example, using an interferometer[54,55]).

An early proposal for achieving quantum computing with anyons was based on fractional quantum Hall states;[56,57] more recent proposals exploit exotic properties of topological superconductors and topological insulators.[58–64] In most such proposals, the anyon braiding by itself is not sufficient for universal quantum computing, but can be supplemented by unprotected (and possibly quite noisy) nontopological operations to realize a universal gate set.[65] Indeed, in some cases[64] braiding of anyons can be modeled faithfully by a time-dependent free-fermion Hamiltonian; therefore, the nonuniversality of braiding operations follows from the observation that free-fermion systems can be simulated classically, together with the presumption that efficient classical simulations of general quantum circuits are impossible.

Since the error rate is suppressed by the energy gap for anyon pair creation, and does not improve as the system size increases, we may anticipate that for very large-scale applications topological quantum computing will need to be supplemented by "standard" methods of quantum error correction. However, if topological protection enforces a very low gate error rate, the overhead cost of using quantum error-correcting codes may be relatively modest.

Classical information in a ferromagnet is protected "passively," because memory errors occur only when the system surmounts an energy barrier whose height increases sharply with system size. Could there be topologically ordered quantum systems that likewise store quantum information passively, providing a mechanism for a "self-correcting" quantum memory?[66] Models realizing this vision are known in four spatial dimensions.[40,67,68] A recently discovered three-dimensional quantum model has a barrier height increasing logarithmically with system size, but for this model the storage time is bounded above, and declines once the system grows beyond an optimal size.[69,70]

9. Quantum Computing vs. Quantum Simulation

One of the most important applications for quantum computing will be simulating highly entangled matter such as quantum antiferromagnets, exotic superconductors, complex biomolecules, bulk nuclear matter, and spacetime near singularities. A general purpose quantum computer could function as a "digital" quantum simulator, in contrast to "analog" quantum simulators based on customizable systems of (for example) ultracold atoms or molecules. The goal of either digital or analog quantum simulation should be achieving quantum supremacy, i.e. learning about quantum phenomena that cannot be accurately simulated using classical systems. In particular, we hope to discover new and previously unsuspected phenomena, rather than just validate or refute predictions made by theorists.

A universal quantum computer will be highly adaptable, capable of simulating efficiently any reasonable physical system, while analog quantum simulators have intrinsic limitations. In particular, it is not clear to what degree the classical hardness hinges on the accuracy of the simulation, and present day quantum simulators, unlike the universal quantum computers of the future, are not fault tolerant. On the other hand, analog quantum simulators may be able to probe, at least qualitatively, exotic quantum phenomena that are sufficiently robust and universal as to be studied without tuning the Hamiltonian precisely. Furthermore, since the characteristic imperfections in analog quantum simulations vary from one experimental platform to another, obtaining compatible results using distinct simulation methods will boost confidence in the results.

10. Conclusions and Questions

I have emphasized the goal of quantum supremacy (super-classical behavior of controllable quantum systems) as the driving force behind the quest for a quantum computer, and the idea of quantum error correction as the basis for our hope that scalable quantum computing will be achievable. To focus the talk, I have neglected other deeply engaging themes of quantum information science, such as quantum cryptography and the capacities of quantum channels. Also, I have not discussed the impressive progress in building quantum hardware, a topic covered by other rapporteurs. I'll conclude by raising a few questions posed or suggested in the preceding sections.

Regarding quantum supremacy, might we already have persuasive evidence that Nature performs tasks going beyond what can be simulated efficiently by classical computers? For example, there are many mathematical questions we cannot answer concerning strongly correlated materials and complex molecules, yet Nature provides answers; have we failed because these problems are intrinsically hard classically, or because of our lack of cleverness so far?

Is quantum simulation (e.g. with cold atoms and molecules) a feasible path to quantum supremacy? Or will the difficulty of controlling these systems precisely prevent us from performing super-classical tasks?

How can we best achieve quantum supremacy with the relatively small systems that may be experimentally accessible fairly soon, systems with of order 100 qubits? In contemplating this issue we should keep in mind that such systems may be too small to allow full blown quantum error correction, but also on the other hand that a super-classical device need not be capable of general purpose quantum computing.

Regarding quantum error correction, what near-term experiments studying noise in quantum hardware will strengthen the case that scalable fault-tolerant quantum computing is feasible? What pitfalls might thwart progress as the number of physical qubits scales up?

Do the observed properties of topologically ordered media such as fractional quantum Hall systems and topological superconductors already provide strong evidence that highly robust quantum error-correcting codes are physically realizable? How much more persuasive will this evidence become if and when the exotic statistics of nonabelian anyons can be confirmed directly?

Which is a more promising path toward scalable quantum computing: topological quantum computing with nonabelian anyons, or fault-tolerance based on standard qubits and quantum error-correcting codes? Will the distinction between these two approaches fade as hardware advances?

Can a quantum memory, like a classical one, be self-correcting, with storage time increasing as the system grows? Can quantum information protected by self-correcting systems be processed efficiently and reliably?

How might quantum computers change the world? Predictions are never easy, but it would be especially presumptuous to believe that our limited classical minds can divine the future course of quantum information science. Attaining quantum supremacy and exploring its consequences will be among the great challenges facing 21st century science, and our imaginations are poorly equipped to envision the scientific rewards of manipulating highly entangled quantum states, or the potential benefits of advanced quantum technologies. As we rise to the call of the entanglement frontier, we should expect the unexpected.

Acknowledgments

I am grateful to the organizers for the opportunity to attend this exciting meeting. This work was supported in part by NSF grant PHY-0803371, DOE grant DE-FG03-92-ER40701, and NSA/ARO grant W911NF-09-1-0442.

References

1. D. Poulin, A. Qarry, R. Somma, and F. Verstraete, Quantum simulation of time-dependent Hamiltonians and the convenient illusion of Hilbert space, *Phys. Rev. Lett.* **106**, 170501 (2011).
2. P. W. Shor, Algorithms for quantum computation: discrete logarithms and factoring, *Proc. 35th Symp. Foundations of Computer Science*, 124-134 (1994).
3. M. H. Freedman, M. Larsen, and Z. Wang, A modular functor which is universal for universal quantum computation, *Comm. Math. Phys.* **227**, 605-622 (2002).
4. D. Aharonov, V. Jones, and Z. Landau, A polynomial quantum algorithm for approximating the Jones polynomial, *Algorithmica* **55**, 395-421 (2009).
5. G. Alagic, S. P. Jordan, R. König, and B. W. Reichardt, Estimating Turaev-Viro three-manifold invariants is universal for quantum computation, *Phys. Rev. A* **82**, 040302 (2010).
6. A. W. Harrow, A. Hassidim, and S. Lloyd, Quantum algorithm for linear systems of equations, *Phys. Rev. Lett.* **103**, 150502 (2009).
7. L. K. Grover, A fast quantum mechanical algorithm for database search, STOC '96: *Proceedings of the 28th Ann. ACM Symp. Theory of Computing* (1996).
8. E. Farhi, J. Goldstone, and S. Gutmann, A quantum algorithm for the Hamiltonian NAND tree, *Theory of Computing* **4**, 169-190 (2008).
9. C. H. Bennett, E. Bernstein, G. Brassard, and U. Vazirani, Strengths and weaknesses of quantum computing, *SIAM J. Computing*, **26**, 1510-1523 (1997).
10. R. P. Feynman, Simulating physics with computers, *Int. J. Theoret. Phys.* **21**, 467-488 (1982).
11. G. Vidal, Efficient classical simulation of slightly entangled quantum computations, *Phys. Rev. Lett.* **91**, 147902 (2003).

12. R. Jozsa, On the simulation of quantum circuits, arXiv:quant-ph/0603163 (2006).

13. M. Cramer, M. B. Plenio, S. T. Flammia, R. Somma, D. Gross, S. D. Bartlett, O. Landon-Cardinal, D. Poulin, and Y.-K. Liu, Efficient quantum state tomography, *Nature Communi.* **1**, 149 (2010).

14. S. D. Bartlett and B. C. Sanders, Requirement for quantum computation, *J. Mod. Opt.* **50**, 2331-2340 (2003).

15. E. Knill, R. Laflamme, and G. J. Milburn, A scheme for efficient quantum computation with linear optics, *Nature* **409**, 46-52 (2001).

16. D. Gottesman, A. Kitaev, and J. Preskill, Encoding a qubit in an oscillator, *Phys. Rev. A* **64**, 012310 (2001).

17. L. G. Valiant, Quantum computers that can be simulated classically in polynomial time, *STOC '01: Proceedings of the 33rd Ann. ACM Symp. Theory of Computing* (2001).

18. B. Terhal and D. P. DiVincenzo, Classical simulation of noninteracting-fermion quantum circuits, *Phys. Rev. A* **65**, 032325 (2002).

19. S. B. Bravyi and A. Yu. Kitaev, Fermionic quantum computation, *Ann. Phys.* **298**, 210-226 (2002).

20. E. Knill and R. Laflamme, Power of one bit of quantum information, *Phys. Rev. Lett.* **81**, 5672-5675 (1998).

21. N. A. Gershenfeld and I. L. Chuang, Bulk spin-resonance quantum computation, *Science* **275**, 350-356 (1997).

22. D. G. Cory, A. F. Fahmy, and T. F. Havel, Ensemble quantum computing by NMR spectroscopy, *Proc. Nat. Acad. Sci.* **95**, 1634-1639 (1997).

23. P. Shor and S. P. Jordan, Estimating Jones polynomials is a complete problem for one clean qubit, *Quant. Inf. Comput.* **8**, 681 (2008).

24. S. P. Jordan and G. Alagic, Approximating the Turaev-Viro invariant of mapping tori is complete for one clean qubit, arXiv:1105.5100 (2011).

25. M. J. Bremner, R. Jozsa, and D. J. Shepherd, Classical simulation of commuting quantum computations implies collapse of the polynomial hierarchy, *Proc. R. Soc. A* **467**, 459-472 (2011).

26. S. Aaronson and A. Arkhipov, The computational complexity of linear optics, arXiv:1011.3245 (2010).

27. S. Lloyd, Universal quantum simulators, *Science* **273**, 1073-1078 (1996).

28. D. Aharonov and A. Ta-Shma, Adiabatic quantum state generation and statistical zero knowledge, *STOC '03: Proceeding of the 35th ACM Symp. Theory of Computing* (2003).

29. A. Yu Kitaev, Quantum measurements and the Abelian stabilizer problem, arXiv:quant-ph/9511026 (1995).

30. A, Aspuru-Guzik, A. D. Dutoi, P. J. Love, and M. Head-Gordon, Simulated quantum computation of molecular energies, *Science* **309**, 1704 (2005).

31. A. Yu. Kitaev, A. Shen, and M. N. Vyalyi, *Classical and Quantum Computation* (American Mathematical Society, 2002).

32. D. Gottesman and S. Irani, The quantum and classical complexity of translationally invariant tiling and Hamiltonian problems, *Proceedings of 50th Symp. Foundations of Computer Science*, 95-105 (2009).

33. E. Farhi, J. Goldstone, S. Gutmann, and M. Sipser, Computation by adiabatic evolution, arXiv:quant-ph/0001106 (2000).

34. I. Kassal, S. P. Jordan, P. J. Love, M. Mohseni, and A. Aspuru-Guzik, Polynomial-time quantum algorithm for the simulation of chemical dynamics, *Proc. Nat. Acad. Sci.* **105**, 18681-18686 (2008).

35. S. P. Jordan, K. S. M. Lee, and J. Preskill, Quantum algorithms for quantum field theories, arXiv:1111.3633 (2011).

36. P. Shor, Scheme for reducing decoherence in quantum memory, *Phys. Rev. A* **52**, R2493-R2496 (1995).

37. A. M. Steane, Error correcting codes in quantum theory, *Phys. Rev. Lett.* **77**, 793-797 (1995).

38. E. Knill and R. Laflamme, Theory of quantum error-correcting codes, *Phys. Rev. A* **55**, 900-911 (1997).

39. A. Yu. Kitaev, Fault-tolerant quantum computation by anyons, *Ann. Phys.* **303**, 2-30 (2003).

40. E. Dennis, A. Kitaev, A. Landahl, and J. Preskill, Topological quantum memory, *J. Math. Phys.* **43**, 4452 (2002).

41. D. Aharonov and M. Ben-Or, Fault-tolerant quantum computation with constant error, *STOC '97: Proceedings of the 29th Ann. ACM Symp. Theory of Computing* (1997).

42. A. Yu. Kitaev, Quantum computations: algorithms and error correction, Russian *Math. Surveys* **52**, 1191-1249 (1997).

43. R. Laflamme, E. Knill, and W. Zurek, Resilient quantum computation: error models and thresholds, *Proc. R. Soc. Lond. A* **454**, 365-384 (1998).

44. P. Aliferis, D. Gottesman, and J. Preskill, Quantum accuracy threshold for concatenated distance-3 codes, *Quant. Inf. Comput.* **6**, 97-165 (2006).

45. B. W. Reichardt, Fault-tolerance threshold for a distance-three quantum code, *Lecture Notes in Computer Science* **4051**, 50-61 (2006).

46. R. Raussendorf and J. Harrington, Fault-tolerant quantum computation with high threshold in two dimensions, *Phys. Rev. Lett.* **98**, 190504 (2007).

47. D. Aharonov, M. Ben-Or, R. Impagliazzo, and N. Nisan, Limitations of noisy reversible computation, arXiv:quant-ph/9611028 (1996).

48. B. Terhal and G. Burkhard, Fault-tolerant quantum computation for local non-Markovian noise, *Phys. Rev. A* **71**, 012336 (2005).

49. D. Aharonov, A. Kitaev, and J. Preskill, Fault-tolerant quantum computation with long-range correlated noise, *Phys. Rev. Lett.* **96**, 050504 (2006).

50. H.-K. Ng and J. Preskill, Fault-tolerant quantum computation versus Gaussian noise, *Phys. Rev. A* **79**, 032318 (2009).

51. L. Viola, E. Knill, and S. Lloyd, Dynamical decoupling of open quantum systems, *Phys. Rev. Lett.* **82**, 2417-2421 (1999).

52. R. W. Ogburn and J. Preskill, Topological quantum computation, *Lecture Notes in Computer Science* **1509**, 341-356 (1999).

53. M. H. Freedman, A. Kitaev, M. J. Larsen, and Z. Wang, Topological quantum computation, *Bull. AMS* **40**, 31-38 (2002).

54. A. Stern and B. I. Halperin, Proposed experiments to probe the non-Abelian nu=5/2 quantum Hall state, *Phys. Rev. Lett.* **96**, 016802 (2006).

55. P. Bonderson, A. Kitaev, and K. Shtengel, Detecting non-Abelian statistics in the nu=5/2 fractional quantum Hall state, *Phys. Rev. Lett.* **96**, 016803 (2006).

56. S. Das Sarma, M. Freedman, and C. Nayak, Topologically protected qubits from a possible non-Abelian fractional quantum Hall state, *Phys. Rev. Lett.* **94**, 166802 (2005).

57. C. Nayak, S. H. Simon, A. Stern, M. Freedman, and S. Das Sarma, Non-Abelian anyons and topological quantum computation, *Rev. Mod. Phys.* **80**, 1083-1159 (2008).

58. A. Yu. Kitaev, Unpaired Majorana fermions in quantum wires, *Physics-Uspekhi* **44**, 131 (2001).

59. L. Fu and C. Kane, Superconducting proximity effect and Majorana fermions at the surface of a topological insulator, *Phys. Rev. Lett.* **100**, 096407 (2008).

60. J. D. Sau, R. M. Lutchyn, S. Tewari, and S. Das Sarma, Generic new platform for topological quantum computation using semiconductor heterostructures, *Phys. Rev. Lett.* **104**, 040502 (2010).

61. J. Alicea, Majorana fermions in a tunable semiconductor device, *Phys. Rev. B* **81**, 125318 (2010).

62. R. M. Lutchyn, J. D. Sau, and S. Das Sarma, Majorana Fermions and a topological phase transition in semiconductor-superconductor heterostructures, *Phys. Rev. Lett.* **105**, 077001 (2010).

63. Y. Oreg, G. Refael, and F. von Oppen, Helical liquids and Majorana bound states in quantum wires, *Phys. Rev. Lett.* **105**, 177002 (2010).

64. J. Alicea, Y. Oreg, G. Refael, F. von Oppen, M. P. A. Fisher, Non-Abelian statistics and topological quantum information processing in 1D wire networks, *Nature Physics* **7**, 412-417 (2011).

65. S. Bravyi and A. Kitaev, Universal quantum computation with ideal Clifford gates and noisy ancillas, *Phys. Rev. A* **71**, 022316 (2005).

66. D. Bacon, Operator quantum error-correcting subsystems for self-correcting quantum memories, *Phys. Rev. A* **73**, 012340 (2006).

67. R. Alicki, M. Horodecki, P. Horodecki, and R. Horodecki, On thermal stability of topological qubit in Kitaev's 4D model, *Open Syst. Inf. Dyn.* **17** (2010).

68. S. Chesi, D. Loss, S. Bravyi, and B. Terhal, Thermodynamic stability criteria for a quantum memory based on stabilizer and subsystem codes, *New J. Phys.* **12**, 025013 (2010).

69. J. Haah, Local stabilizer codes in three dimensions without string logical operators, *Phys. Rev. A* **83**, 042330 (2011).

70. S. Bravyi and J. Haah, Energy landscape of 3D spin Hamiltonians with topological order, *Phys. Rev. Lett.* **107**, 150504 (2011).

Prepared comment

A. Zeilinger: Experiments on the Foundations of Quantum Physics and on Quantum Information

It is evident that experiments on the foundations of quantum mechanics were crucial to give rise to the field of quantum information science.

Immediately after the invention of modern quantum mechanics, discussions about its conceptual implications started. These are most interesting for individual quantum systems. Most significant was the debate between Bohr and Einstein which took place mainly at the Solvay Congresses in 1927 and 1930. Einstein proposed many elegant gedanken experiments, as real experiments with individual particles were not yet possible. Einstein's goal was to defend a realistic world view. Bohr was always able to dismiss Einstein's attacks by taking quantum mechanics completely into account.

Beginning in the 1960s and 1970s, technological progress made experiments with individual particles possible. Interference experiments with matter waves were performed, initially with electrons and neutrons (Fig. 1), the latter being less susceptible to electromagnetic noise and having lower energy.

For neutrons, I mention explicitly the first demonstration of the fact that a 2π rotation of a spin-$^1/_2$ system leads to a phase factor of -1. This is the first demonstration of the phase factor which results from a complete Rabi cycle between two states and which became essential in many quantum computation tasks. Also with neutrons, the first quantitative decoherence experiments with matter waves were performed.

Providing the first realization of individual quantum phenomena, such experiments also led to precision tests of quantum mechanics, e.g. of the linearity of the Schrödinger equation. The relevance of the linearity of the Schrödinger equation stems from the fact that many of the conceptual puzzles, like the Schrödinger cat paradox, stem exactly from quantum superposition, a consequence of the linearity of evolution.

In matter waves, some of the next steps included atomic interferometers of atoms, which laid the groundwork for Bose-Einstein-condensates. Also, interference experiments with fullerenes (Fig. 1) opened up the avenue for realizing quantum phenomena of increasingly large objects. With fullerenes, also the first precision tests of quantum decoherence were performed.

For photons, I would like to mention an experiment by Clauser in the 1970s which, by observing non-classical correlations that cannot be explained by a semi-classical approach, confirmed the photon concept, and the experiment by Grangier and Aspect where they for the first time observed real single photon interference.

Following Bell's theorem, many experiments were performed to confirm the quantum mechanical predictions for entangled states. Today, a local realistic

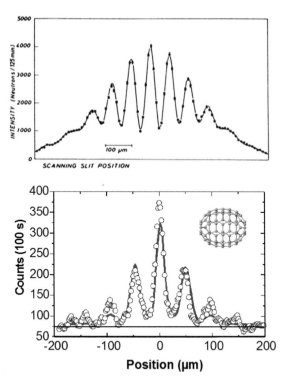

Fig. 1.   Double-slit interference of neutrons (top). The intensities were such that at most one neutron at a time passed through the double-slit assembly. The same statement can be made for the interference of fullerenes (bottom).

world view can only be upheld by taking advantage of so-called loopholes in existing experiments. All loopholes have been closed, but no experiment exists which is completely loophole-free.

There have been other proposals which pointed out the conflict between classical realism and quantum mechanics. Leggett's nonlocal model, which can explain all Bell experiments, is not in conflict with relativity theory. Experiments which look at the complete Bloch sphere of an entangled system have also excluded this nonlocal theory.

Bell, and Kochen and Specker, have shown that for a system of at least dimension 3, one cannot assign non-contextual probabilities to measurement outcomes. Many experiments have been performed about the Kochen-Specker paradox, all confirming quantum mechanics. For a three-state system, this was recently confirmed by Lapkiewicz et al. following a proposal by Klyachko et al. (see Fig. 2). With only five measurements, it is impossible to assign non-contextual values even for pairwise commuting observables.

To the surprise of all early researchers in the field, experiments of this kind started to become useful in the 1990s. Characteristic is the entanglement

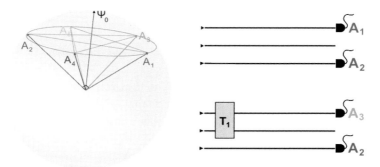

Fig. 2. Testing quantum noncontextuality for a three-state system (according to Klyachko et al., experiment by Lapkiewicz et al.). A spin-1 system in an eigenstate along the central direction is subject to five mutually orthogonal measurements of the square of the spin component. In the experiment, the three-state systems were realized by three propagation modes for individual photons. The probability of detecting the photon in A_2 depends on whether it is measured in the context of A_1 (top) or A_3 (bottom).

of more than two qubits, today called GHZ (Greenberger-Horne-Zeilinger) states. These states, initially interesting because of their nonstatical contradiction with local realism, have now assumed important roles in quantum computation.

Due to enormous technological progress, one can now navigate in Hilbert spaces of dimensions up to the order of 100. This again opens up new experimental possibilities, both in the foundations and in quantum information.

Discussion

A. Aspect Thank you, Anton. Now, let us start the discussion. Do not hope to escape if you do not ask any questions.

X.-G. Wen I have a question for John (Preskill). You have mentioned the idea that unitarity might emerge at large distance. What do you mean by that?

J. Preskill Indeed, on my last slide there was a statement on the list of question: could Nature be fault-tolerant and unitary dynamics emerge at large distances? What I mean is the following. Many quantum error correcting codes have a kind of self-similar structure: if you think of coarse-graining the system, it might be very noisy at short distances, but as you go to longer and longer distances after many coarse-graining steps, the encoded information at the top-level of the hierarchical scheme is very well protected against decoherence. From that point of view, the underlying dynamics is far from unitary but the far infrared dynamics is very nearly unitary, except up to very small corrections. One could ask if that could be a metaphor for what could happen in Nature: if there is some kind of intrinsic decoherence at short distances, could it be that when we integrate out the short distances, we get a very good approximation to unitary dynamics at long distances, like what happens in these quantum error correcting codes? If that it is true, quantum mechanics would reign at intermediate scales. At long scales, where classicality kicks in, there would be decoherence and unitarity would be lost and, similarly, at very short distances where the intrinsic decoherence would be strong.

Y. Aharonov I would like to say something about a new interpretation of quantum mechanics that will perhaps answer some of the questions that where raised here. In classical physics, if you put a boundary condition at a given time, it dictates the entire future and past of the system. In quantum mechanics, this is of course not true because of the uncertainty principle. But in quantum systems if you put two boundary conditions, one in the past and one in the future by making two different kind of experiments, then you can describe the situation in the present by using two wavefunctions instead of one: one wavefunction propagating from the past and the other wavefunction propagating back from the future. If you use the two wavefunctions in your description, new kinds of realities appear. The way to test those realities is with what I call weak measurements and now many laboratories in the world are doing weak measurements and find new values associated with operators that I call weak values. But what is relevant for this discussion here is to look at the wavefunction of the whole universe and assume that this wavefunction has two boundary conditions instead of one. Then we can look at the many-world with all the branches of the many-world and say that the future boundary condition simply selects one of them. This introduces something akin to a collapse, but does not require to add something foreign to quantum mechanics, it is achieved just by adding another boundary condition. Such an approach can completely solve the measurement problem.

A. Aspect Who wants to comment or argue?

T. Leggett Are you saying that if you impose a single initial condition and a single final condition that determines the state uniquely? That sounds improbable.

Y. Aharonov First of all, I have to qualify what I am saying. I can show that it is consistent to put a future boundary condition on systems only if their number of degrees of freedom are extremely large, otherwise it cannot happen. I cannot discuss this in more detail now, but it is a very interesting point. You can show that only what we call macroscopic objects can have a boundary condition that will satisfy the condition that if one macroscopic measurement device gives one answer or another answer, then the future boundary conditions can consistently select one of them. This is only true for macroscopic objects not microscopic ones. It is a very interesting detail that I do not have time to discuss now.

A. Aspect Other comments on this or another question? Igor.

I. Klebanov This is probably a silly question about quantum computing, I'm not an expert in the field. John (Preskill) mentioned that it is good to have a lot of entanglement. But we know that in continuum quantum field theory, the vacuum is a very highly entangled state: if you break space into two regions, the quantum entanglement with respect to these two regions diverges. Could that somehow be used?

J. Preskill In my talk, I described the feasibility of quantum states imaging that we start with a product state and then build-up entanglement with quantum circuits. But, as you ask, because the vacuum of quantum field theory is already highly entangled, do not we somehow have this resource for free? I do not know. In the case of a weakly coupled quantum field theory, I think that we can argue that the vacuum is something that is easy to prepare — a gaussian state is easy to prepare. So, when we have an algorithm for making the ground state of a free field theory, then we can simulate turning on adiabatically the coupling like in the old quantum field theory textbooks to make the vacuum. What would prevent such an algorithm from succeeding would be that the adiabatic evolution fails because the gap gets small, as in some kind of quantum phase transition. But if we can reach the vacuum that we are interested in by starting with a state that is nearly gaussian and turning on the coupling without coming to a quantum phase transition, then this would fit within my description of what a feasible state is. Potentially, though, Nature provides us with something that cannot be described that way. In the case of QCD, for example, we probably know how to create the vacuum state by an efficient algorithm. Nevertheless, it is possible that Nature has been kind enough to give us a powerful quantum resource for free that we could not have created by ourselves.

A. Aspect Now there are many questions. David, you were first.

D. Gross I have a question to John (Preskill) about entanglement. Usually people introduce entanglement with a two-spin system and then when there are many degrees of freedom they divide the system into two parts, so that entanglement is well-defined. But of course quantum states do not consist of only two degrees of freedom, but many of them, if not an infinity. Do we have a good definition of entanglement in this case and do we have even the beginnings of some efficient classification of quantum states in their complexity?

J. Preskill We have a definition of entanglement for many-particle quantum states. It is only a non-trivial definition in the case of mixed states, where it can be subtle to say what is entangled and what it is not. If a mixed state can be expressed as a mixture of product states, then we say that it is separable and if it cannot, it is entangled. Now the question is, can we distinguish between two-part versus three-part versus four-part entanglement, and so on? Yes, one can do that, at least at the level of systems with a few parts. One could say, for instance, that this system has the property that it has three parts, but if you trace out any one of the parts, what remains is separable, so the entanglement in a sense is really three-part entanglement. But do we have classification of many part quantum entanglement for pure states? No, far from it. It is very complex to describe the different types of many-body entanglement. A more promising approach maybe is what we will possibly hear about in the quantum matter session through different ways of classifying different phases of matter which have some universal significance that are stable with respect to small deformations.

A. Aspect Michael Berry has a question.

M. Berry This is a follow-up to the question to John (Preskill) about the vacuum. There is a quantum entanglement which is massive, much more familiar, much more elementary, and which you get for free. This is the entanglement between identical particles, bosons or fermions. When I mention this to my quantum information colleagues, they agree that these particles are entangled but argue that we cannot exploit it, and thus that it is not a resource and therefore is not interesting. Do you agree with that?

J. Preskill I do not agree. In fact in the example that I gave of linear optics, the source of the hardness of simulating that system lies precisely in the indistinguishable particle statistics. It is actually kind of funny, because linear optics with fermions is easy to simulate. Normally, we think that fermionic problems are harder and bosonic problems are the easy ones. But in this case, for computation complexity reasons having to do with determinants being easy to compute and permanents, which are determinants without the minus signs, being hard to compute, it is really the bosons which are hard to simulate. It is the indistinguishable particle statistics, which is at the source of that.

W. Zurek I would like to make one minor addition to the question that was just discussed. When one talks about entanglement, one sweeps under the rug a tremendous complexity of different sorts of correlations, and only very few of them may end-up being useful. For example decoherence arises because of entanglement. So simply having entanglement is not good enough. You have to have it in the right place and you have to be able to control it in very specific ways and makes sure that it stays where it is supposed to. Related to that, I would like to ask a question that has to do with what we actually know about the resources which are essential for quantum computation. For instance, there are tasks that seem to be sped-up in a way that is substantial in systems where entanglement is not present — I'm thinking about the one-qubit quantum computer.

J. Preskill Right. I had a slide which I cut to get under 25 minutes which was about the one-qubit quantum computer. It is an interesting example which illustrates that it is subtle to understand what the source of the power of quantum computing is. This is a model in which in the initial state, there is one qubit in a pure state and N qubits which are in a maximally mixed state. We then perform a quantum circuit, a sequence of unitary gates, and finally do a readout at the end. While this model looks like it should be very easy to simulate classically, we know efficient algorithms that can be run on it, which solve problems for which there is no corresponding classical efficient algorithm. Although, this model does not seem to be as powerful as universal quantum computing, as we cannot run all quantum algorithms in it, it seems to go beyond what we can do classically. But I do not know if it is quite correct to say that there is no entanglement in the system. There is just a little bit of it in the sense that one pure qubit gets distributed among many. The point is that when we consider mixed states, separability versus entanglement is not necessarily the right criterion for being hard to simulate.

T. Leggett I wonder if I could make a comment which was originally meant to be on Mike g's talk. It also relates to a point which John (Preskill) brought up. I certainly agree of course that no quantum system is ever isolated from its environment and that very often the consequences of this can be quite spectacular as in the case of Hyperion. But when one looks at the history of discussions about the viability of observing a macroscopic quantum coherence, then you find that throughout the last decades up to 2000, the effect of decoherence was almost always grossly overestimated. I think that a lot of the part of the reason for that overestimation was that although it is true that all systems are entangled with their environment, an awful lot of that entanglement is actually adiabatic in Nature. That connects with John (Preskill)'s point that possibly you could have a lot of entanglement at high energies and short distances, but nevertheless when you come to look at the behavior in the macroscopic limit it could be much more unitary than what you might have wanted to believe originally.

S. Das Sarma John (Preskill), probably you had to cut this part out, but in your talk you did not emphasize the fact that there are ways of doing topological quantum computation where you do not have to do quantum error correction at all as a matter of principle. The anyons that you talked about are all from SU(2) level 2 conformal field theories. But of course there is nothing that prevents us from having in Nature one of the condensed matter systems which at low-energy and long wavelength obey SU(2) level 3. In such a case, we would have parafermions, which can do universal quantum computation without any error-correction. My own feeling is that this will probably be the ultimate solution. There are candidates in fractional quantum hall effect, which may or may not be SU(2) level 3. Also one can take SU(2) level 2 anyons like in these Majorana fermions and you can do dynamical topology change, which again does away with the need of circuit-level qubits. I am very worried about circuit-level qubits because we have been working on it for 12 years and we are hearing it is just around the corner, but we still have only a couple of qubits, so I think that the road to the future is error-correction free, fault-tolerant topological quantum computation.

J. Preskill Sankar (Das Sarma) is correctly chiding me for talking about topological quantum computation only in the special case where the braiding of anyons alone is not sufficient for a universal quantum computing. I did not mention the possibility of universal braiding, which might be realized by some of the fractional quantum Hall states. In that case, by the sort of arguments that I gave, we have to be talking about a system of interacting fermions, not free fermions, to get a system which would be hard to simulate classically. As far as whether we would call this truly scalable quantum computing, I do not quite agree, because again the system would be protected by the energy gap. Although we could make the error rate very small by operating at a temperature small compared to the gap, strictly speaking that is not scalable, there will be some limit to how large the circuit would be able to perform before the errors would swamp the system. That is what I meant when I said that we would have to add at some stage, if we wanted to do a very big computation, more conventional error-correction.

X.-G. Wen There was some discussion about large entanglement as a resource. Here I just want to make a comment that there is a notion of long-range entanglement, which is important. This topological quantum computation really requires a long-range entanglement and that is what makes it work.

F. Englert I have a question related to the point that Aharonov made. It goes a little bit backwards because I think that your statement should have been done before in the course of the theme. My question is the following. It seems to many people, and I share that opinion, that if one wants to include classical physics in quantum mechanics as a particular case, the many-world picture is unavoidable. I want to see clearly if what you propose with the two boundaries can definitely get rid of this point?

Y. Aharonov The answer is yes.

B. Halperin This is a comment or caution partly in response to Sankar (Das Sarma)'s comment and John (Preskill)'s statement. It is of course true in principle that if we had a topological quantum computer based on level 3 anyons you could do universal quantum computing. It is also true that if the temperature were low enough, we would not have to worry about thermal excitations. But there are other constraints. For example, these things are only protected if the particles are far enough apart, that is things falls off exponentially if you separate the particles, but it is exponential distance divided by what? Typically these distances are a few microns or they could be, maybe they are smaller but we are very far from getting to the range where this exponential is e to the minus 30. I mean if you want protection of 10 to the 10th and microscopic scales after all are fractions of a nano-second, you would like to work for seconds then you need enormous factors there. So that's a big difficulty. Low temperatures, of course, are also a difficulty. If you have a gap you could imagine getting down to very low temperatures, but then there is also non-thermal noise that can cause excitations that may be harder to filter out. So there are many things other than temperature which may cause one to need error-correction. If we get there at all.

W. Phillips This question is for John (Preskill). On your last slide, if I understood correctly, you talked about the fact that a general quantum computer still would not be as powerful as Nature for doing certain quantum problems. I want to explore that idea a little more. We are usually told that if we had a general purpose quantum computer, we could do quantum problems efficiently. But what you are telling us is that maybe it is not quite as efficient as Nature. So what does Nature have that a quantum computer does not have?

J. Preskill I phrased it as a question. What I had in mind is the following. We do know the following things. If you have a universal quantum computer, then you can simulate the dynamics of a local many-body Hamiltonian. I also think that we know how to simulate the dynamics of a quantum field theory modulo the question that Igor brought up — whether we can efficiently prepare the ground state, which is an issue related to whether you encounter a quantum phase transition. But you might ask what about quantum gravity? We have this intriguing hint, AdS/CFT, which says that we can describe quantum gravity using quantum field theory. This suggests that a quantum computer would be able to study quantum gravity. But there are limitations to that statement. We only understand this duality in the case of asymptotically AdS space-times, not flat space-time or the de Sitter space-time that we live in. So maybe a quantum computer of the conventional type — a quantum circuit — is not capable of simulating the dynamics of gravity under some circumstances. If that is true then the quantum gravity computer, which in principle we can realize in Nature by harnessing quantum gravity, would be more powerful. That would be interesting and a good justification for pursuing research into quantum gravity.

A. Aspect   With this general statement, it is time to close this afternoon session. Before closing, I would like to make a biased, totally biased conclusion. I was struck that at two moments, in the second and third parts, we understood that what we need is something to test experimentally. We need something like a Bell inequality to test the classical versus quantum boundary. As expressed by John (Preskill) we need a kind of test to know whether Nature provides all possible states or whether it is restricted. As an experimentalist, I am looking forward with a great interest to some kind of theorem providing the possibility for a test, because at the end of the day Nature is our judge. With this, I thank everybody for making this session lively and exciting.

Session 3

Control of Quantum Systems

Chair: *Peter Zoller*, University of Innsbruck, Austria
Rapporteurs: *Ignacio Cirac*, Max-Planck Institüt für Quantenoptik and *Steven Girvin*, Yale University, USA
Scientific secretary: *Serge Massar*, Université Libre de Bruxelles, Belgium

Rapporteur talk by I. Cirac: Quantum Computing and Simulation with Atoms and Photons

Abstract

This article reviews how interactions between atoms can be engineered by means of external fields, and how those interactions may be used to build quantum information processing devices and simulators. It also describes some of the obstacles one has to overcome in order to build such devices, as well as their potential impact in other branches of Physics.

1. Introduction

Progress in atomic physics allows us today to control and manipulate atoms and ions at an unprecedented level. They can be isolated from the environment by holding them in electric, magnetic or optical traps, and cooled down to extremely low temperatures. Then, we can control their motion and manipulate their internal structure (electronic and spin states) with the help of external forces, as well as interrogate them for extremely long times. Once we have reached such a level of control, an obvious question is: what can we do with them? In this article we give some of the options that are being considered at the moment; in particular, those related to the fields of quantum information and simulation.

Since three decades ago, quantum systems can be controlled at the single particle level. Single atoms or ions (or few of them) can be isolated using magnetic or optical traps, and electric traps, respectively. Those experiments have been (and are being) used to make extraordinarily precise measurements of atomic properties, which in turn have given rise to extraordinary tests of fundamental theories. Furthermore, given the fact that all atoms under such isolation conditions behave in exactly the same way, one could build with them the most precise clocks, something that has many potential applications. In all those experiments it was crucial that atoms do not interact with each other, since this would disturb their properties and lead to errors. In the last years, however, it has been possible to control and even tailor such interactions. For instance, nowadays it is possible to tune the s-wave scattering length between two atoms by employing external magnetic fields (via Feschbach resonances), so that at low temperatures they can attract or repulse each other as desired. Or one can manipulate the internal state of ions so that the interaction strength between two of them is different if they are in different hyperfine states.

Theoretical progress in atomic physics and quantum optics has gone hand-in-hand with the experimental achievements. Theories to accurately describe atomic experiments are well established by now. Simple models exist (like the Jaynes-Cummings model and versions thereof) to describe the interaction of isolated atoms with light, or the effects of light on the motion of a trapped ion. This has generated many ideas on how to create intriguing quantum states of light or in the motion of such particles. Furthermore, new methods have been developed to account for the coupling of atomic systems with all different kinds of environments (via master equations, input-output formalisms, and numerical simulations), or to study the evolution of a quantum system in the presence of continuous monitoring. This enables us to describe many observed phenomena, as well as to predict to what extent an experiment will achieve its goal if atoms are not completely isolated. In theoretical research, the consideration of atom-atom interactions has widen the scope of research as well. First of all, controlled interactions lead to entanglement, one of the key ingredients of several applications in the context of quantum information science. Thus, theoretical research has concentrated in finding ways of using atoms to build quantum processing devices and avoid decoherence, as well as in some more fundamental aspects (like the characterization of multiparticle entanglement), in close collaboration with quantum information theorists. Besides that, theorist working in AMO physics have teamed up with condensed matter scientists and apply their techniques to describe many-body atomic systems. All this has opened up a very wide spectrum of opportunities in the field of AMO physics, and defined new interdisciplinary areas of research.

In general, one may distinguish two approaches to control interactions in atomic systems. The first one, what we will call bottom-up, consists of considering one atom, then two, three, etc, so that the degree of control is kept. This is typically the approach that is required in quantum information, as one has to increase the complexity of experiments little by little until everything is controlled and thus one can scale it up. The second one, the top-down, tries to get control of collective states of many atoms (perhaps millions or billions), even though they cannot be individually addressed or manipulated. This kind of approach may be useful to learn about many-body quantum systems, where collective properties emerge even though one cannot access each individual system, and is the basis of quantum simulation. In the next sections we will address those approaches separately.

2. Bottom-up Approach: Quantum Information

Sets of atoms or ions can be stored in different kinds of traps. Using lasers one can change their internal levels. For instance, denoting by $|0\rangle$ and $|1\rangle$ two different hyperfine levels of the electronic ground state manifold, one may induce arbitrary single-qubit gates $|i\rangle \rightarrow u_{i0}|0\rangle + u_{i1}|1\rangle$, where the u's form a 2×2 unitary matrix. By employing atom-atom interactions one can also change the state of pairs of atoms, and induce, for instance, a so-called controlled-NOT gate, defined as: $|i, j\rangle \rightarrow |i, i \oplus j\rangle$, where $\oplus$ denotes addition modulo 2. By concatenating single-qubit and controlled-NOT gates one can implement any unitary operation acting on a set of qubits. Note that the controlled-NOT gate may be replaced by any other gate acting on two qubits (which is able to entangle them). This, together with the ability to prepare a pure state (e.g. the $|0, 0, ...\rangle$) and to measure individual qubits (in the qubit basis, for instance), constitute the necessary requirements to build a quantum computer.

While single-qubit gates are rather easy to implement in atomic systems using lasers, gates involving two or more atoms are much harder. The reason is that one has to let atoms interact, but still isolate them so that they do not interact with anything else, in order to avoid decoherence. There exist several techniques to achieve this task, and Fig. 1 describes some of them.

The physical system over which the highest control has been gained in recent years is trapped ions.[1,2] They can be stored for days in ion traps, and coherence (ie superpositions of $|0\rangle$ and $|1\rangle$) can be kept for several minutes (and there is no reason why this could not be extended to much longer times). They can be prepared in pure states, their internal state can be efficiently detected, and single qubit gates can be carried out in time scales of the order of microseconds. All this can be performed with errors of the order of 0.1%. Two-qubit gates have also been demonstrated for more than ten years, and fidelities of the order of 99% have also been achieved in this case. Those experiments typically involve 2-4 ions, and demonstrate proof-of-principle ideas. Remarkably, up to 14 ions have recently been entangled.

Neutral atoms have also been controlled, but not at the level of trapped ions. Nevertheless, different kinds of two-qubit gates have been demonstrated using Rydberg atoms, cold collisions, and cavities. Atoms and ions at remote places (separated by few meters) have also been entangled, and basic protocols (like teleportation) have been successfully implemented.

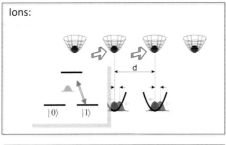

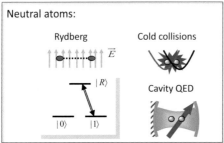

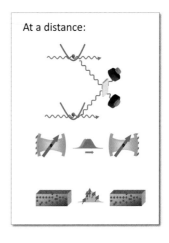

Fig. 1.   Different ways of inducing gates between atoms using interactions. Ions: they are pushed with a laser depending on their internal state. The distance and thus Coulomb energy depends on internal states, resulting in phase shifts that implement a gate. Neutral atoms: (a) A laser promotes them to a Rydberg state; there, in the presence of an external electric field, dipole-dipole interactions are strong enough to induce the desired gate; (b) Cold collisions respecting internal states $|0\rangle$ and $|1\rangle$ but inducing phase shifts may also be used; (c) Gates can also be generated by exchanging photons in a cavity (which is required so that photons do not escape away); At a distance: Atoms may interact at a distance by interchanging photons, or by producing photons which are somewhere else detected, giving rise to an interference.

While atoms and ions are most suitable to store quantum states for relatively long times, photons are better suited to transfer quantum information.[3,4] Typically, two polarization states of a single traveling mode are identified as the states forming a qubit. Coherent superpositions can be sent over distances exceeding 100km. One may use optical fibers to direct the photons, although photon absorption limits the distance over which they can travel. Using postselection (i.e. only counting the photons that arrive) one can still send quantum states over distances beyond 100km, although at a very slow rate. Entangled photons can be produced by non-linear crystals with a fidelity exceeding 99%. In fact, the first demonstration of teleportation was implemented using entangled photon pairs. Since then, basic quantum algorithms with up to 8 photons have been demonstrated.

Methods to transfer quantum states from atoms to photons and back have been put forward and even demonstrated, both with atoms in cavities and with atomic ensembles. Fidelities (i.e. the probability of success for the whole process) are still below 80% in the best cases.

All those experiments open up a wide variety of possibilities. First of all, they provide us with new and powerful tests of Quantum Mechanics. For instance, Bell inequalities (as predicted by any local hidden variable theory) have been violated with massless (photons) and massive (ion) particles, although a full loophole-free test is still pending. Entanglement and superposition has been created with different kinds of atoms, at various distances, and under different conditions. Furthermore, precision measurements using atoms and molecules may shed some light in more fundamental questions, like the change of the hyperfine constant α with time, or the electric and magnetic properties of elementary particles.

Atoms and photons are also well suited for some applications in the field of quantum information. The latter are used in quantum communication, where one employs quantum superpositions to send information either more efficiently or more securely. The first may be used as quantum memories; that is, to store quantum states for long times, a requirement of several applications in the field. Furthermore, one may even think about building small prototypes of quantum computers using them.

Building a quantum computer that outperforms classical devices is a holy grail, but a daunting challenge. It is clear by now that fault tolerant error correction will be required for such a venture, as otherwise errors accumulate and eliminate all advantages offered by quantum mechanics. This technique involves an overhead, since one will have to store a logical qubit in many atoms. One may estimate that in order to build a fully operating quantum computer, several hundred thousands of qubits will be required. They will have to be accessed in parallel, with an error per basic step (quantum logic gate) of the order of 0.1%. Even though this last requirement has been achieved with a couple of trapped ions, it is crucial that this error does not grow with the size of the system. This scaling problem is the major obstacle to build a quantum computer with ions, atoms, or nuclear spins in molecules. It is hard to imagine at the moment how this technological challenge can be achieved with hundred thousand ions or atoms. Nevertheless, those systems are extremely useful to deepen our knowledge about how quantum gates can be implemented, or how the effects of decoherence can be minimized, a knowledge that will be required if we are to build quantum computers with any other system. Let us also note that atomic systems have been proposed for topological quantum computing, a different way of protecting against decoherence, although it is not clear at this stage (at least for atomic systems) if this would offer any advantage with respect to standard methods.

Another application is quantum key distribution, where the idea is to use quantum states to distribute a secret number among different people, so that they can securely communicate using public channels. This is typically achieved using photons, and bit rates well beyond MHz have been demonstrated. Commercial devices are already available, although their cost and versatility cannot compete yet with classical systems. Furthermore, in order to achieve full security a quantum memory will probably be required in order to store quantum states and perform measurements when the device is completely disconnected from the exterior world (to avoid hacking attacks). To enhance the range of applicability, it will also be required to extend quantum communication to longer distances. This can be achieved by using satellites or quantum repeaters. Other security applications, like quantum money or credit cards, require to store superpositions for very long times (days or months), well beyond today's capabilities. In any case, it seems clear that methods to efficiently transfer quantum states from photons to atoms and back, as well as to store superpositions in atoms, will be the key to many of those applications. Another issue which must be considered is the versatility of the devices. Most of the experiments with atoms take place at very low temperatures, and require very complex infrastructures, impossible to carry along or to include in a small device. Ideally, one would like to build compact and portable devices for all those applications. Thus, one should look for systems operating at room temperature and not requiring extreme conditions (like ultra-high vacuum). Two atomic systems stand out in this context: color centers in solids and atomic ensembles. The first are atom–like impurities (like Nitrogen-Vacancy centers in diamond) which have nuclear spins in their vicinities with which they interact. One can purify the state using light, and control the spins using RF or microwave fields, together with the hyperfine interaction. If one could prolong the life time of the nuclear spins (which are typically already very well isolated from the environment), and improve the fidelity of manipulation, this would certainly constitute an important technology for quantum information. The second one consists of atoms at room temperature in a cell which is specially coated to keep the coherence in the collision of the atoms with the cell walls. The high atomic density allows them to absorb photons with probability close to one even at room temperature, and to emit them again after a short time (of the order of milliseconds). If one could manipulate the atoms while they store the quantum information and improve the input-output fidelity, they would provide a basic tool in the context of quantum communication.

3. Top-down Approach: Quantum Simulations

Billions of atoms can be trapped in magnetic or optical traps and cooled to nanokelvin temperatures.[5] Depending on the parity of their elementary components (protons, electrons and neutrons), atoms obey bosonic or fermionic statistics. At those temperatures, atomic bosons condensate and tend to occupy the same quantum state, whereas fermions occupy different single particle states. The achievement of Bose-Einstein condensation in 1995 opened up new lines of research in atomic physics, trying to observe both many-body and coherent quantum phenomena, and making contact with other branches of physics.

Atomic gases can be described by an effective quantum field theory at sufficiently low temperatures, with Hamiltonians of the form

$$H = \int d^3x \Psi_\sigma(x)^\dagger [-\nabla^2 + V_e(x)] \Psi_\sigma(x) + u_{\sigma_i} \int d^3x \Psi_{\sigma_1}(x)^\dagger \Psi_{\sigma_2}(x)^\dagger \Psi_{\sigma_3}(x) \Psi_{\sigma_4}(x).$$
$$(1)$$

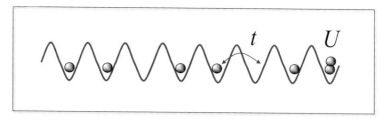

Fig. 2. Atoms in optical lattices can be described in terms of Hubbard models. At low temperatures and densities, they occupy the lowest Bloch band. They can tunnel from one potential well
to the next one, and they strongly interact if they meet in a well. The ratio between the tunneling
and interaction energies, t/U, may be adjusted by choosing the laser intensity (i.e. the depth of
the potentials).

Here $\Psi_\sigma(x)$ is a (bosonic or fermionic) field operator for atoms in internal state
σ, and a sum over the σ's is understood. The first term describes the kinetic and
potential energy of single atoms, whereas the second one describes interactions. V_e
denotes an external potential (created by electric, optical, or magnetic fields), which
can be adjusted. The constants u_σ are proportional to the scattering length, and
can also be adjusted using external fields.

A key element in experiments with cold atomic clouds is that one can tune the
interactions and make the atoms strongly attract, repel, or no interact at all. This
has allowed, for instance, to study the BCS-BEC crossover: fermionic atoms with
two internal states and attractive interactions form Cooper pairs, which get more
and more correlated as the interaction strength grows, until they form molecules
which Bose-Einstein condense when the attractions become repulsive. Another way
of increasing the interaction strength is by confining the atoms in an optical lattice,
which is a periodic potential V_e created by counterpropagating laser fields (in 1, 2,
or 3 spatial dimensions), see Fig. 2. Atoms meeting in one potential well experience
an interaction which now scales as the inverse of the volume they occupy, and
thus by increasing the laser intensity the interaction strength U increases, and the
tunneling energy t decreases. This has led to the observation of the Mott insulator-
superfluid transition with cold atoms: for small values of t/U atoms tend to localize
in individual lattice sites (i.e. they are in a product state with one atom per site),
whereas in the opposite limit atoms tend to delocalize and form a Bose-Einstein
condensate in the lowest Bloch band.

Ultracold atoms can be described by simple field theories whose parameters can
be externally adjusted. One can have bosons, fermions or mixtures thereof, as well
as different internal states. All this makes them ideal as a laboratory to study many-
body quantum systems. In particular, by tuning the different parameters one can
end up with effective theories which imitate those found in other fields of physics. In
this sense, they can be employed as quantum simulators, a term coined by Feynman
thirty years ago,[6] who anticipated that in order to study complex quantum systems

one would have to use another system that can be engineered to evolve according to the Hamiltonian one would like to analyze.

Atoms in optical lattices are specially well suited to study basic models of condensed matter physics.[7] Just by cooling them down and loading them into a lattice, they implement so called Hubbard models, which are described by Hamiltonians of the type

$$H = -t \sum_{<i,j>} a_{i,\sigma}^\dagger a_{j,\sigma} + \sum_{i,\sigma,\sigma'} u_{\sigma,\sigma'} a_{i,\sigma}^\dagger a_{j,\sigma'}^\dagger a_{j,\sigma'} a_{i,\sigma} \tag{2}$$

Here, $a_{i,\sigma}$ annihilates an atom at site i in the internal state σ, the sum in i and j is restricted to nearest neighboring lattice sites, and t and u are parameters which can be externally adjusted. This Hamiltonian can be easily obtained starting from (1), writing a periodic potential V_e, projecting on the first Bloch band, and neglecting small non–local terms. This model is related to many strongly correlated phenomena in solid state systems, some of which are not fully understood. Ultracold atoms may thus help us to analyze and better understand those models. Moreover, using external lasers one can make the the parameters t and U site- and internal state-dependent, as well as time-dependent; and even complex, which breaks up time-reversal symmetry. Thus, even though atoms are neutral and have small magnetic moments, they can simulate the action of strong electric and magnetic fields. One can also form different geometries using proper laser configurations, like square, triangular, hexagonal, or Kagome lattices, and adjust the density. As mentioned above, one can have bosonic or fermionic atoms, or mixtures thereof. With them one can simulate the physics of electrons interacting with phonons, whose role is played by fermionic and bosonic atoms, respectively. Furthermore, if atoms strongly interact with each other and for a prescribed density and low temperature, they will form a Mott phase where each atom occupies a different site. Virtual tunneling between adjacent sites will give rise to an effective interaction which will depend on the internal atomic states. With this mechanism, one would be able to simulate magnetic systems, where the role of spins is taken by the internal atomic levels, and the spin-spin interaction can be adjusted using external fields. In fact, the possibility of adjusting the interaction parameters u should allow us to emulate a variety of strongly correlated phenomena. All these interactions will be short range, since the basic mechanism can be described in terms of the contact interaction appearing in (1). There are several possibilities in order to simulate Hamiltonians with long-range interactions. For instance, one can take atoms with large magnetic moment, so that dipole-dipole interactions are sufficiently strong. Or one can partially excite Rydberg states using lasers, in order to enhance electric dipole-dipole terms. One can get even stronger interactions if one uses polar molecules instead of atoms.

The actual experimental situation with atoms in optical lattices is very promising. Many of the basic ingredients for quantum simulations have already been demonstrated. At the moment, one of the major obstacles is related to the temperature requirements imposed by some of the simulations. Even though very low temperatures (down to hundred of pico K) have been achieved, one or two order of magnitude lower temperatures are required in order to enter interesting regimes which are presently not theoretically understood. Given that atomic interactions are extremely weak as compared to those between electrons, much lower temperatures are required in order for the same phenomena to appear. Nevertheless, experimental progress in the field is continuous and many new ideas and techniques will likely soon allow experimentalists to simulate interesting models.

One may wonder at which stage a quantum simulator will indeed help us to solve scientific problems which cannot be attacked with classical computer simulations. This is hard to tell, since there exist very powerful theoretical methods to describe many-body quantum systems that are also progressing dramatically in the last years. At the moment, however, it seems that the description of dynamics in strongly correlated system is extremely complex using classical computers, since one has to simulate extraordinarily entangled states violating the so-called area law (which states that in the grounds state of a local Hamiltonian the entanglement entropy between a region and the rest scales like the boundary of that region, and not as its volume). With atoms in optical lattices, quantum simulation of dynamics should be, in contrast, relatively simple.

Another relevant system for quantum simulators is that of trapped ions. When loaded in an electric or magnetic trap, they crystallize forming different structures. With the help of microtraps, one should be also able to create arbitrary geometries of ion crystals. Then, the motion of the ions can be described in terms of a bosonic field (phonons), which can be made to interact with the internal states of each ion by using lasers, giving rise to similar models as the ones proposed with atoms in optical lattices. The advantage of this system is that ions interact more strongly and can be more easily detected, although its main disadvantage is that one can only simulate bosonic (and/or spin) systems with them, at least in a relatively simple way. So far, experiments with few ions have been carried out, although in principle one could use already existing traps containing thousand of them for quantum simulations.

One may think about going beyond condensed matter systems and simulate relativistic quantum field theories with cold atoms as well. Even though this may sound impossible, as atoms are very heavy so that relativistic effects in their motion are practically absent, one may use a different approach. For instance, if one has a Bose-Einstein condensate, one may rewrite $\Psi(x) \to \varphi(x) + \delta\Psi(x)$, where φ is a classical field (describing the condensate wavefunction), and $\delta\Psi$ reflects the small quantum fluctuations (phonons). Replacing this in Eq. (1), expanding up to second order in $\delta\Psi$, and taking the Fourier transform, one obtains a linear dispersion relation for low energies, as it is well known from Bogoliubov theory. Thus, phonons behave as relativistic particles in a curved space whose metric depends on the condensate wavefunctions. Furthermore, if one has several internal states, the effective Hamitonian may contain terms resembling pair creation. This occurs, for instance, if one has spin 1 particles and most of the atoms are initially in the $m = 0$ state. The collision term in (1) then would give rise to terms of the form $\varphi^2 \Psi_1^\dagger \Psi_{-1}^\dagger$, something which resembles some inflation model (pre-heating). Thus, it seems that quantum simulation may also become interesting through their connection to cosmological and relativistic models.

A more ambitious goal is to simulate interacting field theories with atoms, as they appear in high-energy physics. Mixtures of bosonic and fermionic atoms, as well as molecules, are described by Hamiltonians of the form

$$ H = \int \Psi_\sigma^\dagger [-\nabla^2 + V_e] \Psi_\sigma + u \int (\Phi_\mu^\dagger \Psi_\sigma \Psi_{\sigma'} + h.c.) + v \int \Phi_\sigma^\dagger \Phi_{\sigma'}^\dagger \Phi_\sigma \Phi_{\sigma'} + \dots \quad (3) $$

The molecular field Φ_μ may be viewed as a gauge field A_μ which mediates interactions between other fermionic fields Ψ_σ. May be using the flexibility offered by atomic systems, together with lattices (to give rise to the proper dispersion relations), and different bosonic and fermionic atomic and molecular systems, one may engineer QED-like or even QCD-like models. An even more challenging task would be to simulate molecular systems, as they appear in quantum chemistry. One can imagine fermionic atoms playing the role of electrons, and some external potential replacing the position of the nuclei. For that, one would need to figure out how to induce Coulomb-like interactions among the electrons, something which requires new ideas. Nevertheless, if this is possible some day, quantum simulations would not only have an extraordinary impact in other areas of physics, but also in other branches of science.

4. Conclusions

In this contribution we have given a superficial overview of the theoretical and experimental situation in AMO physics in the context of quantum information and simulation. The progress experienced during the last 30 years allows us to control, manipulate, and measure single or few atoms and photons, as well as ultracold atomic clouds. Control of atomic interactions has been achieved in the last fifteen years and has opened up whole new lines of research. On the one hand, the basic principles of quantum information devices (like quantum computers, repeaters or memories) have been demonstrated. It is not clear at the moment how long it will still take to make those devices competitive with respect to their classical counterparts, or if solid state quantum systems will prove better suited for this purpose. But nevertheless, atomic systems are helping us to test quantum mechanics, learn about how to avoid (or even use) decoherence, implement error correction procedures, or use basic interactions to efficient prepare quantum states. All this knowledge has already proven very useful in current experiments with solid state devices, including nano-mechanical oscillators, superconducting qubits and resonant cavities, or quantum dots. Apart from that, it seems that the degree of control acquired so far will also have intriguing applications in precision measurements and sensing. On the other hand, the ability to engineer the interactions with cold atomic gases, as well as the possibility of dealing with fermionic, bosonic atoms with different internal (spin) states, and even molecules, has led to a new field of research, quantum simulation. The degree of control at the moment is not enough to simulate many interesting physical systems, but given the progress and the activity in the field one would expect that this system will give rise to very intriguing applications. For the moment, it has established a solid link among different areas in physics, like AMO and condensed matter physics; and other branches of science, like computer science and high energy physics are or will likely be joining this effort. The main goal is to tame the quantum world of atoms, molecules and photons, and engineer the interactions. This will surely have outstanding implications in other disciplines and, certainly, as it has always happened in the past when we got access to new laws of physics, may lead to many surprises.

References

1. J. I. Cirac, L. M. Duan, and P. Zoller, in Experimental quantum computation and information, *Proc. Int. School of Physics "Enrico Fermi"*, Course CXLVIII, p. 263, edited by F. Di Martini and C. Monroe (IOS Press, Amsterdam, 2002). arXiv:quant-ph/0405030
2. R. Blatt and D. J. Wineland, *Nature* **453**, 1008 (2008).
3. N. Gisin, G. Ribordy, W. Tittel, and H. Zbinden *Rev. Mod. Phys.* **74**, 145 (2002).
4. J.-W. Pan, Z.-B. Chen, C.-Y. Lu, H. Weinfurter, A. Zeilinger, and Marek Zukowski, *Rev. Mod. Phys.* (to be published), arXiv:0805.2853
5. I. Bloch, J. Dalibard, and W. Zwerger *Rev. Mod. Phys.* **80**, 885 (2008). arXiv:0704.3011
6. R. P. Feynman, *Int. J. Theoretical Phys.*, **21**, 467 (1982).
7. I. Buluta, F. Nori, *Science*, **326**, 108 (2009).

Prepared comments

A. Zeilinger: Quantum Communication and Computation with Entangled Photons

Entanglement has a very unusual history. Maybe this is best signified by the number of citations of the original 1935 Einstein-Podolsky-Rosen (EPR) paper[a] according to the ISI Web of Knowledge. Both the English name "entanglement" and the German name "Verschränkung" were coined by Schrödinger in the same year. Immediately after its appearance, the EPR paper received only very few citations, among them, two by Schrödinger and one by Bohr. Then, the field lay dormant until the 1960s. A first strong rise of the citations was triggered by John Bell's discovery in 1964 that the EPR concept of local realism implies a conflict with quantum mechanics. This led to significant experimental and theoretical activity, for example, the first experiment by Freedman and Clauser and the elegant experiments by the Aspect group. Another rise of citations of the EPR paper happened in the 1990s. It was due to the discovery that entanglement is a cornerstone concept in quantum information science, particularly quantum computation, quantum teleportation and some forms of quantum cryptography.

Photons are generally accepted to be the ideal carriers of information in a future quantum internet. Here, in quantum communication, I would like to comment on two concepts, on entanglement-based quantum cryptography and on quantum teleportation. Quantum teleportation, together with entanglement swapping, is considered to be the ideal means of communication between future quantum computers. Entanglement swapping is the extension of teleportation to teleporting entangled states themselves. Entanglement swapping is very interesting, as it allows two photons which do not share any common past to become entangled. This confirms that entanglement is not just a consequence of conservation principles. Rather, two systems are entangled when all information about their individual identity is irrevocably erased. From the application point of view, entanglement swapping with its generalization to quantum repeaters as proposed by Duan, Zoller and Cirac is very important, because in principle, it can lead to connecting distant quantum computers.

Quantum cryptography exploits the fact that a general unknown quantum state can neither be observed without disturbing it nor can it be perfectly cloned. Therefore, an eavesdropper intercepting the quantum channel which is used to establish a secret key between two parties, Alice and Bob, can easily be detected. A quantum cryptography system which is built on entangled states in such a high quality that it allows loophole-free tests

---

[a] Albert Einstein, Boris Podolsky and Nathan Rosen, Can quantum-mechanical description of physical reality be considered complete? *Phys. Rev.* **41**, 777 (15 May 1935).

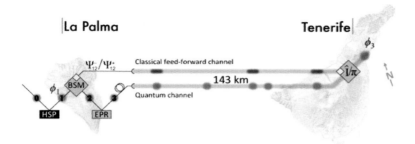

Fig. 1. Quantum teleportation over 143 kilometers between two Canary Islands.

of Bell inequalities provides unconditional security. Therefore, providing a loophole-free Bell test is not only of fundamental importance, but also of practical significance.

Other challenges in quantum cryptography include to increase distances and data rates. Therefore we need very ambitious research on sources for single and entangled photons, on detectors and on long-distance quantum communication. There have been network tests of quantum cryptography systems, most notably in Boston, Vienna and Tokyo. For long-distance quantum communication and quantum cryptography, the method of choice at present is to use satellites.

We have heard the interesting challenges and possibilities for quantum computation in Ignacio Cirac's talk. A disadvantage for photons is that photons do not like to interact. Therefore, to build a routinely usable controlled-not-gate directly for interacting photons is a daunting task. Knill, Laflamme and Milburn suggested to use the nonlinearity of quantum measurement to provide the effective interaction between two photons. Then, a quantum algorithm is implemented as a sequence of measurements on a suitable entangled state. As Raussendorf and Briegel showed, the randomness of the individual measurement can be corrected by feed-forward such that future measurements depend on earlier measurement results. That way, it is possible to build a universal deterministic quantum computer. Such systems have been used in various laboratories to implement fundamental quantum computation primitives such as search algorithms and factoring.

Another approach for photonic quantum communication is coherent photon conversion, where strong classical beams can enhance the nonlinear interaction between quantum states.

Most recently, measurement-based quantum computation has been shown to provide a solution to the long-standing challenge of blind computation. In blind quantum computation, the operator of a central quantum server, say, in cloud computing, not only has no access to the data used by a client,

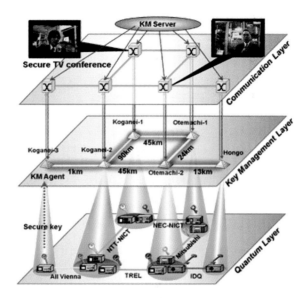

Fig. 2.   Layout of the quantum cryptography network test in Tokyo (Sasaki *et al.*).

but also the calculation procedure itself remains completely unknown to its operator.

A photonic quantum computer would provide extremely high speed, and it would ideally match to photonic quantum communication systems. The technical challenges do not appear to be unsurmountable.

W. Ketterle: From Strongly Interacting Atomic Systems to Optical Lattices

In my comment, I discuss the frontiers of cold atom science. Ultracold atoms are the building blocks to realize novel Hamiltonians, or in other words, to explore Hilbert space. Hilbert space is vast, and so far we have explored only a small section of it. Using quantum control and cooling we will advance further.

For more than a decade, cold atoms represented weakly interacting systems. Laser cooling achieved temperatures typically around 100 μK in a classical gas, far away from quantum degeneracy. Evaporative cooling led to Bose-Einstein condensation at temperatures typically around 100 nK. Bose-Einstein condensates are weakly interacting many-body systems described by mean field physics through the Gross-Pitaevskii equation. The next milestone was the cooling of fermions to quantum degeneracy. Strong interactions led to pairing and fermionic superfluidity. The exploration of the BEC-BCS crossover was a major achievement. These systems show strong pair correlations, but the pairs can still be described by a mean field approximation. Correlations become stronger when the kinetic energy is small compared to the interaction energy. Strongly correlated matter has been created in optical lattices which reduces the kinetic energy, and trough Feshbach resonances which enhance the interaction energy. Strong correlations in optical lattices lead to Mott insulator physics.

All these developments focused on the external degree of freedom, motion. Superfluidity is coherent motion, Mott insulator physics is the suppression of motion by interactions. To go beyond motion is one of the frontiers of cold atom science, and this involves spin ordering in the form of quantum magnetism.

One of the major goals is the realization of the low-temperature phases of the fermionic Hubbard model. Many people regard this model to be the minimum model for high T_c superconductors. So the goal is to observe d-wave superfluidity by hole doping the well known antiferromagnetic phase of the Fermi-Hubbard model at half filling.

A major challenge is the temperature requirement. The Hubbard model is usually parameterized by two parameters: the tunneling or hopping matrix element t, and U, the on-site interaction between two atoms occupying the same lattice site. So far, the physics which has been explored involves particle hole excitations at energy U, and direct first order tunneling at energy scale t. However, antiferromagnetic ordering is caused by exchange interactions, or second order tunneling which has a rate t^2/U. This is usually on the order of 100 picokelvin. Such temperatures have been achieved in my group in proof of principle experiments through adiabatic cooling, but not in a situation where spin ordering could have occurred. In the Fermi-Hubbard model, d-wave superfluidity occurs at even lower temperatures than the Neel temperature at half doping. Therefore, an intermediate goal for the realization of d-wave superfluidity is antiferromagnetic ordering at half filling.

Besides pursuing novel cooling schemes to reach lower temperatures, there are at least three possibilities to raise the phase transition temperature. (1) Use light atoms (lithium) which tunnel faster due to their smaller mass. (2) Use stronger coupling than second-order tunneling in the form of electrostatic interactions. This can be realized with Rydberg admixtures, or polar molecules which interact via strong electric dipole moments. (3) A third possibility is to realize magnetic ordering not with spins, but in the density sector. This was recently accomplished at Harvard, where an Ising model was realized where spin up and spin down correspond to different occupation numbers (zero or two) on each lattice site.

Let me briefly mention two other frontiers of cold atom science. One is precision many body physics. Usually, the Hamiltonian for a cold atoms system is exactly known since the interactions in cold atomic gases are short range. Therefore, precision calculations of transition temperatures and other thermodynamic quantities are possible, and can be directly compared to experiments. It is unprecedented to have calculations for a superfluid of strongly interacting fermions at the level of a few %, but this has become possible now due to advances in quantum Monte Carlo simulations. These methods have been validated through experiments which have reached a similar level of precision.

Finally, another frontier is quantum dynamics. I mentioned above that the slow speed of tunneling can be challenge (since phase transitions have a very low temperature), but it is also a blessing, because it means that systems don't equilibrate quickly. Therefore, it is possible to prepare cold atoms systems far away from equilibrium and study such states and their dynamics.

D. Wineland: Experimental Prospects of Quantum Computing with Trapped Ions

Peter Shor's introduction of a quantum-mechanics-based algorithm for efficient number factoring was followed by a dramatic increase of activity in the field of quantum information science. The possible realization of general-purpose quantum information processing (QIP) is now explored in many settings, including condensed-matter, atomic, and optical systems. Trapped atomic ions have proven to be a useful system in which to study the required elements for such a device. Ions are interesting, in part, because qubits based on their internal states can have very long coherence times, in some cases exceeding ten minutes.

Ignacio Cirac and Peter Zoller were the first to propose a general scheme for quantum computing, which was based on trapped ions.[b] Due to their mutual Coulomb repulsion, cold ions that are held in a potential well naturally form into arrays of spatially separated qubits. With the use of focused laser beams, this spatial separation enables selective qubit addressing, coherent manipulations, and high-fidelity qubit-state readout with state-dependent laser scattering. A single-qubit rotation on an individual ion is implemented by applying a focused laser beam or beams onto that ion (Fig. 1). In addition, because the ions are coupled through their mutual Coulomb interaction, their combined motion is best described by normal modes. In general, the motion of each mode is shared amongst all the ions; therefore, a selected mode can act as a data bus for transferring information between ions. A logic gate between two selected ions is ideally implemented by first freezing out the motion of the ions (putting all modes in their ground states) with laser cooling. Referring to Fig. 1, laser beam 1 then transfers the internal

---

[b] J. I. Cirac and P. Zoller, *Phys. Rev. Lett.* **74**, 4091 (1995).

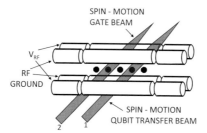

SPIN - MOTION
GATE BEAM

V_{RF}

RF

GROUND

SPIN - MOTION
QUBIT TRANSFER BEAM

2 1

Fig. 1. Scheme for trapped-ion quantum computation proposed by Cirac and Zoller. Quadrupolar electrodes are configured to produce a linear array of trapped ion qubits (filled black circles). Two diagonally opposite rods support an RF potential to produce a ponderomotive pseudopotential transverse to the trap's (horizontal) axis. Static potentials applied to the end electrode segments confine ions along the axis. Ideally, all motional modes are laser-cooled to the ground state before logic operations are implemented. The quantized modes of motion can be used as a data bus to share information between ion qubits that are selected by focused laser beams (see text).

qubit state of one ion onto the qubit formed by the ground and first excited state of a particular mode of motion. Laser beam 2 then performs a logic gate between the motion qubit and a second selected ion qubit. Finally, the initial transfer step on the first ion is reversed, restoring the motion to the ground state and the first qubit to its original state. Overall, these operations implement a logic gate between the internal qubit states of the two selected ions. Although more streamlined versions of the gates have been realized[c] the basic idea that information can be transferred through the motional modes remains. With these tools, simple algorithms have been demonstrated.

To scale to more complex operations, a way to handle large numbers of qubits must be found and logic gate errors must be reduced. For simplicity of construction when scaling to larger numbers, the three-dimensional trap electrode structure shown schematically in Fig. 1 can be transformed into a structure where all the electrodes lie in a plane and the ions are trapped above the plane.[d] For this construction, we can take advantage of established techniques used for micro-fabrication that have been developed in the microelectronics industry.

Current operation errors are significantly above those required for fault tolerance (error probability per gate $< 10^{-4}$), and efforts towards scaling to a large system are only beginning. Solving these problems will involve significant technical challenges, but straightforward solutions are being pursued. In the meantime, some of the basic ideas of QIP are starting to be applied to metrology, for example in atomic spectroscopy.[e]

[c] R. Blatt and D. J. Wineland, *Nature* **453**, 1008 (2008).
[d] J. M. Amimi *et al.*, *New J. Phys.* **12**, 033031 (2010).
[e] P. O. Schmidt *et al.*, *Science* **309**, 749 (2005).

In the early 1980's, Richard Feynman proposed that one quantum system might be used to efficiently simulate the dynamics of other quantum systems of interest. This is now pursued by many laboratories. In one approach, the available interactions in a quantum processor are used to simulate certain classes of physical problems. For trapped ions, it has been possible to use the interactions employed in the various multi-qubit logic gates to simulate other processes of interest such as phase transitions in a quantum magnet.[f] A more general approach is to use a series of simple logic operations in short time steps to simulate arbitrary Hamiltonians. The basic features of this "digital quantum simulation" have been demonstrated by the Innsbruck group.[g]

As trapped-ion experiments are refined, we might be able to provide more stringent tests of certain quantum phenomena. For example, with antici-pated technical improvements, QIP systems will become larger, more com-plex, and more entangled. This will press issues such as the measurement problem and fundamental decoherence, and may enable the possibility of realizing situations like Schrödinger's cat with a macrocscopic sample of qubits.

---

[f]See for example, R. Islam *et al.*, *Nature Commun.* **2**, 377 (2011).
[g]B. Lanyon *et al.*, *Science* **334**, 57 (2011).

Discussion

J. Hartle From the point of view of testing quantum mechanics, as we look over these very interesting experiments, which one provides the most sensitive test of the principle of superposition?

I. Cirac There are several systems with which one can test the principle of superposition, in particular photon and ions. With photons you can test it by looking at interferences and using single photon detection. Or in the case of ions creating a superposition, waiting for half an hour (like Dave Wineland was saying before), and then check whether you will see interferences.

A. Zeilinger There has been a recent beautiful experiment by Gregor Weihs where he tested the superposition principle by looking at a situation where you have, not two slits, but three slits. You have superpositions of three amplitudes. And he tested whether there are any terms which go beyond what you would just get by application of the Born rule. This was a very recent experiment. And many years ago – and I must apologize since this was our experiment – we did an experiment with neutrons where we tested the linearity of the Schrödinger equation. This kind of experiment would be much harder with atoms, since neutrons don't like to interact with the environment, so they are much better isolated.

J. Hartle I remember the neutron experiments. I was curious to know what would be the best test in atomic physics.

A. Zeilinger The best – with photons – I think is this recent experiment by Weihs.

A. Leggett This question concerns the use of optical lattice systems to simulate the Hubbard model. I think undoubtedly that an analog solution of the Hubbard model is of enormous interest in many body physics. But I would comment that not everyone, by any means, would agree that the Hubbard model is a good starting point to discuss high temperature superconductivity. One of the things it most obviously does not have is the long range Coulomb interaction which, for all sorts of reasons, one would think would be much more important in two dimensions than in three. So my question is whether there is some prospect of doing an optical lattice simulation which does incorporate some part of the long range Coulomb interaction?

W. Ketterle I can offer only an indirect answer: That's why people are interested in atoms which have a Rydberg mixture because the high polarisability of Rydberg atoms means that they have long range interactions. And the same motivation is responsible for some of the advances with polar molecules because they also have long range interactions. But that is in the making: nobody has used this yet in optical lattices.

S. Dimopoulos We heard a lot about this interesting interface between atomic physics and other fields like condensed matter and quantum computation. One other exciting interface is the interface of atomic physics and gravity. In particular Jason Hogan at Stanford has a ten meter interferometer, as we speak. I'm told he is testing the equivalence principle by dropping Rubidium 85 and 87 atoms "a la Galileo". By comparing their motion he can test the equivalence principle to 15 decimals, which is a factor of 300 better than the best measurements that come from lunar laser ranging.

There is also the possibility, by having two ten meter atom interferometers, to build a gravity wave detector. There are proposals to that effect and Hogan is very interested in them. They are considering putting such interferometers in space. The idea is that you could have gravity wave detectors which are far smaller than LISA, simply because atoms move much slower than light. One could thus measure similar time scales and similar frequencies as LISA. The prospects are also good for studying Einstein's theory, not by studying the motion of stars and planets, but by studying the motion of atoms in gravitational field of the earth with the exquisite precision that atomic physics provides.

So I think it is a very fertile field for the future, the interaction of atoms and gravity. And of course there also are applications of these principles for studying basic physics. I am not talking about analog simulations, but the decades old program of measuring say electric dipole moments. So does anybody have a comment on the interface between atoms and gravity?

W. Phillips I want to comment about using atom interferometers to detect gravity waves. There was a bit of excitement a few years ago with a proposal to do exactly what is suggested here, by making a more compact gravity wave detector. The motivation was that because atoms are themselves acted upon more strongly by gravity than photons, that the gravity detector would be more sensitive. Well we did an analysis of that with some colleagues from the university of Maryland – some general relativity colleagues – and I participated as the atomic physics resource. We came to the conclusion that such gravity wave detectors are no more sensitive than optical detectors. My own naive way of thinking of this is that what you are trying to do when you measure gravity waves is measure the change in the metric, and atoms don't do that job any better than photons do. There is a place for atom interferometry in gravity wave detection, and it is basically using an atom interferometer as a very fancy mirror for an optical interferometer. By doing it the right way you can cancel out certain kinds of common mode motion in the mirrors, which can make the gravity wave detectors more robust. And therefore the hope that using an atom interferometer as a gravity wave detector in the naive way is apparently not right.

However, since it is proved that atoms are better sensors of inertial forces, such as rotation, than is a similar optical interferometer, there is still some hope that there might be a way to use that feature of the atoms. Just briefly, if you wanted to build a Sagnac interferometer – something that could detect rotation –, an atom interferometer of the same design is going to be something like 10^{10} or 10^{11} times more sensitive than an optical interferometer with the same number of photons as you have atoms, which of course is never the case. But making that into a gravity wave detector, well no one has figured out how to do that. But I think that would be a wonderful thing if somebody could do it.

If I can have one more minute I will comment on the question of using atom quantum computation techniques to address fundamental quantum mechanics, as asked by Jim. One of the things I found very exciting, and I would like to hear what people here think about it, is a proposal that using a general holographic principle – saying that if you know everything that happens on the surface of a volume, then you know everything that is happening inside as well – you could claim that the total amount of information available in the universe is limited by the surface of the universe divided into bits that are the size of a Planck length, but assuming that that information is classical. And then you would find that the total amount of information would not be sufficient to describe the entanglement of 400 qubits. Now nobody has ever entangled 400 qubits, but if you did, then you would basically disprove this idea. Now 400 qubits is a lot bigger than anything anybody has done so far, but its not by any means outside the possibilities we might imagine, even within the lifetime of the youngest people in the audience. And I would really be interested in knowing whether this is a reasonable idea that is worth refuting. I would love to refute it because I think it is silly, because why should you think that the information should be restricted to classical information when the world is quantum. But I would love to hear what other people think.

J. Maldacena I think that the bounds that are often discussed are bounds from quantum information. And in situations like gauge theory, or gravity duality, its really bounds on quantum information that you are testing. So the area of some volume is giving you the number of qubits of information.

W. Phillips Well the claim was that if you thought of it as classical bits instead of quantum bits, then you got a reasonable limit. If you think of it as qubits, then the amount of information is plenty for all practical purposes.

W. Zurek I want to come back to the question that Jim (Hartle) raised, and which Anton (Zeilinger) started to answer. He did not mention one of his most important experiments, which was his fullerenes which tested the superposition principle probably better than anything to date. But there is more in the offing. There are very interesting proposals, both by the chairman of the session and by the rapporteur, to put systems of the size of a virus in superposition. And there are bold people like Dick Bouwmeester who are pushing it even further, and trying to check if gravity may even at some point cause decoherence.

But I really have question for Wolfang (Ketterle). Remember that a dozen years ago we were hiking together, and I was trying to talk you into doing quantum phase transitions in a Bose-Einstein condensate. And you said that was far into the future. But from what you have been saying today, it seems like you are there. You can go through the phase transition, and look at what happens at the other end, for instance how many topological defects, how many vortex lines will be created as you go through. Is that something that is really on the horizon?

W. Ketterle Yes. That is correct. People are doing quantum quenches. They do rapid cooling of systems. Freeze in vorticity. Check for the development of vortices as a function of quench time. And in terms of quantum phase transitions, specially the superfluid to Mott insulator phase transition, quantum criticality has been examined. So in that sense the atomic systems are slow enough that we can do some quenches and study the dynamic behavior at quantum phase transitions. That is already happening, and we will see much more within the near future.

M. Berry Concerning this question of superposition, and testing superposition, it is true that the modern experiments, as described, test superposition to a high degree of accuracy. But I think we should not forget that there were experiments in the 50s and 60s with molecular beams where, in the scattering of atoms and molecules, you did see the interference between different classical trajectories which emerge in the same direction in the differential scattering cross section. And there, you can see several levels of interference. You can see fringes on different scales: three or more classical paths interfering exactly as one would expect from the principle of superposition. So its a pretty firm and substantial achievement.

D. Gross In the same domain, in particle physics I am always amused by these tests of the superposition principle such as K regeneration and neutrinos interfering after traveling over cosmic distances. To put it a different way: to test the superposition principle, one would like to know what a violation would mean, what is the measure of the accuracy of the test. For the equivalence principle we know what that means. What is the measure of a violation of the superposition principle? What is an alternative model which has a parameter which you could bound by these experiments? I mean, everything we do is quantum physics: we are testing the superposition principle daily.

S. Das Sarma Coming back to the question of quantum phase transitions in many body Hamiltonians, I want to make a point which is known to experts but which has not come up in the discussion. The Mott to superfluid transition for the Bose-Hubbard model corresponds to a bosonic Hamiltonian. And for bosonic Hamiltonians, just regular digital simulation is very good, because you don't have a sign problem. What you are doing in the laboratory is very good. Its a physical realization. But its not a much bigger system than what we can do with simulations. So, its very good, it verifies what we know spectacularly. We see dynamics. But even dynamics can be simulated for bosonic systems very well. For fermions we have seen

BCS-BEC crossover in the laboratory. But that is also a mean field physics. So again theory understands it very well. We are going to see it in all its glory. But I would say, to make a provocative statement, I myself did not learn anything from it because it was something I expected. Mean field is something I can do.

Where we are going to make a real breakthrough is when we do something like these fermionic systems, which is what Wolfgang was emphasizing. Because of the sign problem, digital simulation is completely helpless there. That is what we are waiting for. That is where we will learn something. Even it if it is the Hubbard model, whether it solves high T_c or not is irrelevant because its a very important model problem that we would like to know the answer to, and digital simulation does not give us the answer.

G. 't Hooft May I just add a remark to the question of testing linearity and the superposition principle. I think you must realize that such tests are very similar to testing the validity of probability theory, where probabilities add up and multiply the way we learn from the laws of logic. I think when you ask that question, you realize that all depends on the setup of the experiment. And the same when you talk about linearity and superposition: what do you mean when you talk about these concepts? I don't think that by itself you can quantify it, as David (Gross) was asking.

D. Wineland I think that in this game of trying to establish who has the best test of the superposition principle, the rules are not well established. Are two level systems interesting, or should it be entangled systems? Is the length scale important? Is the number of constituents, the number of degrees of freedom, important? I at least don't have any guidelines. I would be interesting to hear what people think is the most important.

J. Hartle To chime in on this question, one simple measure is to ask if a superposition decays, and on what time scale does it decay? But I agree with these comments also that it is not a very well defined question.

A. Aspect About the last question, about how superpositions decay, I would like to recall an old historical discussion of the paper of Einstein, Podolsky and Rosen. There was the hypothesis that entanglement would be a property at short distances that would decay when the members of the pair go far away. Now we know, at least for photons, that that is not the case. In our experiment we already put detectors at 13 meters, then Anton put them at thousands of meters, hundreds of kilometers. I don't know whether it is an answer to Jim, but it is a piece of an answer. Entanglement survives at long distances with photons.

A. Zeilinger Just to try to answer a little the question of our conference chair. I fully agree that it is very difficult to define what you mean by a test, and there have been many. The two tests which I mentioned actually had some quantitative statements, some quantitative alternatives. One was a very beautiful non linear variant of the Schrödinger equation invented by

Iwo Bialynicki-Birula (he is one of the grand old men of polish physics, as you know). He had very good reasons – at least at that time – why his proposal would not have been discovered in any experiments yet. So there was a quantitative statement. And in the experiment by Weihs its the same: there was an explicit model and they could put numbers on it.

Your comment, Dave, is actually the reason why I did not mention the fullerene experiment. They showed superposition, but we did not put any number on any alternative theory.

The general problem is that luckily quantum mechanics is so tight that it is very difficult to invent reasonable tests.

T. Leggett I agree 100percent with Dave Wineland's point that the criteria for what is interesting to test in a superposition experiment are not well defined. However it does seem to me that one possible criteria of interest is how close are the states which are to be superposed to being what we would all recognize as being distinct at the macroscopic level. And what is very interesting I think is that some of the recent experiments on SQUIDS, by any reasonable criteria, are pretty close to involving states which are as macroscopically distinct as those which could be distinguished by the naked eye. So I think that is one possible criteria, but by no means the only one.

W. Phillips I want to comment on this question of macroscopic superpositions, because I agree that this is certainly one of the frontiers, to create macroscopic superpositions and see whether they decay. It is certainly a different kind of question than creating microscopic superpositions, and seeing whether they decay. And decay, well of course they will decay, but will they decay because of something we understand, or because of something we don't yet understand. That's the really interesting part.

So I want to talk about the SQUIDS. Because on the face of it, these SQUID circuits appear to be macroscopic superpositions: they involve roughly a gazillion electrons in different states. On the other hand, they represent a distance in flux quanta of 1. Its the superposition of, say, a state of 1 flux quantum and a state of zero flux quanta. So the question arises: is this really a macroscopic superposition? One of the things in this regard which I don't understand very well is that Birgitta Whaley at Berkeley has an explicit calculation asking how many single particles, or single pairs, would you need to change in order to go from one state to the other. And that number, as she calculates, is extremely modest. But I see that you want to comment now.

T. Leggett Well I actually have an unpublished note in response to Birgitta Whaley in which I compare that situation with the situation of observing the smallest dust particle that you can see with the unaided naked eye moving by its diameter in one second. The number in that case is much smaller than the number that Brigitta has calculated. Its very surprising but its true.

W. Phillips But that then means that these are not particularly macroscopic superpositions.

T. Leggett Well I think that is a matter of definitions.

P. Zoller Well Sankar had raised a question, and I think there were some comments related to his question. Essentially what he had said is that if you take bosons in equilibrium, then you can do path integral Monte Carlo, and we have efficient classical codes to do these things. So where is the borderline between quantum simulations and when classical simulations are possible.

S. Sachdev I wanted to come back to Wolfgang's (Ketterle) presentation where he was mentioning quantum magnetism being realized by cold atoms. But even the subject of quantum magnetism is extremely vast, and there are many interesting possibilities of ground states of magnets. And there is a lot of very interesting routes to realizing more complex magnetic states, some of which could not be realized in a classical digital computer. I think that may be a shorter route to achieving the challenge posed by Sankar (Das Sarma), than going all the way to fermions.

Rapporteur talk by S. Girvin: Quantum Machines: Coherent Control of Mesoscopic Solid-State Systems

Abstract

Once the exclusive province of atomic physics, coherent control of quantum states is rapidly becoming routine in certain mesocopic solid-state systems. Remarkable experimental progress in coupling light to the motion of matter through feeble radiation pressure forces has at long last 'put the mechanics into quantum mechanics', allowing mechanical motion to be cooled to its quantum ground state and permitting controlled production of enormous non-linear optical effects. Superconducting electrical circuits can now be used as novel non-linear quantum optical systems for microwave photons. Through clever quantum engineering and design, the phase coherence time of 'artificial atoms' constructed from superconducting Josephson junctions has risen nearly five orders of magnitude in the last decade. We stand at the threshold of a new era of modular 'quantum machines' which can be assembled to carry out computational and signal processing tasks that are impossible on classical devices.

1. Introduction

A quantum machine is a device whose degrees of freedom are intrinsically quantum mechanical. While we do not yet fully understand the properties of such devices, there is great hope (and some mathematical proof) that such machines will have novel capabilities that are impossible to realize on classical hardware. One might think that quantum machines were first constructed long ago. For example, the laser and the transistor would seem to rely on quantum physics for their operation. It is clear that the frequency of a laser cannot be computed without the quantum theory that predicts the excitation energies of the atoms in the laser. Similarly the optimal bias voltage of a bipolar transistor depends on the electronic band gap energy of the material from which it is made. Nevertheless, it is only the particular values of the operating *parameters* of these machines that are determined by quantum physics. Once we know the values of these parameters, we see that these are classical machines because their degrees of freedom are purely classical. Indeed the light output from a laser is special because it is exactly like the RF output of the classical oscillator that powers a radio station's antenna. Similarly, the currents and voltages in an ordinary transistor circuit need not be treated as non-commuting quantum operators in order to understand the operation of the circuit. Its degrees of freedom are, for all intents and purposes, classical.

2. Superconducting Qubits and Quantum Microwave Circuits

Experimental progress over the last decade in creating and controlling quantum coherence in superconducting electrical circuits has been truly remarkable. Simple quantum machines have already been built using superconducting circuits which can manipulate and measure the states of individual microwave quanta,[1-6] entangle two[7,8] and three qubits,[9,10] run simple quantum algorithms[11,12] and perform rudimentary quantum error correction.[13]

The quantum electrodynamics of superconducting microwave circuits[14-18] has been dubbed 'circuit QED' by analogy to cavity QED in quantum optics. In this quantum optics approach to microwave circuits, the superconducting qubits play the role of artificial atoms whose properties can be engineered. Despite being large enough to be visible to the naked eye (1mm in size and comprising trillions of Al atoms), these artificial atoms have essentially a single degree of freedom (collective charge oscillations) and hence a very simple discrete set of quantized energy levels which are as well understood as those of the prototypical single-electron atom, hydrogen. The reduction of such a large many-electron system to a single degree freedom follows from two key features of the physics: the superconductivity which gaps out particle-hole pair excitations in the electron gas of the metal and the strength and long-range nature of the Coulomb interaction which gaps out the remaining short-wavelength collective oscillations.

It has proven possible to put these artificial atoms into coherent superpositions of different quantum states so that they can act as quantum bits. Through clever engineering, the coherence times of such superposition states have risen nearly five orders of magnitude from nanoseconds for the first superconducting qubit created in 1999[19] to 50-100 microseconds today using so-called 'transmon' qubits.[20,21] This 'Schoelkopf's Law' for the exponential growth of coherence time is illustrated in Fig. 1. Future improved qubit designs, microwave circuit designs, and materials improvements should allow this trend to continue unabated. In addition to being a potentially powerful engineering architecture for building a quantum computer, the 'circuit QED' paradigm opens up for us a novel new regime to study ultra-strong coupling between 'atoms' and individual microwave photons.[22]

2.1. *Quantum Limited Amplification, Measurement and Feedback*

In the early days of quantum physics, theorists struggling to understand the role of measurements in the quantum theory were forced to argue over the implications of various 'gedanken' experiments. The rapid progress in quantum microwave circuits and qubits now allows these fundamental measurement experiments to be carried out and theorists are forced to confront actual data. The formalisms developed to understand quantum noise and measurements in quantum optics have been largely unknown to condensed matter physics but there is now growing interdisciplinary communication on the subject.[23] The two quadratures of a microwave signal

$$\hat{X} \equiv \frac{\hat{a} + \hat{a}^\dagger}{2} \tag{1}$$

$$\hat{Y} \equiv -i\frac{\hat{a} - \hat{a}^\dagger}{2} \tag{2}$$

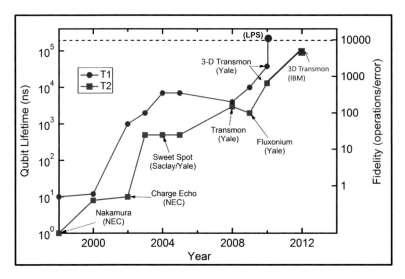

Fig. 1. 'Schoelkopf's Law' illustrating the steady exponential rise in superconducting phase qubit energy relaxation times T_1 and phase coherence times T_2. The scale on the right-hand side shows the number of operations that can be carried out in one coherence time assuming a typical gate operation time of 20ns.

are canonically conjugate. To preserve the canonical commutation relations an amplifier that amplifies one quadrature by a factor G must necessarily *de-amplify* the other quadrature by the same factor[23]

$$\hat{X}_{\text{out}} = G\hat{X}_{\text{in}} \tag{3}$$

$$\hat{Y}_{\text{out}} = G^{-1}\hat{Y}_{\text{in}}. \tag{4}$$

Such 'phase sensitive' amplification is sometimes useful and appropriate. If however we want to build an ordinary 'phase preserving' amplifier which amplifies both quadratures equally, we are forced by the rules of quantum mechanics to entangle our signal with another signal channel (known as the idler signal)

$$\hat{a}_{\text{out}} = G\hat{a}_{\text{in}} + \sqrt{G^2 - 1}\,\hat{b}_{\text{in}}^{\dagger} \tag{5}$$

$$\hat{b}_{\text{out}} = G\hat{b}_{\text{in}} + \sqrt{G^2 - 1}\,\hat{a}_{\text{in}}^{\dagger}. \tag{6}$$

This familiar form is nothing more than a Bogoljubov transformation between the two input modes $\hat{a}_{\text{in}}, \hat{b}_{\text{in}}$ and the two output modes.[23] Even if the amplifier is operated with no signal input at the idler port, the amplified vacuum noise from this extra port necessarily adds noise to the output. Every amplifier, even if quantum mechanically perfect, lowers the signal to noise ratio! This added noise is a consequence of the Heisenberg uncertainty principle and cannot be avoided. There is a long history of parametric amplifiers based on Josephson junctions, but only in the last few years have they become both practical and able to operate close to this quantum limit of minimum added noise.[24-27]

Now that robust and practical quantum-limited amplifiers are available, the world of quantum signal transduction and quantum feedback control[28-30] is open to us. Classical control theory is now well-developed and is ubiquitous in modern technology, playing an essential role in guaranteeing the stability of complex systems ranging from transistor amplifiers to chemical manufacturing plants. Quantum feedback is more subtle because observation of a quantum system changes it due to the back-action of the measurement.[23] An interesting recent example is the stabilization of photon number (Fock) states in a microwave cavity through repeated measurements using Rydberg atoms.[31] The 'natural' state of a driven microwave cavity is a coherent state which is a superposition of different photon number states. By controlling the classical drive on the cavity based on information fed back from measurements of atoms passing through the cavity, the Paris group was able to stabilize the cavity in a state with a definite (pre-selected) photon number. Such Fock states are useful resources in quantum communication and continuous variable quantum computation. Korotkov[32] has proposed a quantum feedback scheme for superconducting qubits to produce persistent Rabi oscillations and this has recently been successfully implemented by the Berkeley group.[33] Quantum feedback, control and autonomous error correction is a new area for mesoscopic systems which is now poised to take off.

2.2. *Future Directions for Superconducting Qubits*

The rate of progress in the realization of quantum microwave circuits over the past decade has been truly remarkable and represents both progress towards building novel quantum machines and realizing non-linear quantum optics in a novel strong-coupling regime. Circuit QED is much more than atomic physics with wires. We have a set of modular elements which are readily connected together. Hence, we have the opportunity to assemble large scale structures from these quantum building blocks and do some real quantum engineering. Further progress will require scaling up both the number of qubits and resonators and continuing to advance coherence times and gate fidelities. As the number of qubits grows, it will be important to increase the on-off ratio of the couplings among them. Simply detuning them from each other will probably not be sufficient and interference between two coupling channels which can null out the net coupling will likely be needed. Houck et al. have developed a novel tunable transmon qubit structure in which the vacuum Rabi coupling can be tuned over a wide range using magnetic flux to control the interference between two internal modes of the qubit.[34-36]

Another exciting direction involves using multiple physical qubits to realize individual logical qubits that overcome the difficulties of maintaining stable transition frequencies. In particular, the possibility of topological protection[37-41] is beginning to be explored in superconducting qubits.[42] The central idea is that qubits are constructed in which the ground and excited states are degenerate and this degeneracy is robust against local variations in Hamiltonian parameters. Even if the energy levels are not exactly degenerate, it would be very useful to have a qubit with a "Lambda" energy level scheme, that is, two nearly degenerate levels that can be coupled via stimulated Raman pulses through a third level. This would be advantageous both as a robust qubit and for purposes of fundamental quantum optics studies. It seems reasonably certain that this cannot be achieved without applied magnetic flux to frustrate the Josephson couplings (as in a flux qubit or in the fluxonium qubit). Indeed the fluxonium qubit[43] may turn out to be quite useful as a Lambda system.

The development of large resonator/qubit arrays will be interesting not only as a quantum computation architecture[44] but also for fundamental quantum optics purposes including quantum simulations of many-body bosonic systems. An array of resonators each containing a qubit that induces a Kerr nonlinearity will be a realization of the boson Hubbard model[45] which exhibits both superfluid and Mott insulator phases. There is now a burgeoning interest in seeing 'quantum phase transitions of light'.[46–67] The possibility has even been suggested of realizing the fractional quantum Hall effect for bosons by creating a synthetic magnetic field seen by the photons as they hop on a lattice of resonators.[65] Fig. (2) shows a particularly interesting proposed example of a simple honeycomb lattice of coplanar waveguide resonators, each of which contains a superconducting qubit. The polariton excitations live on the bonds of this lattice or equivalently on the sites of the dual Kagome lattice. The Kagome lattice[68–70] is a great importance in the study of frustrated spin systems in part because of the flat energy band dispersion shown in the inset of Fig. (2). The dimensionless measure of how strong the correlations are in a bosonic many-body system like this is the ratio of the Hubbard U (self-Kerr) to the bandwidth. Hence a flat band automatically leads to strong correlations. Experimental evidence[71] for the flat bands has been seen in low-disorder kagome arrays of several hundred resonators and experiments which include interactions induced by the addition of qubits can be expected in the near future.

Since the transmon qubit is itself an anharmonic oscillator, it might be easiest to simply use a lattice of coupled transmons to realize the boson Hubbard model (if one is content with only negative values of the Kerr coefficient). The advantage of using a lattice of resonators is that their resonance frequencies can be closely matched to a single fixed value. The Kerr coefficient induced by coupling each resonator to an off-resonant qubit will have some variation due to variations in qubit transition frequencies, but this disorder in the Hubbard U will be more tolerable than disorder in the photon 'site energies.' A further advantage is that if the qubits are tuned to be degenerate or nearly degenerate with the resonators, the upper and lower 'polariton' excitations of the combined system have negative and positive Kerr coefficients respectively, allowing one to explore the Bose-Hubbard model with both attractive and repulsive boson-boson interactions. Just as cold atom systems

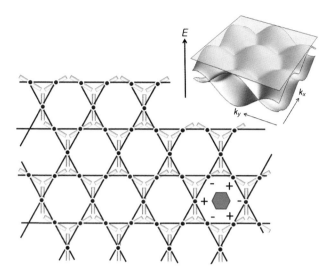

Fig. 2. Honeycomb lattice of coplanar waveguide microwave resonators (blue bars), each coupled to a superconducting qubit (red dots). The resulting 'polariton' excitations live on the bonds of the lattice or equivalently on the sites of the dual Kagome lattice (whose bonds are the red lines). Inset: bandstructure of the Kagome lattice which contains three bands corresponding to the three sites per unit cell of the lattice. One of the bands is completely flat (dispersionless) meaning the excitations are localized in space and cannot propagate. One of the localized modes is illustrated on one of the hexagons in the lower right. Notice the perfect destructive interference between the positive and negative amplitudes when the excitation attempts to hop to an adjacent site outside the given hexagon. This is the source of the localization.

are now used to simulate condensed matter models, so we may be able to use photon/polaritons as interacting strongly correlated bosons, which can be probed, measured and controlled in ways that are impossible in ordinary condensed matter.

The demands placed on classical computers to design and simulate even small numbers of qubits and resonators is already enormous. As the size of our quantum machines grows, the modeling complexity will grow exponentially. Similarly the experimental measurements and process tomography needed to verify the operation of larger quantum machines will become extremely challenging. We will have to develop calibration and verification protocols that can reliably vet each segment of a quantum processor without the luxury of complete end-to-end process tomography. Today, classical computers are sufficiently complex that we cannot design the next generation by hand. We must use the current generation of computers to design the next. Eventually this might be true of quantum machines. However, one wonders if we might soon find ourselves in a situation where both existing quantum machines and classical computers are not powerful enough to be used to design and model the next generation of quantum machines. How will we cross *that* desert?

3. Quantum Optomechanics

Another exciting area which is rapidly developing is optomechanics[72–74] in which feeble radiation pressure forces are harnessed to control the state of mechanical motion of both microscopic and macroscopic objects. The radiation pressure forces are enhanced by using high-finesse optical resonators to increase the circulating power and the optomechanical coupling is achieved through the parametric dependence of the cavity frequency ω_R on the position x of a deformable mechanical element. The Hamiltonian has the generic form

$$H = \omega_R(x)a^\dagger a + \omega_M b^\dagger b \tag{7}$$

where assume we can focus on only a single optical mode and a single mechanical mode, and a and b are respectively the lowering operators for the photon and phonon number for these modes. Recalling that x is a quantum variable $x \equiv x_{ZPF}(b+b^\dagger)$, and assuming its fluctuations are small, we can Taylor expand the parametric coupling to obtain

$$H = \left[\omega_R(0) + \frac{\partial \omega_R}{\partial x}x_{ZPF}(b+b^\dagger) + \frac{1}{2}\frac{\partial^2 \omega_R}{\partial x^2}x_{ZPF}^2(b+b^\dagger)^2 + \ldots\right]a^\dagger a + \omega_M b^\dagger b. \tag{8}$$

We see that the parametric coupling leads to 3-wave and 4-wave mixing terms between the optical and mechanical degrees of freedom. By driving the cavity red-detuned from $\omega_R(0)$ by an amount equal to the mechanical frequency ω_M, one can achieve cavity-assisted laser cooling of the mechanical motion. Conversely for blue detuning one can achieve parametric amplification of mechanical motion which can lead to self-oscillations.[72] This effect has also been used in the microwave domain to achieve low-noise parametric amplification of microwave signals[75] with the mechanical mode playing the role of the idler channel.

The fact that the Hamiltonian is not quadratic in field operators means that we are dealing with a new kind of non-linear optical medium. Photon-photon interactions are created by the fact that the radiation pressure from the first photon displaces the mechanical system which in turn changes the optical properties of the resonator as seen by the second photon. Such effects can in principle be used to produce squeezing, and to switch light with light in an all-optical communications network. This form of mechanically-induced optical non-linearity may prove very advantageous compared to the weak electro-optic response which produces ordinary non-linear optical effects in most materials.

One of the holy grails in the field is detection of quantum jumps in the phonon number. Using quadratic parametric coupling, one can take advantage of the fact that the slowly varying part of $(b+b^\dagger)^2 = 2\hat{n}+1$ contains the phonon number, it is possible in principle to see quantum jumps, but it will be technically challenging.[73] Another major goal of the field is to see the quantum shot noise of the optical radiation pressure (or even better, the radiation pressure of a single photon). Recent remarkable experiments[76,77] with a flexible membrane capacitor coupled to a microwave resonator have achieved mechanical ground-state cooling and the strong coupling regime in which the mechanical and optical modes hybridize with coupling stronger than the damping but only for the case where the resonator is strongly driven (i.e. not at the single photon level). The strong coupling limit opens up the prospect of coherent swapping operations between the mechanical state and the microwave state and hence remote state transfer between distant mechanical systems via exchange of microwave photons. Remote entanglement via photon measurements may also be possible.[78] Progress is also being made hybrid systems in which mechanical motion is coupled to superconducting qubits[79] or to trapped atoms.[80,81]

4. Summary

We are at the beginning of a Golden Age for coherent quantum control of mesoscopic systems. Coherence times for superconducting qubits are growing exponentially at a rate of nearly an order-of-magnitude every two years. We are entering a new era of quantum-limited amplification and detection, quantum feedback control, quantum state transfer, and quantum error correction for systems of superconducting qubits, and mechanical and microwave oscillators. We are exploring and testing quantum mechanics in entirely new regimes and may someday be able to use quantum machines to create entirely new technologies for the 21st century.

Acknowledgments

The author is grateful to his experimental colleagues and collaborators Michel Devoret, Rob Schoelkopf, and Jack Harris for numerous discussions and collaborations on optomechanics and circuit QED. The discussion of superconducting qubits presented here is a brief summary based in part on a much more extensive review of the subject which the author presented at the 2011 Les Houches Summer School on Quantum Machines.[18] The author's research is supported by the National Science Foundation, the Army Research Office and IARPA.

References

1. A. A. Houck, D. I. Schuster, J. M. Gambetta, J. A. Schreier, B. R. Johnson, J. M. Chow, L. Frunzio, J. Majer, M. H. Devoret, S. M. Girvin and R. J. Schoelkopf, *Nature* **449**, 328 (2007).
2. M. Hofheinz, E. M. Weig, M. Ansmann, R. C. Bialczak, E. Lucero, M. Neeley, A. D. O'Connell, H. Wang, J. M. Martinis and A. N. Cleland, *Nature* **454**, 310 (2008).
3. M. Hofheinz, H. Wang, M. Ansmann, R. C. Bialczak, E. Lucero, M. Neeley, A. D. O'Connell, D. Sank, J. Wenner, J. M. Martinis and A. N. Cleland, *Nature* **459**, 546 (2009).
4. B. R. Johnson, M. D. Reed, A. A. Houck, D. I. Schuster, L. S. Bishop, E. Ginossar, J. M. Gambetta, L. DiCarlo, L. Frunzio, S. Girvin and R. Schoelkopf, *Nature Physics* **6**, 663 (2010).
5. M. Mariantoni, H. Wang, R. C. Bialczak, E. Lucero, M. Neeley, A. O'Connell, D. Sank, M. Weides, J. Wenner, T. Yamamoto, Y. Yin, J. Zhao, J. M. Martinis and A. Cleland, *Nature Physics* **7**, 287 (2011).

6. H. Wang, M. Mariantoni, R. C. Bialczak, M. Lenander, E. Lucero, M. Neeley, A. O'Connell, D. Sank, M. Weides, J. Wenner, T. Yamamoto, Y. Yin, J. Zhao, J. M. Martinis and A. N. Cleland, *Phys. Rev. Lett.* **106**, p. 060401 (2011).
7. M. Ansmann, H. Wang, R. C. Bialczak, M. Hofheinz, E. Lucero, M. Neeley, A. D. O'Connell, D. Sank, M. Weides, J. Wenner, A. N. Cleland and J. M. Martinis, *Nature* **461**, 504 (2009).
8. J. M. Chow, L. DiCarlo, J. M. Gambetta, A. Nunnenkamp, L. S. Bishop, L. Frunzio, M. H. Devoret, S. M. Girvin and R. J. Schoelkopf, *Phys. Rev. A* **81**, p. 062325 (2010).
9. M. Neeley, R. C. Bialczak, M. Lenander, E. Lucero, M. Mariantoni, A. D. O'Connell, D. Sank, H. Wang, M. Weides, J. Wenner, Y. Yin, T. Yamamoto, A. N. Cleland and J. M. Martinis, *Nature* **467**, p. 570 (2010).
10. L. DiCarlo, M. Reed, L. Sun, B. Johnson, J. Chow, J. Gambetta, L. Frunzio, S. Girvin, M. Devoret and R. Schoelkopf, *Nature* **467**, 574 (2010).
11. L. DiCarlo, J. M. Chow, J. M. Gambetta, L. S. Bishop, B. R. Johnson, D. I. Schuster, J. Majer, A. Blais, L. Frunzio, S. M. Girvin and R. J. Schoelkopf, *Nature* **460**, 240 (2009).
12. M. Mariantoni, H. Wang, T. Yamamoto, M. Neeley, R. C. Bialczak, Y. Chen, M. Lenander, E. Lucero, A. D. O'Connell, D. Sank, M. Weides, J. Wenner, Y. Yin, J. Zhao, A. N. Korotkov, A. N. Cleland and J. M. Martinis, *Science* **334**, p. 61 (2011).
13. M. D. Reed, L. DiCarlo, S. E. Nigg, L. Sun, L. Frunzio, S. M. Girvin and R. J. Schoelkopf, *Nature* **482**, 382 (2012).
14. M. Devoret, *Quantum Fluctuations* (Elsevier, Amsterdam, 1997), pp. 351–385.
15. A. Blais, R.-S. Huang, A. Wallraff, S. M. Girvin and R. J. Schoelkopf, *Phys. Rev. A* **69**, p. 062320 (2004).
16. A. Wallraff, D. I. Schuster, A. Blais, L. Frunzio, R.-S. Huang, J. Majer, S. Kumar, S. M. Girvin and R. J. Schoelkopf, *Nature* **431**, 162 (2004).
17. R. Schoelkopf and S. Girvin, *Nature* **451**, p. 664 (2008).
18. S. M. Girvin, Circuit QED: Superconducting qubits coupled to microwave photons, *Proc. 2011 Les Houches Summer School on Quantum Machines* (Oxford University Press, 2012).
19. Y. Nakamura, Y. A. Pashkin and J. S. Tsai, *Nature* **398**, 786 (1999).
20. H. Paik, D. I. Schuster, L. S. Bishop, G. Kirchmair, G. Catelani, A. P. Sears, B. R. Johnson, M. J. Reagor, L. Frunzio, L. I. Glazman, S. M. Girvin, M. H. Devoret and R. J. Schoelkopf, *Phys. Rev. Lett.* **107**, p. 240501 (2011).
21. C. Rigetti, S. Poletto, J. M. Gambetta, B. L. T. Plourde, J. M. Chow, A. D. Corcoles, J. A. Smolin, S. T. Merkel, J. R. Rozen, G. A. Keefe, M. B. Rothwell, M. B. Ketchen and M. Steffen, arXiv:1202.5533 (2012).
22. M. Devoret, S. Girvin and R. Schoelkopf, *Annalen der Physik* **16**, 767 (2007).
23. A. A. Clerk, M. H. Devoret, S. M. Girvin, F. Marquardt and R. J. Schoelkopf, *Rev. Mod. Phys.* **82**, 1155 (2010), (Longer version with pedagogical appendices available at: arXiv.org:0810.4729).
24. N. Bergeal, R. Vijay, V. E. Manucharyan, I. Siddiqi, R. J. Schoelkopf, S. M. Girvin and M. H. Devoret, *Nature* **465**, 64 (2010).
25. N. Bergeal, F. Schackert, M. Metcalfe, R. Vijay, V. E. Manucharyan, L. Frunzio, D. Prober, R. J. Schoelkopf, S. M. Girvin and M. H. Devoret, *Nature Physics* **6**, 296 (2010).
26. M. A. Castellanos-Beltran, K. D. Irwin, G. C. Hilton, L. R. Vale and K. W. Lehnert, *Nature Physics* **4**, 929 (2008).
27. N. Roch, E. Flurin, F. Nguyen, P. Morfin, P. Campagne-Ibarcq, M. H. Devoret and B. Huard, *Phys. Rev. Lett.* (in press), arXiv:1202.1315 (2012).

28. H. Mabuchi, *Phys. Rev. A* **78**, p. 032323 (2008).
29. A. Pechen, C. Brif, R. Wu, R. Chakrabarti and H. Rabitz, *Phys. Rev. A* **82**, p. 030101 (2010).
30. J. Kerckhoff, H. I. Nurdin, D. S. Pavlichin and H. Mabuchi, *Phys. Rev. Lett.* **105**, p. 040502 (2010).
31. C. Sayrin, I. Dotsenko, X. Zhou, B. Peaudecerf, T. Rybarczyk, S. Gleyzes, P. Rouchon, M. Mirrahimi, H. Amini, M. Brune, J.-M. Raimond and S. Haroche, *Nature* **477**, 73 (2011).
32. A. N. Korotkov, Quantum Bayesian approach to circuit QED measurement, *Proc. 2011 Les Houches Summer School on Quantum Machines* (Oxford University Press, 2012), p. arXiv:1111.4016.
33. R. Vijay, C. Macklin, D. H. Slichter, S. Weber, K. Murch, R. Naik, A. N. Korotkov, and I. Siddiqi, *Bull. Am. Phys. Soc.*, p. H29.00009 (2012).
34. J. M. Gambetta, A. A. Houck and A. Blais, *Phys. Rev. Lett.* **106**, p. 030502 (2011).
35. S. J. Srinivasan, A. J. Hoffman, J. M. Gambetta and A. A. Houck, *Phys. Rev. Lett.* **106**, p. 083601 (2011).
36. A. J. Hoffman, S. J. Srinivasan, J. M. Gambetta and A. A. Houck, *Phys. Rev. B* **84**, p. 184515 (2011).
37. A. Kitaev, *Ann. Phys.* **303**, 2 (2003).
38. L. B. Ioffe and M. V. Feigelman, *Phys. Rev. B* **66**, p. 224503 (2002).
39. L. B. Ioffe, M. V. Feigel'man, A. Ioselevich, D. Ivanov, M. Troyer and G. Blatter, *Nature* **415**, 503 (2002).
40. B. Doucot, M. V. Feigelman and L. B. Ioffe, *Phys. Rev. Lett.* **90**, p. 107003 (2003).
41. B. Doucot, M. V. Feigelman, L. B. Ioffe and A. S. Ioselevich, *Phys. Rev. B* **71**, p. 024505 (2005).
42. S. Gladchenko, D. Olaya, E. Dupont-Ferrier, B. Doucot, L. B. Ioffe and M. E. Gershenson, *Nature Physics* **5**, 48 (2009).
43. V. E. Manucharyan, J. Koch, L. Glazman and M. Devoret, *Science* **326**, 113 (2009).
44. F. Helmer, M. Mariantoni, A. G. Fowler, J. von Delft, E. Solano and F. Marquardt, *EPL* **85**, p. 50007 (2009).
45. M. P. A. Fisher, P. B. Weichman, G. Grinstein and D. S. Fisher, *Phys. Rev. B* **40**, 546 (1989).
46. A. D. Greentree, C. Tahan, J. H. Cole and L. C. L. Hollenberg, *Nature Physics* **2**, 856 (2006).
47. F. Illuminati, *Nature Physics* **2**, 803 (2006).
48. M. J. Hartmann and M. B. Plenio, *Phys. Rev. Lett.* **99**, p. 103601 (2007).
49. T. C. Jarrett, A. Olaya-Castro and N. F. Johnson, *EPL* **77**, p. 34001 (2007).
50. D. Rossini and R. Fazio, *Phys. Rev. Lett.* **99**, p. 186401 (2007).
51. M. J. Hartmann, F. G. S. L. Brandao and M. B. Plenio, *NJP* **10**, p. 033011 (2008).
52. M. I. Makin, J. H. Cole, C. Tahan, L. C. L. Hollenberg and A. D. Greentree, *Phys. Rev. A* **77**, p. 053819 (2008).
53. M. Aichhorn, M. Hohenadler, C. Tahan and P. B. Littlewood, *Phys. Rev. Lett.* **100**, p. 216401 (2008).
54. J. Cho, D. G. Angelakis and S. Bose, *Phys. Rev. A* **78**, p. 062338 (2008).
55. J. Cho, D. G. Angelakis and S. Bose, *Phys. Rev. Lett.* **101**, p. 246809 (2008).
56. J. Zhao, A. W. Sandvik and K. Ueda, Insulator to superfluid transition in coupled photonic cavities in two dimensions, arXiv:0806.3603, (2008).
57. N. Na, S. Utsunomiya, L. Tian and Y. Yamamoto, *Phys. Rev. A* **77**, p. 031803 (2008).
58. S.-C. Lei and R.-K. Lee, *Phys. Rev. A* **77**, p. 033827 (2008).
59. A.-C. Ji, Q. Sun, X. C. Xie and W. M. Liu, *Phys. Rev. Lett.* **102**, p. 023602 (2009).

60. J. Hartmann, F. Brandao and M. Plenio, *Laser & Photon. Rev.* **2**, 527 (2009).
61. M. Grochol, *Phys. Rev. B* **79**, p. 205306 (2009).
62. D. Dalidovich and M. P. Kennett, *Phys. Rev. A* **79**, p. 053611 (2009).
63. I. Carusotto, D. Gerace, H. E. Tureci, S. D. Liberato, C. Ciuti and A. Imamoglu, *Phys. Rev. Lett.* **103**, p. 033601 (2009).
64. S. Schmidt and G. Blatter, *Phys. Rev. Lett.* **103**, p. 086403 (2009).
65. J. Koch and K. Le Hur, *Phys. Rev. A* **80**, p. 023811 (2009).
66. M. Lieb and M. Hartmann, *New J. Phys.* **12**, p. 093031 (2010).
67. S. Schmidt, D. Gerace, A. A. Houck, G. Blatter and H. E. Türeci, *Phys. Rev. B* **82**, p. 100508 (2010).
68. J. T. Chalker and J. F. G. Eastmond, *Phys. Rev. B* **46**, 14201 (1992).
69. L. Balents, M. P. A. Fisher and S. M. Girvin, *Phys. Rev. B* **65**, p. 224412 (2002).
70. P. Sindzingre and C. Lhuillier, *EPL* **88**, p. 27009 (2009).
71. D. Underwood, W. Shanks, A. Hoffman, J. Koch and A. Houck, *Bull. Am. Phys. Soc.*, p. B29.00003 (2012).
72. F. Marquardt and S. M. Girvin, *Physics* **2**, p. 40 (2009).
73. J. D. Thompson, B. M. Zwickl, A. M. Jayich, F. Marquardt, S. M. Girvin and J. G. E. Harris, *Nature* **452**, 72 (2008).
74. M. Poot and H. S. J. van der Zant, *Phys. Rep.* **511**, 273 (2012).
75. F. Massel, T. T. Heikkilä, J.-M. Pirkkalainen, S. U. Cho, H. Saloniemi, P. J. Hakonen and M. A. Sillanpää, *Nature* **480**, 351 (2011).
76. J. D. Teufel, T. Donner, D. Li, J. W. Harlow, M. S. Allman, K. Cicak, A. J. Sirois, J. D. Whittaker, K. W. Lehnert and R. W. Simmonds, *Nature* **475**, 359 (2011).
77. J. D. Teufel, D. Li, M. S. Allman, K. Cicak, A. J. Sirois, J. D. Whittaker and R. W. Simmonds, *Nature* **471**, 204 (2011).
78. K. Børkje, A. Nunnenkamp and S. M. Girvin, *Phys. Rev. Lett.* **107**, p. 123601 (2011).
79. A. D. O'Connell, M. Hofheinz, M. Ansmann, R. C. Bialczak, M. Lenander, E. Lucero, M. Neeley, D. Sank, H. Wang, M. Weides, J. Wenner, J. M. Martinis and A. N. Cleland, *Nature* **464**, 697 (2010).
80. K. Hammerer, M. Wallquist, C. Genes, M. Ludwig, F. Marquardt, P. Treutlein, P. Zoller, J. Ye and H. J. Kimble, *Phys. Rev. Lett.* **103**, p. 063005 (2009).
81. S. Camerer, M. Korppi, A. Jöckel, D. Hunger, T. W. Hänsch and P. Treutlein, *Phys. Rev. Lett.* **107**, p. 223001 (2011).

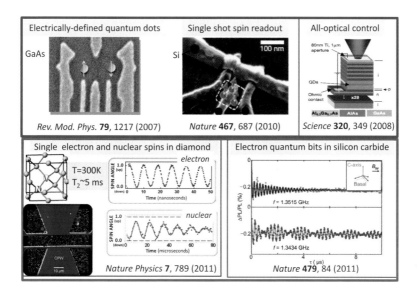

Fig. 1.   Quantum control of single spins in semiconductors: physics and materials science.

Prepared comment

D. Awschalom: Experiments on the Foundations of Quantum Physics and on Quantum Information

Those of us working in semiconductor physics live in a relatively dirty world – especially dirty when exploring quantum mechanics, where a myriad of interactions conspire to destroy coherent states (impurities, phonons, etcetera). While the challenges loom large, many of us have preferred a strategy of measurement rather than calculation.

This has led to a series of true surprises over the past decade regarding the robustness of quantum states based on the spins of electrons and nuclei in the solid state. For example, using relatively simple optical techniques, it was possible to create ensembles of coherent electron spin states that persisted far longer than expected, could be rapidly manipulated, and could be transported hundreds of microns with traditional electrical gating techniques. And while most of the experiments took place at liquid helium temperatures, some materials enabled these ensemble of states to survive to room temperature, albeit with shortened coherent times.

Going to lower dimensions with higher quantum confinement energies, such as using colloidal quantum dots, ensembles of isolated spins were prepared and measured. In general, many of the experiments were (and are) driven by the intimate connection between spin and photons driven by well-defined selection rules in the solid state, enabling quantum communication.

Encouraged by the success of atomic physics, the community forged ahead with the goal of localizing single electron and nuclear spins for quantum information processing and metrology. While technologically challenging, the last few years have seen success with complex carefully microfabricated devices that include electrically gated quantum dots in GaAs, single spin Si transistors, and with semiconductor heterostructures.

Recently, a different approach has emerged: the idea of abandoning perfection in the solid state, and embracing historically-annoying defects in materials (many of us survive by abandoning perfection and embracing our defects). That is, instead or removing defects in solid state structures, create them as needed for quantum processing of wavefunctions.

In many cases, these defects strongly trap individual electron spins, enabling room temperature initialization, measurement, and control of single quantum states: a powerful example is the N-V center in diamond, with Rabi oscillations from a single electron spin and a single nuclear spin. Gigahertz control, single-shot readout, and even schemes – like atomic physics – that exploit Landau-Zener interactions enable coherent transfer to the adjacent nuclear spin, creating a prototype single nuclear spin memory. And it is scalable.

This opens the door to using theoretical screening techniques to identify new materials for quantum information processing with many new possibilities for the field. For example, in Santa Barbara, based on density functional calculations, over a dozen new systems were predicted, the first of which is silicon carbide (SiC).

This is particularly interesting, as SiC is a common industrial material used to build high power electronics, JFETS, and MEMs sensors, with large scale wafers readily available relatively inexpensive. And with a great deal of industrial interest and experience in microfabrication, it is a system that could enable hybrid classical-quantum technologies.

And recently, divacancy defects were observed at room temperature with coherent electron spin properties similar to diamond, but in a material emitting photons within the telecom region of the spectrum.

So, my message is that the quantum mechanical properties of individual particles in semiconductors may offer very exciting opportunities for future information processing, integrating computation and communication in unique ways. And that the field is moving very rapidly, and perhaps in somewhat unexpected directions: it is a powerful example of the interplay between physics and materials science driving research.

Discussion

G. Horowitz I have a question for David (Awschalom). Is it known what is special about diamond and Silicon carbide to make these materials work so well?

D. Awschalom We have some pretty good guesses about it: they are very robust lattices, so they are quite immune to phonon effects such as scattering information; they have a pretty high bandgap, approximately 3 or more eV, so compared to kT the energy scale is very favourable; and in many of these systems the density of nuclear spins is quite low. So in some of these materials the density of nuclear spins is a few parts per billion, and you directly see the effect on the decoherence. In addition there are spin preserving optical transitions that allow to initialize these quantum states with very high fidelity actually. I should say that DFT calculations have shown there about 20 materials that are predicted to work like this.

W. Phillips My question is for Steve (Girvin) about the transmon qubits. You were saying that there is pretty good evidence that spontaneous emission is an important feature in the decay of the qubits. So I was hoping you would say a little bit more about that. I assume what you are talking about is that if you put the transmon in an eigenstate it decays by spontaneous emission as opposed to putting into some sort of a superposition which would correspond to an actual current in the transmon, whereupon it would radiate classically. So I was wondering if you could say a little bit about that in connection with the way in which we have thought traditionally about atoms because there is this point that if you put an atom into a superposition state it has a classically oscillating dipole moment and you expect it radiates classically. If you put it into an eigenstate of course you don't, but it does because – well some people would say because of the vacuum fluctuations. Can you tell that story for transmons?

S. Girvin Sure. So there is this Purcell effect with which, if you have an atom in a cavity, you can enhance the spontaneous emission rate by having the peak of the density of states in the cavity, or suppress it by having the cavity filter out the vacuum fluctuations at that frequency. And Schoelkopf has done an experiment in which you put the qubit exactly in its excited state and then measure that indeed one photon is radiated, but by doing a homodyne measurement – trying all the different phases of the homodyne measurement – you see zero coherence. There is no average phase. But if you put the transmon in a coherent superposition of up and down with a certain phase you can see that phase put into the electromagnetic field and the photon. The electromagnetic field is in a superposition of zero and one photon with that exact phase. So people tend to think of spontaneous emission as an incoherent process but what they really mean is that if you start with the atom definitely excited you will have no definite phase of the photon that comes out. But it is perfectly coherent transfer of that phase information, if it is there, to the photon in the decay process.

W. Phillips So what I am wondering is does the magnitude, that is the rate, at which this thing radiates, does it agree with what you predict? I mean presumably you can calculate exactly what its coupling is to the vacuum.

S. Girvin There is no difference. The probability of getting the photon out is just proportional to the fraction of the state which is excited. And the rate, as far as we can tell, is indistinguishable.

B. Halperin Presumably the issue is how much photons are going out of the cavity mode you want. That would be one issue of decay. But presumably you know that that is small in your case.

S. Girvin There are different kinds of resonators and two dimensional circuits, and more recently these three dimensional ones. We do have concerns that some of the finite lifetime is due to poor microwave hygiene. That is there may be modes that should not be there, but are, and have low Q, and take away some of the probability. So that is one of the motivations for using these 3D cavities which have a cleaner mode structure.

B. Halperin The other question I wanted to ask. You mentioned this scheme – you went through it very quickly – of a ring with a magnet trying to introduce a synthetic gauge field with a little ring. Can you say a little bit more about what you need to achieve that?

S. Girvin Photons are neutral. You need some way for them to acquire some kind of charge which is sensitive to some kind of gauge field. So by having this ring, and not having a bosonic particle hole symmetry – we think of the Cooper pairs as bosons – by breaking that particle-hole symmetry and by having a magnetic flux through the ring you can show that there is a kind of Hall effect, or circulation effect, which will give a net phase to the hopping matrix element of the photon that passes through that object to the next resonator.

B. Halperin But to break the particle hole symmetry you must actually have a superconductor that is very far from ...

S. Girvin You must have a very small capacitance and the discreteness of the charges on each of the islands is crucial. I am talking about charge asymmetry in the Bose-Hubbard model describing the superconducting islands, not microscopic fermion particle-hole asymmetry about the fermi level. I am happy to go into details later. Its sort of a technical issue.

B. Altshuler Steve (Girvin), I just want you to comment. As far as I understood you need about two orders of magnitude in coherence time for a single qubit. Can you comment about scalability? Because we heard from Ignacio Cirac that for atomic qubits this is the main problem. Do you think that these qubits will be more scalable?

S. Girvin Well the thing that we tell the funding agencies is that we can actually see our qubits. They are glued down where we put them. They stay there. The mode matching of the optical fields is very simple, because they are just passing through wires. There are a lot of reasons why in principle it may be good for scaling. But in theory, theory and practice are the same, but in practice they are different.

P. Zoller So let us move on to the next discussion contribution. You heard from Ignacio (Cirac) this morning that Bill (Phillips) is now doing high energy physics. I think the background is that atoms have no charge, so if you want to do interesting things like quantum Hall effect, you have to do artificial magnetic fields. And this is what he will talk about.

Prepared comments

W. Phillips: Quantum Engineered Synthetic Fields

As has been pointed out elsewhere during this meeting, neutral atoms as well as ions are good simulators of Hamiltonians that may be interesting in condensed matter physics. (This is important because such Hamiltonians may be calculationally intractable.) I also want to emphasize that one can use atoms to make quantum systems that are not analogous to condensed matter systems. These are new systems, interesting in their own right. The Bose-Hubbard model is an example that is not analogous to the typical condensed matter system. The Bosonic fractional Hall effect is another example that is not an analog to a condensed matter system, but is certainly interesting in its own right.

But there is an important problem: neutral atoms are neutral. If you want to simulate the effects of magnetic fields, which are essential to Hall effects and quantum Hall effects, then you have to do something else and that is where synthetic fields come in. As shown in the figure, we write the Hamiltonian for a free particle as $H = p^2/2m$, and the usual way in which we include a magnetic field is to take $H = \frac{(\vec{p}-q\vec{A})^2}{2m}$ where q is the charge and the magnetic field is the curl of $\vec{A}$. How are we going to get a Hamiltonian that looks like that for neutral atoms? Traditionally one of the ways that people have done it is to take your neutral atom system and rotate it. Why does rotation mimic a magnetic field? If you go into the rotating frame, there is a Coriolis force and that Coriolis force has the form $\vec{\Omega} \times \vec{v}$, which looks just like the Lorentz force $\vec{v} \times \vec{B}$, which is responsible for a lot of the interesting things that happen when you put electrons in a magnetic field. Unfortunately, this rotation trick has never worked well enough that you could get close to the quantum Hall regime. The reasons for this failure are rather technical. Nevertheless, rotation has produced lots of good results, like vortex lattices in cold gases. But it doesn't look like rotation is going to get to the quantum Hall regime.

Fortunately, there is an alternative. This idea for making a synthetic magnetic field is the brain child of Ian Spielman. A number of other people have considered other, similar approaches to making synthetic fields, but I am going to talk about what Ian has done because Ian is in my group, and because his approach has been so successful. So look again at the Hamiltonian of a charged particle in a magnetic vector potential. The expression $(\vec{p} - q\vec{A})^2/2m$ looks like the energy-momentum dispersion curve of a free particle, except that the minimum of the curve is shifted away from $p = 0$ to some non-zero momentum that is given by the vector potential. So how do we do that? We use light. We couple together different spin states of an atom. One of the great things about atoms is that they have these internal spin states, and we can couple them together using laser fields.

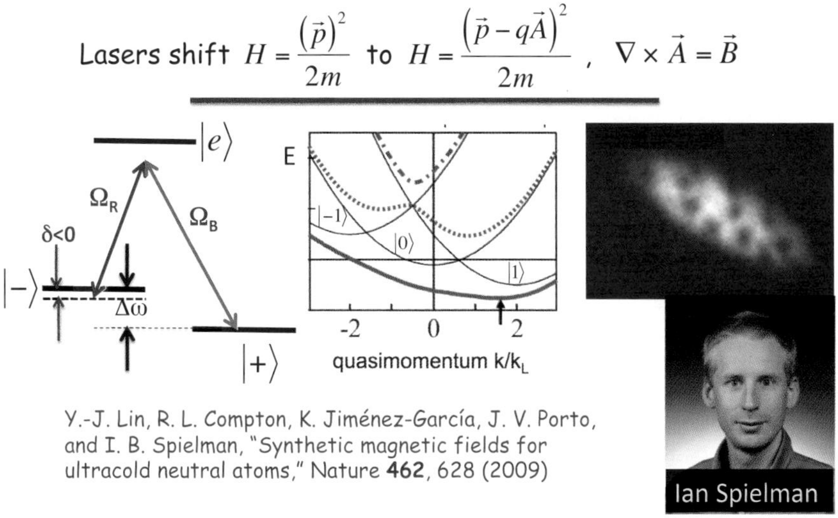

Lasers shift $H = \dfrac{(\vec{p})^2}{2m}$ to $H = \dfrac{(\vec{p} - q\vec{A})^2}{2m}$, $\nabla \times \vec{A} = \vec{B}$

Y.-J. Lin, R. L. Compton, K. Jiménez-García, J. V. Porto, and I. B. Spielman, "Synthetic magnetic fields for ultracold neutral atoms," Nature **462**, 628 (2009)

Ian Spielman

Fig. 1. Synthetic magnetic fields.

The energy level diagram on the left side of the figure shows that one spin state, the state I am calling $| - \rangle$ is coupled to the state called $| + \rangle$ through a two photon process indicated by the red and blue arrows. A little further to the right is an energy level diagram of the actual system we use, rubidium in its electronic ground state, and in the $F = 1$ hyperfine state. There are three spin states and they are coupled together by similar laser beams. Now look at the thin parabolic lines that are labeled $| - \rangle$, $| 0 \rangle$, and $| + \rangle$. These are the three free-particle states that correspond to the three spin states. We couple them together and they produce the thick red line, which has its minimum shifted away from zero-momentum. Then we use a gradient magnetic field to give this shift a spatial gradient (a curl) and what results is a synthetic magnetic field. You can see the results of that in the photograph on the right, just above the picture of Ian Spielman. The gas contains vortices because of the synthetic magnetic field. This entry of vortices into the neutral-atom gas is just as if one had applied a magnetic field to a charged system. We are hoping this will lead to an atomic, bosonic analog of the quantum Hall effect.

S. Das Sarma: Topological Phases in Cold Atomic Systems

I will discuss topological phases here. The paradigm for describing phases of matter is the Landau-Ginzburg-Wilson theory, which uses the central concepts of an order parameter and the underlying symmetry to describe and distinguish various phases of matter. This applies to classical phases and quantum phases, and the theory (with its appropriate quantum generalization) works equally well for the temperature-driven thermodynamic (classical) phase transition and a tuning parameter (in the Hamiltonian) driven zero-temperature quantum phase transition. Spontaneous breaking of some underlying symmetry as described by an effective field theory is the key concept, leading to powerful theoretical ideas such as universality, fixed points, relevant or irrelevant or marginal operators dominating the theory of phases of matter over the last fifty years. The theory is beautiful bringing particle physics and condensed matter physics communities together in a synergistic manner using the powerful language of effective field theories in real condensed matter systems, and allows calculation of various quantities through the well-established tools of the renormalization group technique. The theory has enjoyed great triumphs in describing classical phases of matter (solid/fluid, magnet/non-magnet, etc.) as well as quantum phase transitions (Anderson localization, Mott transition, quantum magnetism, quantum superfluids, Kondo effect, etc.).

There are, however, striking examples of quantum phases (most notably, integer and fractional quantum Hall states) where this symmetry and order parameter based description of matter (or more specifically, of the ground state of a many-body system) fails miserably. There is no symmetry which is broken spontaneously in a fractional quantum Hall state and there is no order parameter to describe the ground state. The system is in a topological phase of matter with a unique ground state quantum degeneracy and a gap protecting this degeneracy. The nontrivial topology is best understood by imagining the system in a special geometry (e.g. on a torus), and the key is a unique topological ground state quantum degeneracy which does not follow from any obvious symmetry in the Hamiltonian. The situation is in some vague qualitative sense the inverse of the spontaneous symmetry breaking (where the ground state wavefunction has less symmetry than the Hamiltonian); in a topological phase of matter, the ground state wavefunction has a hidden topological (and hence, nonlocal) invariance which emerges only at long wavelengths and low energies, which is not present in the parent Hamitonian.

Recently, it has become clear that topology, like symmetry, is a powerful way to distinguish quantum phases of matter with the ground state having a quantum degeneracy which is protected by a gap and cannot be destroyed by local operators. The gapped topological system is a quantum insulator (e.g. quantum Hall states), which is fundamentally different from a trivial band insulator (although a band insulator can, in principle, also be thought as a topological system, albeit a trivial one), and cannot be connected to a band insulator adiabatically by tuning a parameter in the Hamiltonian. The only way to go from a topological system to a nontopological (trivial) system is through a topological quantum phase transition.

Typically, the gapped topological system has gapless modes at the surface or the edge of the system, and in some situations the low-lying excitations of the system are anyons (e.g. 1/3 fractional quantum Hall state) or even non-Abelian anyons (e.g. 5/2 fractional quantum Hall state). Often, the topological phase can be described by an underlying topological quantum field theory (TQFT) although given an interacting microscopic Hamiltonian, it is not easy to derive the corresponding TQFT. The noninteracting topological phases (e.g. 2D integer quantum Hall, 3D topological insulator, topological superconductor as in a chiral or helical p-wave superconductor) are relatively easy to deal with, and a great deal of current activities in the subject have concentrated on the topological classification of such states.

Since it is relatively straightforward to write down theoretical noninteracting model Hamiltonians (often, but not always, involving spin-orbit coupling) whose ground states have some topological phases, cold atomic gas systems have emerged as a powerful experimental platforms to study topological phases since it is often possible to create designer (essentially noninteracting) Hamiltonians in laboratory-based cold atomic systems. The subject is still in its infancy, but there is great excitement that quantum Hall states and topological superconductivity (i.e. chiral p-wave superfluids) could possibly be realized soon in cold atomic gases. If that happens, then it may become possible to directly study abstract ideas such as Chern numbers (characterizing topological phases) and Majorana fermions (which arise naturally as zero energy modes in chiral topological superconductors) in the laboratory. The solid state systems are problematic (unless nature just provides a topological phase for free as happened with quantum Hall states) in this respect because the applicability of model noninteracting Hamiltonians to specific laboratory solid state systems is always uncertain, but there has been impressive recent advance in realizing topological superconductivity and Majorana fermions in superconductor-semiconductor hybrid structures and topological insulators in strongly spin-orbit coupled materials. A particularly important technological motivation for pursuing topological systems is that certain kinds of topological systems (e.g. topological superconductors with Majorana fermions) have non-Abelian anyons occurring naturally in the system, which can be effectively utilized as robust topologically-protected quantum memory for use in quantum computation.

Discussion

P. Zoller Thank you Sankar. Let me ask you a question. In atomic physics the gaps that you would expect for these kinds of things are probably very very small. And so do you see applications of these ideas you talk about topological quantum computing in the context of atoms.

S. Das Sarma I would say that topological quantum computation is far in the future, but seeing a non abelian particle in Nature would be extremely exciting. Yes the gap is very small. But so is your temperature. So the important thing is your gap divided by temperature. And that is not so bad.

H. Nicolai I have a question about synthetic gauge fields. As I could see this was about electromagnetism. But do you think you will be able to synthesize non abelian gauge fields with this idea?

W. Phillips Yes.

L. Randall I have a question for Bill (Phillips) as well. To get a gauge field is more than introducing this particular interaction. There are a lot of other constraints on your system. There are a lot of other terms which should not appear, or should have definite relations. So how does your system, or any system of this sort, guarantee that?

W. Phillips The electromagnetic case is admittedly trivial, although interesting. What we have done does not rely very much on the fact that it is a gauge field. I mean it just does the electromagnetic analog. I mean the electromagnetic field is a gauge field. That is not very important for the kind of things that you are thinking of. So to expand a little bit on the previous question. I was a little bit flip in saying "yes" we could make a non abelian gauge field. The idea there is that you would use the spin degree of freedom of the atoms to create, not only effective fields, but states that are superpositions of different spin states. And because you have introduced the spin degree of freedom into this then you will get states and operators that have tensor character and therefore have the non abelian feature to them. So that is the direction in which that is going. Now in answer to your question about how we engineer things so that it has all the right properties. Well I don't know that we will get everything that we want to be exactly analogous to the kinds of gauge fields that everybody wants. But we are hoping we will get something that will be interesting.

D. Gross Can you also have gauge fields dynamical?

W. Phillips Ha! Well I am guessing by that you mean could I have everything Maxwell's equations gives me. For example for the electromagnetic gauge fields. And the answer is, as far as we can tell, no. What we are creating are basically static fields, slowly varying fields. We are not going to get radiation out of this kind of field. So we can produce magnetic fields. We can produce in fact synthetic electric fields by having a time derivative of the magnetic vector potential. That produces a synthetic electric field, and we have seen the effects of that, mainly because its just neat to do. I don't think there is much application of making that kind of synthetic field.

D. Gross I think we would all be more comfortable if you used the word "vector field", as opposed to "gauge field".

W. Phillips OK. I am happy to do that.

D. Gross And also the term "background" in this case.

W. Phillips Yes. So what we make is the analog of the vector magnetic potential. We see the effects of its curl being a synthetic magnetic field, we see the effects of its time derivative being a synthetic electric field. What we are not seeing is radiation fields. And I don't believe we would be able to see things that are equivalent to a radiation field. At least we have not seen how we would do that. So I think the answer to the question: would there be dynamics – if that is what you mean by dynamics? No.

P. Zoller Bill, there are some ideas to make these quantities dynamical. But I guess maybe Ignacio wanted to comment on that.

I. Cirac Yes exactly. I mentioned in my talk that in principle you could now have at the same time atomic fermions and bosons, and if they interact appropriately then you will have an interaction theory of bosons and fermions. And you can also put a constraint such as Gauss law just by putting terms in your Hamiltonian. Then in the end you can have an effective theory that corresponds to QED in which there is a gauge field that is dynamical. This is not what happened in your experiment. But this in principle could be done with cold atoms.

G. Gibbons So would you extend that hope to the gravitational case?

I. Cirac Well I don't know how to do it. I mentioned that you could also simulate some gravitational effects, but in that case the gravitational field is external, so it does not fulfill any Einstein equation. Of course in the case of cold atoms there is a feedback on your condensate which will change your metric, also dynamically, but it does not do it in the way that would correspond to Einstein's equations. I don't know how to do that.

G. Gibbons Continuing. Is it foreseeable that you can also cook up gauge theories that don't exist such as anomalous gauge theories?

W. Phillips I don't know, maybe Ignacio (Cirac) does?

I. Cirac In principle, if you give me your Hamiltonian, and your Hamiltonian contains up to two particle interactions, bosons and fermions, then there is a way of simulating it. Of using external fields in your atomic systems in such a way that your evolution of your atoms will be governed by your Hamiltonian. That is in theory. It is very far in the future because at present experiments cannot do that. But in principle one could think of having something like that. I don't know this gauge theories that don't exist. But write the Hamiltonian, and we will tell you the tool box that you should use.

P. Zoller Maybe we should cut the discussion here. The conclusion is that Ignacio (Cirac) said that atomic physicists know how to do a lot things that do not exist in Nature. That's a good conclusion here. Let's move on to the next contribution.

Prepared comments

B. Altshuler: Quantum Disorder

A. Aspect: Anderson Localization of Ultra Cold Atoms in an Optical Disorder: a Quantum Simulator

Anderson localization (AL) was proposed more than 50 years ago to under-stand how disorder can lead to the total cancellation of electron conduction in certain materials. It is a purely quantum, one-particle effect, which can be interpreted as due to interference between the various amplitudes associated with the scattering paths of a matter wave propagating among impurities. According to the celebrated scaling theory, AL dramatically depends on the dimension, and, in the three-dimensional (3D) case, a mobility edge is pre-dicted. It is an energy threshold separating localized states, which decay to zero at infinity and correspond to insulators, from extended states, which correspond to conductors. However, determining the precise value of the mobility edge, and the corresponding critical behavior around it, remains a challenge for microscopic theory, numerical simulations, and experiments.

Figure 1(a) describes schematically an experiment that has allowed us to directly observe Anderson localization of ultracold atoms in a disordered optical potential.[a] A dilute Bose-Einstein condensate (BEC) of ultracold ^{87}Rb atoms, initially trapped by the red-detuned crossed laser beams, is released until the interaction energy between atoms is negligible, and then suddenly submitted to a repulsive disordered potential. This potential is due to the optical speckle field produced by two crossed, blue-detuned, wide coherent laser beams along the x- and z- axes, which pass through diffusive plates and are focused onto the atoms. The (paramagnetic) atoms are suspended against gravity by a magnetic field gradient (produced by the yellow coils), and the expansion of the atomic cloud can be observed for times as long as 6 s. This observation is realized with a EMCCD camera that images the fluorescence produced by a resonant probe, and yields the atomic column density $\tilde{n}(y, z)$ resulting from integration along the x-axis.

Figure 1(b) is a false color representation of a realization of the disordered potential in the $x = 0$, $y = 0$, and $z = 0$ planes. It is well known that the complex electric field in a laser speckle is a gaussian random process, so that one can easily calculate the autocorrelation function of the light intensity (i.e. of the potential) which is a fourth order correlation function of the complex electric field. The result is shown on Fig. 1(c). The correlation

---

[a]F. Jendrzejewski, A. Bernard, K. Müller, P. Cheinet, V. Josse, M. Piraud, L. Pezzé, L. Sanchez-Palencia, A. Aspect, and P. Bouyer, Three-dimensional localization of ultracold atoms in an optical disordered potential, *Nature Physics* **8** (2012) 398. See also that article for references.

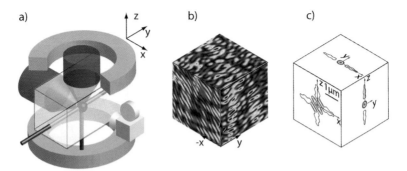

Fig. 1.   Experiment. (a) Experimental setup (see text for details). (b) False color representation of a realization of the disordered potential in $x = 0$, $y = 0$, and $z = 0$ planes. (c) Plots of the 3D autocorrelation function of the disordered potential in $x = 0$, $y = 0$, and $z = 0$ planes (the equal level lines represent levels separated by 14% of the maximum value).

radii, along the main axes (axis y and the two bisecting lines of $x - z$), are 0.11 μm, 0.27 μm and 0.08 μm, which means our disorder is a genuine 3D disorder, although it is not isotropic.

Figure 2(a) shows two typical examples of our observations of the expansion of the atomic cloud, for two different average amplitudes V_R of the disorder. When the disorder amplitude is small enough that the mobility edge is below all atoms energies, there is no localization, and one observes a standard diffusive behavior. In contrast, when the mobility edge is large enough to permit localization of a significant fraction of the atoms, the cloud is composed of two parts: a non evolving localized fraction corresponding to atoms with an energy below the mobility edge, and a diffusing fraction corresponding to atoms with an energy above the mobility edge. This statement is supported by a phenomenological analysis, assuming that the observed profiles are the sum of two contributions: (i) a steady localized part that is the replica of the initial profile $\tilde{n}_i(y, z)$; (ii) a diffusive expanding part $\tilde{n}_D(y, z, t)$, whose contribution at the center decays with time towards zero. More precisely, we decompose the observed column density as

$$\tilde{n}(y, z, t) = f_{\text{loc}} \times \tilde{n}_i(y, z) + \tilde{n}_D(y, z, t), \tag{1}$$

and we check on Fig. 2(b) that the mean squared radii Δu^2 of the column density profiles ($u \in \{y, z\}$) vary with time as $\Delta u(t)^2 = \Delta u(t_i)^2 + 2\langle D^u \rangle (t - t_i)$. Figure 2(c) shows that the column density at the center tends asymptotically towards a finite value, which is determined by a fit to the function $\tilde{n}(0, 0, t)/\tilde{n}_i(0, 0) = A + B(t - t_i)^{-1}$, where A refers to the localized part. The $(t - t_i)^{-1}$ evolution of the column density at the center is expected for a diffusive behavior when the size of the initial profile is negligible. It results from the integration over one dimension of the $(t - t_i)^{-3/2}$ evolution expected for the 3D density at the origin. The constant A of the fit is then

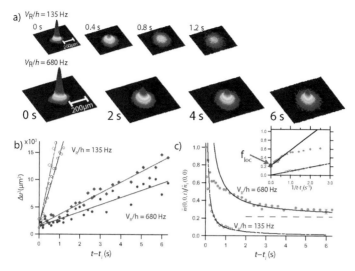

Fig. 2. Evolution of the atomic cloud for two different amplitudes of the disorder. (a) Column density in the $y-z$ plane, at various delays after application of the disorder. As shown in Fig. 2(b), the expanding parts have a diffusive behavior in both cases. (b) Time evolution of the mean squared radii along y (blue) and z (red) of the column density profiles. The anisotropy of the disorder, visible in Fig. 1(b), is reflected on the diffusion coefficients. (c) Evolution of the column density at the center, and determination of its asymptotic value, yielding the localized fraction f_{loc} (asymptotic value of the fitted black solid line, see text).

interpreted as the localized fraction of atoms, f_{loc}. It is found equal to 22% for $V_{\mathrm{R}}/h = 680$ Hz, and 1% for $V_{\mathrm{R}}/h = 135$ Hz.

Although we have no direct proof that the localized component is due to Anderson localization, we stress that we do not know any other possibility to understand its existence. Indeed, the classical percolation threshold in our disorder is less than $10^{-2}V_{\mathrm{R}}$, so that no atom has an energy below that threshold. Moreover, the correlation energy of our disorder $E_{\mathrm{R}} = \hbar^2/m\sigma_{\mathrm{R}}^2$ ($E_{\mathrm{R}}/h \simeq 6.5$ kHz for an average correlation radius $\sigma_{\mathrm{R}} \simeq 0.13$ μm) is larger than the disorder amplitudes V_{R} used in the experiment. We are thus in the quantum disorder regime, in which local minima of the disordered potential do not support bound states, eliminating the possibility of quantum trapping in individual local minima.

The experiment can be improved significantly by loading atoms in the disorder with a well defined controlled energy, in contrast with the present situation where the sudden ramping of the disorder leads to a broad energy distribution. It will then be possible to measure precisely the mobility edge, and to determine the critical exponents around the mobility edge. Since there are only approximate theories –still under debate– for calculating these quantities, we will have a way to test these theories. Moreover, it will be possible to introduce controlled interactions between the atoms, and to address the difficult (many body) problem of Anderson localization in the presence of interactions. From this point of view, our system can be considered a quantum simulator, in the spirit of Feynman definition,[b] since any simulation of its evolution on a classical computer seems definitely out of reach.

---

[b]R.P. Feynman, Simulating physics with computers, *Int. J. Theoret. Phys.* (1982), 21, 467.

Discussion

S. Das Sarma I have to respond to Alain (Aspect). The point is what Ignacio (Cirac) was telling during the break. Surely there are bosonic problems which are hard to simulate on a digital computer. No question about it. But since there is no sign problem, all we need is a factor of 10, or factor of 20 increase in digital simulation to be able to incorporate disorder. But when you come to fermions, because of the sign problem, its an exponential difficulty. So suppose you get a computer to simulate something. You add one electron or two electrons, and you need something which is 100 times larger. So there is a qualitative difference between the two. So I completely agree with you that, if you have bosons and disorder, Anderson localization is an interesting problem. But if you have fermions it is qualitatively different. That is the point I was making.

A. Aspect This is very interesting. Because of course we also plan to put fermions in our experiment. But I want to understand how difficult is the problem when we introduce interactions between bosons?

P. Zoller There was one remark by Polyakov.

A. Polyakov I just want to point an interesting analogy. There is the phenomenon of quark confinement. And in some cases it can be viewed as the result of instantons, so you have random flashes of gauge fields in space-time and they are random, and quarks cannot propagate. I just want to point out this analogy which holds, on a technical level, with Anderson localization.

M. Berry This is a question for Boris (Altshuler). You made the analogy between KAM theorem and Anderson localization. But bearing in mind that KAM is a theorem concerned with chaos and non chaos in non linear equations, and that Anderson localization is, at its origin, a phenomenon that occurs with linear equations, did you mean this analogy as a serious precise one, or just roughly?

B. Altshuler I think it is neither precise nor rough. I do believe that if we take the problem that arises in the classical limit what you will get when you quantize it is a problem which is localized in the space of the quantum number, since each point in the space of the quantum number is actually the invariant tori. At the same time, quantum systems are more resistant to chaos than classical, and that is why certain problems that are chaotic in the classical limit are known to be localized in second space in quantum. So this is the relation between the two.

A. Leggett In the condensed matter context there is a very interesting prediction that if you take a set of fermions and put them in a lattice in a disordered medium, sufficiently disordered that they undergo Anderson localization, and if you then switch on a weak interaction they will then become immediately superconducting without ever going through the normal phase on the way. As far as I know it is controversial whether that has ever been seen in real condensed matter systems. I was wondering what are the prospects for testing that in optical lattices?

A. Aspect I think the prospects are good. In principle we have the tools for doing that. All these experiments are awfully difficult, but all the tools are available. Its a matter of investment, of time, etc...

G. Parisi Let's come to the beautiful experiment on localization. I understand that somebody like Alain (Aspect) would wish to have the theory tell him exactly where is the threshold for localization, which amount of particles remain localized, and so on. But I think this is a very difficult question for theorists. Indeed if we have to compute the transition temperature of a ferromagnet like iron, this is not easy. What would be much more important is to get the critical exponent with a correct fit. And you already mentioned there is one critical exponent which is related to dimensional diffusion and I think that the most important from the theoretical point of view would be to get the precise understanding of this exponent.

B. Altshuler Just one phrase. It looks like there is indeed a difference of mentality between the communities of condensed matter and atomic physics. I fully agree with Giorgio (Parisi) that in condensed matter we never worry about calculating things like transition temperature. What we aim to understand is critical behaviour. At the same time people in atomic physics want more because they have a much cleaner system. But I do agree that this critical behaviour would be indeed something which would really increase our understanding and will make us happy.

A. Aspect This is also on the to do list. We think we can study critical exponents. And I would like to cite an experiment which has been done in Lille. What they do is not exactly Anderson localization. It is rather 1D dynamical localization, i.e. localization along one direction in the momentum space. But it has been shown by theorists that if one uses 3 different frequencies, there is a mapping between the 3D Anderson localization problem and that problem. And in that situation they have been able indeed to measure critical exponents. The problem is that in 3D as far as I understand, the mapping is not rigorous, so we do not know if it is an exact of approximate mapping of the 3D Anderson localization problem. That is just to say that, yes, cold atoms should allow us to determine critical exponents.

P. Zoller Thank you Alain. That is a very nice conclusion of this session. Let us thank everybody who contributed.

Session 4

Quantum Condensed Matter

Chair: *Bertrand Halperin*, Harvard University, USA
Rapporteur: *Subir Sachdev*, Harvard University, USA
Scientific secretaries: *Nathan Goldman*, Université Libre de Bruxelles, Belgium
and *Jacques Tempere* Universiteit Antwerpen, Belgium

Rapporteur talk by S. Sachdev: The Quantum Phases of Matter

Abstract

I present a selective survey of the phases of quantum matter with varieties of many-particle quantum entanglement. I classify the phases as gapped, conformal, or compressible quantum matter. Gapped quantum matter is illustrated by a simple discussion of the $\mathbb{Z}_2$ spin liquid, and connections are made to topological field theories. I discuss how conformal matter is realized at quantum critical points of realistic lattice models, and make connections to a number of experimental systems. Recent progress in our understanding of compressible quantum phases which are not Fermi liquids is summarized. Finally, I discuss how the strongly-coupled phases of quantum matter may be described by gauge-gravity duality. The structure of the $N_c \to \infty$ limit of $SU(N_c)$ gauge theory, coupled to adjoint fermion matter at non-zero density, suggests aspects of gravitational duals of compressible quantum matter.

1. Introduction

Some of the most stringiest tests and profound consequences of the quantum theory appear in its application to large numbers of electrons in crystals. Sommerfeld and

Bloch's early theory of electronic motion in metals treated the electrons as largely independent particles moving in the periodic potential created by the crystalline background. The basic principles were the same as in Schrödinger's theory of atomic structure: the electrons occupy 'orbitals' obtained by solving the single particle Schrödinger equation, and mainly feel each other via Pauli's exclusion principle. Extensions of this theory have since led to a remarkably complete and quantitative understanding of most common metals, superconductors, and insulators.

In the past thirty years, the application of quantum theory to many particle physics has entered a new terrain. It has become clear that many phases of quantum matter cannot be described by extensions of the one-particle theory, and new paradigms of the quantum behavior of many particles are needed. In an influential early paper,[1] Einstein, Podolsky, and Rosen (EPR) emphasized that the quantum theory implied non-local correlations between states of well separated electrons which they found unpalatable. Bell later showed[2] that such non-local correlations could not be obtained in any classical hidden variable theory. Today, it is common to refer to such non-local EPR correlations as *quantum entanglement*. Many varieties of entanglement play a fundamental role in the structure of the phases of quantum matter, and it is often long-ranged. Remarkably, the long-range entanglement appears in the natural state of the many materials at low enough temperatures, and does not require delicate preparation of specific quantum states after protection from environmental perturbations.

The structure of Sommerfeld-Bloch theory of metals is summarized in Fig. 1. The electrons occupy single-particle states labelled by a momentum k below the Fermi energy E_F. The states with energy equal to E_F define a $(d-1)$-dimensional 'Fermi surface' in momentum space (in spatial dimension d), and the low energy excitations across the Fermi surface are responsible for the metallic conduction. When the Fermi energy lies in an energy gap, then the occupied states completely fill a set of bands, and there is an energy gap to all electronic excitations: this defines a band insulator, and the band-filling criterion requires that there be an even number of electrons per unit cell. Upon including the effect of electron-electron interactions, which can be quite large, both the metal and the band insulator remain adiabatically connected to the states of free electrons illustrated in Fig. 1. Finally, in the Bardeen-Cooper-Schrieffer theory, a superconductor is obtained when the electrons form pairs, and the pairs Bose condense. In this case, the ground state is typically adiabatically connected to the Bose-Einstein condensate of electron pairs, which is a simple product of single boson states.

Here, I will give a selective survey of the phases of quantum matter which cannot be adiabatically connected to free electron states, and which realize the different flavors of many-body quantum entanglement. I will organize the discussion by classifying the states by the nature of their excitation spectrum. Readers interested primarily in strange metals can skip ahead to Sec. 4.

First, in Sec. 2, I will consider phases in which there is a gap to all excitations in the bulk matter (although, there may be gapless excitations along the boundary).

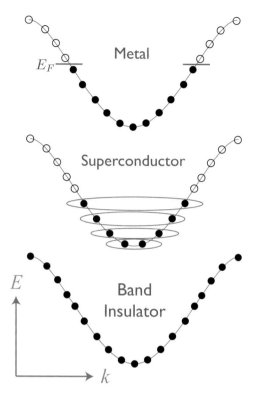

Fig. 1.   Schematic of the phases of matter which can be described by extensions of the independent electron theory.

Despite the absence of low energy excitations, such states can have subtle forms of many body entanglement which are described by topological field theories.

Section 3 will consider states which are gapless, with the zero energy excitations only found at isolated points in the Brillouin zone. Such states often have an excitation spectrum of massless relativistic particles, with the role of the velocity of light being played by a smaller velocity associated with the lattice Hamiltonian. Moreover, many such states are described by a quantum field theory which is invariant under conformal transformations of spacetime, and hence Sec. 3 will describe 'conformal' quantum matter.

Section 4 will turn to 'compressible' quantum matter, in which the density of particles can be varied smoothly by an external chemical potential, without changing the basic characteristics of the phase. All known examples of such phases have zero energy excitations along a $(d-1)$-dimensional surface in momentum space, just as in the free electron metal.[a] However, there is much experimental and theoretical

---

[a]The categories of conformal and compressible matter overlap in $d = 1$. Almost all of our discussion will focus on $d = 2$ and higher.

interest in describing so-called 'strange metals', which are compressible states not smoothly connected to the free electron metal. I will summarize recent theoretical studies of strange metals.

Section 5 will discuss emerging connections between the above studies of the phases of quantum matter and string theory. I will summarize a new perspective on the gravity duals of compressible states.

2. Gapped Quantum Matter

The earliest discussion of the non-trivial phases of gapped quantum matter emerged in studies of Mott insulators. These insulators appear when Coulomb repulsion is the primary impediment to the motion of electrons, rather than the absence of single particle states for band insulators. In a situation where there are an odd number of electrons per unit cell, the independent electron approach necessarily leads to partially filled bands, and hence predicts the presence of a Fermi surface and metallic behavior. However, if the Coulomb repulsion, U, is large compared to the bandwidth, W, then the motion of the charge of the electrons can be sufficiently suppressed to yield a vanishing conductivity in the limit of zero temperature.

For a simple example of a Mott insulator, consider electrons hopping in a single band on the triangular lattice. After the Coulomb repulsion localizes the electron charge, the Hilbert space can be truncated to the quantum states in which there is precisely one electron on each site. This Hilbert space is not trivial because we have not specified the spins of the electrons: indeed the spin degeneracy implies that there are 2^N states in this truncated Hilbert space, in a lattice of N sites. The degeneracy of these spin states is lifted by virtual processes involving charge fluctuations, and these lead to the Hamiltonian of a Heisenberg antiferromagnet

$$\mathcal{H}_{AF} = J \sum_{\langle ij \rangle} \vec{S}_i \cdot \vec{S}_j + \ldots \tag{1}$$

where $\vec{S}_i$ is the $S = 1/2$ spin operator acting on the electron on site i, $J \sim W^2/U > 0$ is the exchange interaction generated by the virtual processes, and the ellipses indicates omitted terms generated at higher orders in a W/U expansion. The exact ground state of $\mathcal{H}_{AF}$ is not known. However, for the truncated Hamiltonian with only nearest neighbor exchange, there is good numerical evidence[3] that the ground state has long-range antiferromagnetic (Néel) order of the type illustrated in Fig. 2. In this state, the spins behave in an essentially classical manner. Each spin has a mean polarization, oriented along the arrows in Fig. 2, and there are quantum spin fluctuations about this average direction. There are gapless spin-wave excitations above this ground state, and so this state is not an example of gapped quantum matter. Furthermore, this ground state is adiabatically connected to the classical state with frozen spins, and so this state does not carry the kind of quantum entanglement we are seeking.

To obtain gapped quantum matter, we have to consider the possibility of a different type of ground state of antiferromagnets in the class described by $\mathcal{H}_{AF}$. This

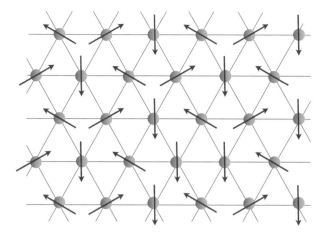

Fig. 2.   Ground state of the Heisenberg antiferromagnet on the triangular lattice with long-range antiferromagnetic order. This state is not an example of gapped quantum matter.

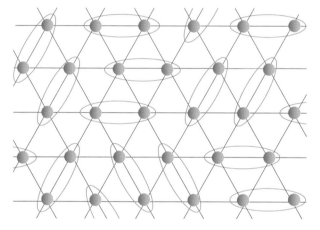

Fig. 3.   A snapshot of the RVB state on the triangular lattice. Each ellipse represents a singlet valence bond, $(|\uparrow\downarrow\rangle - |\downarrow\uparrow\rangle)/\sqrt{2}$. The RVB state is a superposition of all different singlet pairings, of which only one is shown above.

is the resonating valence bond (RVB) state of Fazekas and Anderson,[4] illustrated in Fig. 3. This state is a linear superposition of the very large number of possible singlet pairings between the electrons. It thus generalizes the chemical resonance of π-bonds in the benzene ring to an infinite number of electrons on a lattice. Such a RVB wavefunction was written down early on by Pauling,[5] who proposed it as a theory of a correlated metal. Anderson[6] applied the RVB state to the spin physics of a Mott insulator, and Kivelson *et al.*[7] noted that it exhibits the phenomenon of spin-charge separation: there are *spinon* excitations which carry spin $S = 1/2$ but do not transfer any charge, as shown in Fig. 4.

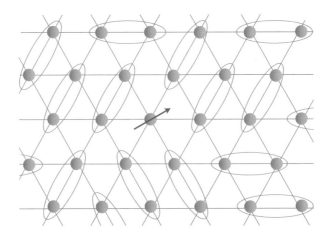

Fig. 4.   A spinon excitation of the RVB state. The spinon carries spin $S = 1/2$ but is electrically neutral.

Our understanding of the physics of RVB states advanced rapidly after the discovery of cuprate high temperature superconductivity in 1986. Baskaran and Anderson[8] pointed out that a natural language for the description of RVB-like states is provided by lattice gauge theory: the constraint on the Hilbert space of one electron per site can be mapped onto the Gauss law constraint of lattice gauge theory. This mapping implies that RVB states can also have neutral, spinless excitations which are the analogs of the 'photon' of gauge theories. However, for this picture of the RVB state to hold, it is required that the gauge theory have a stable deconfined phase in which the spinons can be considered as nearly free particles. Rokhsar and Kivelson[9] described RVB physics in terms of the 'quantum dimer' model, and discovered a remarkable solvable point at which the simplest RVB state, the equal superposition of all nearest-neighbor singlet pairings, was the exact ground state. Fradkin and Kivelson[10] showed that the quantum dimer model on a bipartite lattice was equivalent to a certain compact $U(1)$ lattice gauge theory. However, it remained unclear whether the solvable RVB state was a special critical point, or part of a RVB phase. It was subsequently argued[11] that such $U(1)$ RVB states are generically unstable in $d = 2$ (but not in $d = 3$) to confinement transitions to states in which the valence bonds crystallize in periodic patterns (now called valence bond solids (VBS); see Fig. 12 later). A stable RVB phase first appeared[b] in independent works by Wen,[14] and Read and the author,[15,16] who identified it as a deconfined phase of a discrete $\mathbb{Z}_2$ gauge theory.[17–20] The quantum dimer model on the triangular and kagome lattices provides examples of $\mathbb{Z}_2$ RVB phases,[21] and includes the exactly solvable model as a generic point within the phase.[22–24] The $\mathbb{Z}_2$ RVB state has a

---

[b]I also note the chiral spin liquid,[12,13] which breaks time-reversal symmetry spontaneously. It is closer in spirit to quantum Hall states to be noted later, than to the RVB which breaks no symmetries.

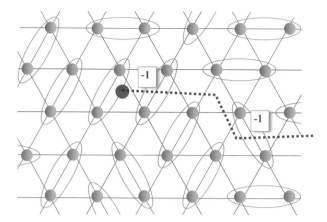

Fig. 5.   The gapped 'vison' excitation of the $\mathbb{Z}_2$ RVB state. This is an excited state whose wavefunction is similar to the RVB ground state of Fig. 3. However, the linear superposition over different valence bond configurations has changes in the signs of the terms: the sign is determined by the parity of the number of valence bonds which intersect a 'branch-cut' (the dashed red line) emanating from the center of the vison. The properties of the vison are independent of the specific location of the branch-cut, which can be viewed as a gauge choice. There is $\mathbb{Z}_2$ gauge flux only in the plaquette indicated by the red circle.

gap to all excitations, and so this is a realization of gapped quantum matter with long-range entanglement. The analog of the photon in this discrete gauge theory is a gapped topological excitation known as a 'vison', and is illustrated in Fig. 5. This is a vortex-like excitation, and can propagate across the antiferromagnet like any point particle. It carries neither spin nor charge and only energy, and so is 'dark matter'.

One of the important consequences of the existence of the vison is that the degeneracy of the RVB state depends upon the topology of the manifold upon which the spins reside: hence it is often stated that the RVB state has 'topological order'. Imagine placing the triangular lattice antiferromagnet on the torus, as shown in Fig. 6. Then we can arrange the branch-cut of the vison so that the $\mathbb{Z}_2$ gauge flux penetrates one of the holes of the torus. For a sufficiently large torus, this gauge flux has negligible effect on the energy, and so leads to a two-fold ground state degeneracy. We can also place the $\mathbb{Z}_2$ in the other hole of the torus, and so the ground state is four-fold degenerate.[7,9,25,26]

This ground state degeneracy of the $\mathbb{Z}_2$ RVB state can be viewed as a reflection of its long-range entanglement. Note that all spin-spin correlation functions decay exponentially fast in the ground state. Nevertheless, there are EPR-type long-range correlations by which the quantum state 'knows' about the global topology of the manifold on which it resides. The non-trivial entanglement is also evident in Kitaev's solvable models[27,28] which realize $\mathbb{Z}_2$ spin liquids.

Another measure of the long-range entanglement is provided by the behavior of the entanglement entropy, S_E. The definition of this quantity is illustrated in

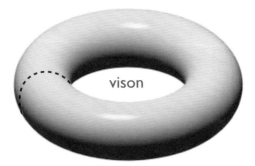

Fig. 6.   Topological ground state degeneracy of the $\mathbb{Z}_2$ RVB state. The triangular lattice antiferromagnet is placed on the surface of the torus. The dashed line is a branch cut as in Fig. 5. The $\mathbb{Z}_2$ gauge flux is now contained in the hole of the torus and so has little influence on the spins.

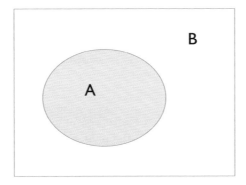

Fig. 7.   The entanglement entropy of region A is defined by tracing over all the degrees of freedom in region B, and computing the von Neumann entropy of the resulting density matrix.

Fig. 7. We divide the triangular lattice antiferromagnet into two spatial regions, A and B. Then we trace over the spins in region B and obtain the density matrix $\rho_A = \text{Tr}_B \rho$, where $\rho = |\Psi\rangle\langle\Psi|$, with $|\Psi\rangle$ the ground state of the full triangular lattice. The entanglement entropy is $S_E = -\text{Tr}\,(\rho_A \ln \rho_A)$. A fundamental feature of the entanglement entropy is that for gapped quantum matter it is expected to obey the 'area law'; for the present two-dimensional quantum system, this is the statement[c]

$$S_E = aP - \gamma \tag{2}$$

where P is the perimeter of the boundary between regions A and B. The constant a depends upon microscopic details of the system under consideration, and is not particularly interesting. Our attention is focused on the value of the offset γ: this is believed to provide a universal characterization of the entanglement of the quantum

---

[c]Eq. (2) is defined in the limit $P \to \infty$, taken for a fixed shape of the region A; with such a limit, there is no ambiguity in the definition of γ.

state. For a band insulator, the entanglement can only depend upon local physics near the boundary,[29] and it is expected that $\gamma = 0$. For the $\mathbb{Z}_2$ RVB state, it was found[30–32] that $\gamma = \ln(2)$: this value of γ is then a signature of the long-range entanglement in this state of gapped quantum matter.

These topological aspects of the $\mathbb{Z}_2$ RVB state can be made more explicit by a mapping of the $\mathbb{Z}_2$ gauge theory to a doubled Chern-Simons gauge theory.[33–37] The latter is a topological field theory, and there is a direct connection between its properties and those of the $\mathbb{Z}_2$ RVB state. Indeed the 4-fold ground state degeneracy on a torus, and the value of the offset γ in the entanglement entropy can also be computed in this Chern-Simons theory.[38,39]

A good candidate for a $\mathbb{Z}_2$ RVB state is the kagome antiferromagnet.[40–43] Two recent numerical studies[44,45] has provided remarkable conclusive evidence for the constant $\gamma = \ln(2)$ in the entanglement entropy. And neutron scattering experiments on such an antiferromagnet display clear signatures of deconfined spinon excitations.[46] There is also compelling evidence for fractionalization and topological order in an easy-axis kagome antiferromagnet.[47–49] And finally, several recent studies[50–53] have argued for a gapped spin liquid in a frustrated square lattice antiferromagnet.

Another large set of widely studied examples of gapped quantum matter states are the quantum Hall states, and the related chiral spin liquids.[12,13] We will not discuss these here, apart from noting their relationship to the $\mathbb{Z}_2$ RVB states. The quantum Hall states do not respect time-reversal invariance and so their topological properties can be described by Chern-Simons theories with a single gauge field (in the simplest cases). Like the $\mathbb{Z}_2$ RVB states, they have ground state degeneracies on a torus, and non-zero values of the entanglement entropy offset γ. The quantum Hall states also generically have gapless edge excitations which play a crucial role in their physical properties, and such gapless states are not present in the simplest $\mathbb{Z}_2$ RVB state we have discussed above.

3. Conformal Quantum Matter

This section considers phases of matter which have gapless excitations at isolated points in the Brillouin zone. A simple recent example is graphene, illustrated in Fig. 8. This has a low energy spectrum of 4 massless Dirac fermions. These fermions interact with the instantaneous Coulomb interaction, which is marginally irrelevant at low energies, and so the Dirac fermions are free. The theory of Dirac fermions is conformally invariant, and so we have a simple realization of a conformal field theory in 2+1 spacetime dimensions: a CFT3. More recently Dirac fermions have also appeared, in both theory and experiment, on the boundary of topological insulators.

However, our primary interest is in strongly-interacting CFTs, which provide realizations of quantum matter with long-range entanglement. One thoroughly studied example is provided by the coupled dimer antiferromagnet, illustrated in Fig. 9. This is described by the nearest-neighbor Heisenberg antiferromagnet in Eq. (1),

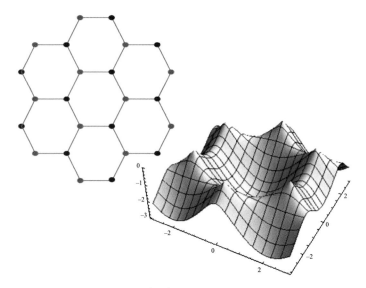

Fig. 8.   The carbon atoms in graphene (top). The π orbitals on the carbon atoms from a half-filled band, the lower half of which is shown (bottom). Notice the Dirac cones at six points in the Brillouin zone. Only two of these points are inequivalent, and there is a two-fold spin degeneracy, and so 4 two-component massless Dirac fermions constitute the low energy spectrum.

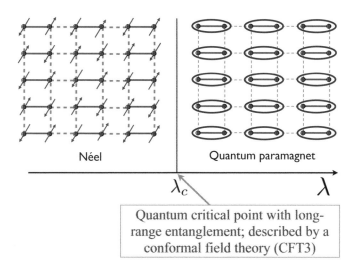

Fig. 9.   The coupled dimer antiferromagnet. The Hamiltonian is as in Eq. (1), with the red bonds of strength J, and the dashed green bonds of strength J/λ ($J > 0$, $\lambda \geq 1$).

but with two values of the exchange interactions, with ratio λ. For large λ, the system decouples into dimers, each of which has a spin-singlet valence bond. This is the quantum paramagnet, which preserves all symmetries of the Hamiltonian and has a gap to all excitations. On the other hand, for λ close to unity, we obtain a

Néel state with long-range antiferromagnetic order, similar to that in Fig. 2. Both these states have short-range entanglement, and are easily understood by adiabatic continuity from the appropriate decoupled limit. However, in between these states is a quantum critical point at $\lambda = \lambda_c$. There is now compelling numerical evidence[54] that this critical point is described by the CFT3 associated with the Wilson-Fisher fixed point of an interacting field theory of a relativistic scalar with 3 components. Thus a simple generic Heisenberg antiferromagnet flows at low energy to a fixed point with not only relativistic, but also conformal, invariance.

A notable feature of this CFT3, and of others below, is that it has long-range entanglement in the same sense as that defined for gapped quantum matter via Eq. (2). The constant γ is non-zero,[55] and is a characteristic universal property of the CFT3 which depends only on the nature of the long-distance geometry, but not its overall scale.

A possible realization of the coupled dimer antiferromagnet critical point in two dimensions is in ref. 56, but detailed measurements of the excitation spectrum are not available. However, TlCuCl$_3$ provides nice realization in three dimensions,[57] as shown in Fig. 10. In this case, the quantum-critical point is described by the theory of the 3-component relativistic scalar in 3+1 dimensions; the quartic interaction term is marginally irrelevant, and so the critical point is a free CFT4. The experiments provide an elegant test of the theory of this quantum critical point, as

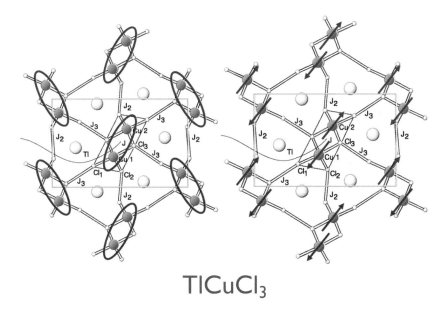

Fig. 10. Quantum phase transition in TlCuCl$_3$ induced by applied pressure. Under ambient pressure, TlCuCl$_3$ is a gapped quantum paramagnet with nearest-neighbor singlet bonds between the $S = 1/2$ spins on the Cu sites (left). Under applied pressure, long-range Néel order appears (right).

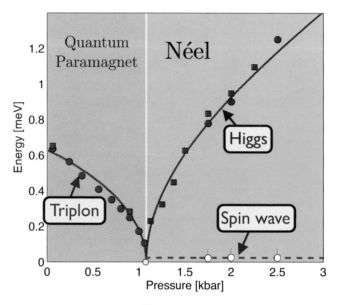

Fig. 11.   Neutron scattering measurements[57] of the excitation spectrum across the quantum phase transition in TlCuCl$_3$. The quantum paramagnet has "triplon" excitations, corresponding to $S = 1$ triplet sets hopping from dimer to dimer. The Néel state has gapless spin-wave excitations associated with the broken symmetry. It also has a Higgs particle excitation, associated with the oscillations in the magnitude of the Néel order parameter.

shown in Fig. 11. The quantum paramagnet has a 'triplon' excitation, which can be interpreted as the oscillation of the scalar field $\vec{\phi}$ about $\vec{\phi} = 0$. The Néel phase has gapless spin-wave excitations, which are the Goldstone modes associated with the broken O(3) symmetry. However, the Néel phase also has an excitation[d] corresponding to oscillations in the magnitude, $|\vec{\phi}|$, which is the Higgs boson, as discussed in refs. 58, 59. Because we are in 3+1 dimensions, we can use mean-field theory to estimate the energies of the excitations on the two sides of the critical point. A simple mean-field analysis of the potential for $\vec{\phi}$ oscillations in a Landau-Ginzburg theory shows that[60]

$$\frac{\text{Higgs energy at pressure } P = P_c + \delta P}{\text{Triplon energy at pressure } P = P_c - \delta P} = \sqrt{2}, \qquad (3)$$

where P_c is the critical pressure in Fig. 11, and δP is a small pressure offset. This ratio is well obeyed[60] by the data in Fig. 11

The quantum antiferromagnet of Fig. 9 is special in that it has two $S = 1/2$ per unit cell: this makes the structure of the quantum paramagnet especially simple, and allows for description of the quantum critical point by focusing on the fluctuations of the Néel order alone. The situation becomes far more complex, with new types

---

[d]The Higgs excitation is damped by its ability to emit gapless spin waves: this damping is marginal in $d = 3$, but much more important in $d = 2$.

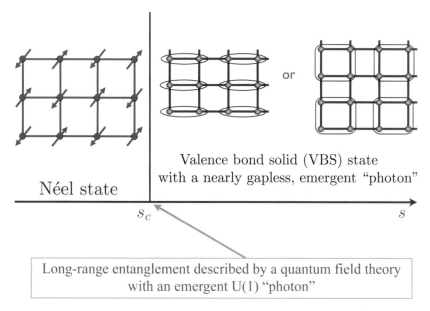

Néel state

Valence bond solid (VBS) state
with a nearly gapless, emergent "photon"

s_c s

Long-range entanglement described by a quantum field theory
with an emergent U(1) "photon"

Fig. 12.   Possible phases of a square lattice antiferromagnet, tuned by additional frustrating interactions which are controlled by the parameter s. The Néel state breaks spin rotation symmetry. The VBS state breaks lattice symmetry by modulating the amplitudes of singlet bonds on the various links of the lattice.

of CFT3s, when we consider models with a single $S = 1/2$ per unit cell. A prominent example is the frustrated square lattice antiferromagnet. With only nearest neighbor interactions, the square lattice antiferromagnet has long-range Néel order, as illustrated on the left side of Fig. 12. After applying additional interactions which destabilize the Néel state, but preserve full square lattice symmetry, certain antiferromagnets exhibit a quantum phase transition to a VBS state which restores spin rotation invariance but breaks lattice symmetries.[61] It has been argued[62,63] that this quantum phase transition is described by a field theory of a non-compact U(1) gauge field coupled to a complex bosonic spinor i.e. a relativistic boson which carries unit charge of the U(1) gauge field and transforms as a $S = 1/2$ fundamental of the global SU(2) spin symmetry (note to particle theorists: here "spin" refers to a global symmetry analogous to flavor symmetry, and so there are no issues with the spin-statistics theorem). Evidence for this proposal has appeared in numerical studies by Sandvik,[64] as illustrated in Fig. 13, which shows remarkable evidence for an 'emergent photon'. It is possible that the experimental system of ref. 56 exhibits a Néel-VBS transition.

A separate question is whether the critical point of this theory of a non-compact U(1) photon coupled to the relativistic boson is described by a CFT3. The existence of such a 'deconfined critical point' has been established by a $1/N$ expansion, in a model in which the global SU(2) symmetry is enlarged to SU(N). Recent numerical

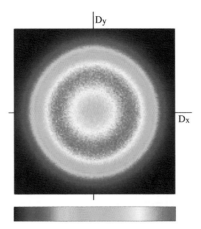

Fig. 13. Results from the studies of a square lattice antiferromagnet by Sandvik.[64] The measurements are at the $s = s_c$ critical point between the Néel and VBS states of Fig. 12. D_x is a measure of the VBS order along the x direction: $D_x = \sum_j (-)^{j_x} \vec{S}_j \cdot \vec{S}_{j+\hat{e}_x}$, and similarly for D_y; here $j \equiv (j_x, j_y)$ labels square lattice sites, and $\hat{e}_x$ is a unit vector in the x direction. The emergent circular symmetry of the distribution of D_x and D_y is evidence for the existence of a gapless scalar field, which is the dual of the emergent U(1) photon.[63,64]

studies[65] also show strong support for the existence of the deconfined critical theory for $N > 4$. The important $N = 2$ case has not been settled, although there is now evidence for a continuous transition with rather slow transients away from the scaling behavior.[66,67]

In the near future, ultracold atoms appear to be a promising arena for experimental studies of CFT3s. Bosonic ^{87}Rb atoms were observed to undergo such a quantum phase transition in an optical lattice,[68] as shown in Fig. 14. The quantum critical point here is described by the same relativistic scalar field theory as that discussed above for the dimer antiferromagnet, but with the field $\vec{\phi}$ now having two components with a global O(2) symmetry linked to the conservation of boson number.[69] This critical point is described by the Wilson-Fisher fixed point in 2+1 dimensions, which realizes a strongly interacting CFT3. Experiments on the superfluid-insulator transition in two dimensions have now been performed,[70] and this opens the way towards a detailed study of the properties of this CFT3. In particular, the single-site resolution available in the latest experiments[71,72] promises detailed information on real time dynamics with detailed spatial information.

These and other experiments demand an understanding of the real time dynamics of CFT3s at non-zero temperatures (T). We sketch the nature of the $T > 0$ phase diagram for the superfluid-insulator transition in Fig. 15. In the blue regions, the long time dynamics are amenable to a classical description: in the "superfluid" region we can use the Gross-Pitaevski non-linear wave equation, while in the "insulator" region we can describe the particle and hole excitations of the insulator using a Boltzmann equation. However, most novel is the pink "Quantum Critical" region,[73]

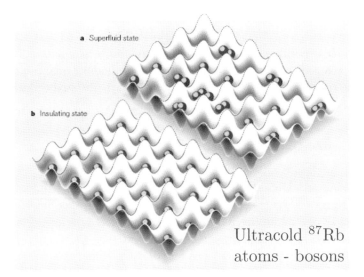

Fig. 14.   Bosons in an optical lattice undergo a superfluid-insulator transition as the depth of the optical lattice is increased, when there is an integer density of bosons per site. The critical theory is described by a relativistic field theory of a complex scalar with short-range self interactions.

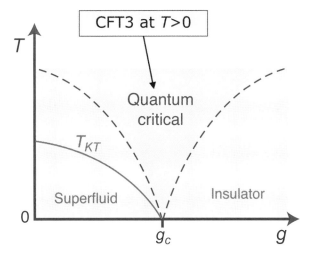

Fig. 15.   Non-zero temperature phase diagram of the superfluid-insulator transition in two spatial dimensions. Quasi-long-range superfluid order is present below the Kosterlitz-Thouless transition temperature T_{KT}. The dashed lines are locations of crossovers.

where classical models cannot apply at the longest characteristic time scales. In fact, in this region, all the characteristic time scales are set by temperature alone, and we have[74,75]

$$\tau = \mathcal{C} \frac{\hbar}{k_B T} \qquad (4)$$

where τ is some appropriately defined relaxation time, and $\mathcal{C}$ is a universal constant characteristic of the CFT3. The computation of $\mathcal{C}$, and related dissipative and transport co-efficients is a challenging task, and is not easily accomplished by the traditional expansion and renormalization group methods of quantum field theory. It is in these questions that the methods of gauge-gravity duality have had some impact, as the author has reviewed elsewhere.[76]

4. Compressible Quantum Matter

As the name implies, compressible states are those whose "density" can be varied freely by tuning an external parameter. Remarkably, there are only a few known examples of states which are compressible at $T = 0$. On the other hand, compressible quantum phases are ubiquitous in intermetallic compounds studied in recent years, and many of their observable properties do not fit into the standard paradigms. So a classification and deeper understanding of the possible compressible phases of quantum matter is of considerable importance.

Let us begin our discussion with a definition of compressible quantum matter.[76,77]

- Consider a continuum, translationally-invariant quantum system with a globally conserved U(1) charge $\mathcal{Q}$ i.e. $\mathcal{Q}$ commutes with the Hamiltonian H. Couple the Hamiltonian to a chemical potential, μ, which is conjugate to $\mathcal{Q}$: so the Hamiltonian changes to $H - \mu\mathcal{Q}$. The ground state of this modified Hamiltonian is compressible if $\langle\mathcal{Q}\rangle$ changes smoothly as a function of μ, with $d\langle\mathcal{Q}\rangle/d\mu$ non-zero.

A similar definition applies to lattice models, but let us restrict our attention to continuum models for simplicity.

Among states which preserve both the translational and global U(1) symmetries, the only traditional condensed matter state which is compressible is the Fermi liquid. This is the state obtained by turning on interactions adiabatically on the Sommerfeld-Bloch state of non-interacting fermions. Note that in our definition of compressible states we have allowed the degrees of freedom to be bosonic or fermionic, but there are no compressible states of bosons which preserve the U(1) symmetry.

One reason for the sparsity of compressible states is that they have to be gapless. Because $\mathcal{Q}$ commutes with H, changing μ will change the ground state only if there are low-lying levels which cross the ground state upon an infinitesimal change in μ. For the gapless states of conformal quantum matter considered in Sec. 3, a scaling argument implies a compressibility $\sim T^{d-1}$. So such states are compressible only in $d = 1$. The known $d = 1$ compressible states are 'Luttinger liquids' or their variants: they have a decoupled massless relativistic scalar with central charge $c = 1$ representing the fluctuations of $\mathcal{Q}$. We will not be interested in such states here.

The key characteristic of the Fermi liquid is the Fermi surface. For interacting electrons, the Fermi surface is defined by a zero of the inverse fermion Green's function

$$G_f^{-1}(|\boldsymbol{k}| = k_F, \omega = 0) = 0. \tag{5}$$

The Green's function is a complex number, and so naively the variation of the single real parameter $|\boldsymbol{k}|$ in Eq. (5) does not guarantee that a solution for k_F is possible. However, we can find k_F by solving for the real part of Eq. (5). In all known cases, we find that the imaginary part of G_f^{-1} also vanishes at this k_F: this happens because k_F is the momentum where the energy of both particle-like and hole-like excitations vanish, and so there are no lower energy excitations for them to decay to.[e]

In a Fermi liquid, the Green's function has a simple pole at the Fermi surface with

$$G_f^{-1} = \omega - v_F q + \mathcal{O}(\omega^2, q^2) \tag{6}$$

where $q = |\boldsymbol{k}| - k_F$ is the minimal distance to the Fermi surface (see Fig. 16), and v_F is the Fermi velocity. The relationship between k_F and the density $\langle \mathcal{Q} \rangle$ in a Fermi liquid is the same as that in the free fermion state: this is the Luttinger relation,

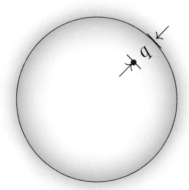

Fig. 16.   The Fermi surface. The fermion at the blue point is momentum $-q$ away from the nearest point on the Fermi surface. There is a sharp quasiparticle pole on the Fermi surface for a Fermi liquid, as in Eq. (6). There are no quasiparticles in non-Fermi liquids, but a continuum of low energy excitations which obey Eq. (8); nevertheless, the position of the Fermi surface is well-defined, and it encloses volume which obeys a Luttinger relation. For the model in Eq. (7), this Fermi surface is *hidden*, because the single fermion Green's function in Eq. (8) is not a gauge-invariant observable.

---

[e]This argument also shows why the equation for the zeros of the Green's function $G_f(|\boldsymbol{k}|, \omega = 0) = 0$ generically has no solution (although they are important for the approach reviewed in ref. 78). In this case the vanishing of the real part has no physical interpretation, and the imaginary part need not vanish at the same $|\boldsymbol{k}|$.

which equates $\langle Q \rangle$ to the momentum-space volume enclosed by the Fermi surface (modulo phase space factors).

Numerous modern materials display metallic, compressible states which are evidently not Fermi liquids. Most commonly, they are associated with metals near the onset of antiferromagnetic long-range order; these materials invariably become superconducting upon cooling in the absence of an applied magnetic field. The onset of antiferromagnetism in metals, and the proximate presence of "high-T_c" superconductivity, was discussed by the author at the Solvay Conference; however, the subject has been reviewed in a separate recent article,[79] and so will not be presented here.

Here, we note another remarkable compressible phase found in the organic insulator $EtMe_3Sb[Pd(dmit)_2]_2$. This has a triangular lattice of $S = 1/2$ spins, as in the antiferromagnet discussed in the beginning of Sec. 2; however, there are expected to be further neighbor ring-exchange interactions beyond the nearest-neighbor term in Eq. (1), and possibly for this reason the ground state does not have antiferromagnetic order, and nor does it appear to be the gapped $\mathbb{Z}_2$ RVB state. Remarkably, the low temperature thermal conductivity of this material is similar to that of a metal,[80] even though the charge transport is that of an insulator: see Fig. 17. Thus

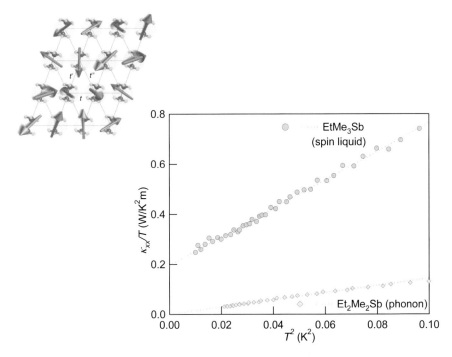

Fig. 17.   From ref. 80. The longitudinal thermal conductivity κ_{xx} as a function of temperature (T) for $EtMe_3Sb[Pd(dmit)_2]_2$, an insulating antiferromagnet of $S = 1/2$ spins on a triangular lattice (sketched at the top). The notable feature is the non-zero value of $\lim_{T \to 0} \kappa_{xx}/T$, which is characteristic of thermal transport of fermions near a Fermi surface.

this material is a charge insulator, but a thermal metal. One possible explanation is that there is a Fermi surface of spinons,[81–86] which would also be consistent with the observed non-zero spin susceptibility. This Fermi sea of spinons realizes a phase of compressible quantum matter, where the conserved charge $\mathcal{Q}$ is identified with the total spin.

Motivated by these and other experiments, we now turn to a discussion of a much-studied realization of compressible quantum matter which is not a Fermi liquid. This is the problem of fermions, ψ, at non-zero density coupled to an Abelian or non-Abelian gauge field, A^a, of a Lie group.[f] We can schematically write the Lagrangian as

$$\mathcal{L} = \psi^\dagger \left(\partial_\tau - i A_\tau^a t^a - \mu h \right) \psi - \frac{1}{2m} \psi^\dagger \left(\boldsymbol{\nabla} - i \boldsymbol{A}^a t^a \right)^2 \psi + \frac{1}{4g^2} F^2 \qquad (7)$$

where F is the field tensor, τ is imaginary time, μ is the chemical potential, t^a are the generators of the gauge group, h is the generator of the conserved charge $\mathcal{Q}$ (h is distinct from, and commutes with, all the t^a), and m is the effective mass. In the application to spin liquids, ψ represents the fermionic spinons, and A^a is the emergent gauge field of a particular RVB state.

Let us summarize the present understanding of the properties of (7) in spatial dimension $d = 2$, obtained by conventional field-theoretic analysis.[87–99] There is a universal, compressible 'non-Fermi liquid' state with a Fermi surface at precisely the same k_F as that given by the free electron value. However, unlike the Fermi liquid, this Fermi surface is *hidden*, and characterized by singular, non-quasiparticle low-energy excitations. It is hidden because the ψ fermion Green's function is not a gauge-invariant quantity, and so is not a physical observable. However, in perturbative theoretic analyses, the ψ Green's function can be computed in a fixed gauge, and this quantity is an important ingredient which determines the singularities of physical observables. In the Coulomb gauge, $\boldsymbol{\nabla} \cdot \boldsymbol{A}^a = 0$, the ψ Green's function has been argued to obey the scaling form[96]

$$G_\psi^{-1} = q^{1-\eta} \Phi(\omega/q^z) \qquad (8)$$

where q is the momentum space distance from the Fermi surface, as indicated in Fig. 16. The function Φ is a scaling function which characterizes the continuum of excitations near the Fermi surface, η is an anomalous dimension, and z is a dynamic critical exponent. The Fermi liquid result clearly corresponds to $\eta = 0$ and $z = 1$, and simple form for Φ. For the present non-Fermi liquid, the exponent η was recently estimated in loop expansions.[96,97] It was also found that $z = 3/2$ to three loops,[96] and it is not known if this is an exact result.[g]

---

[f]It is not clear whether such models apply to EtMe$_3$Sb[Pd(dmit)$_2$]$_2$. The theories of refs. 81–83 have continuous gauge groups, while those of refs. 84–86 have discrete gauge groups; Eq. (7) does not apply to the latter.

[g]In the condensed matter literature, it is often stated that this theory has $z = 3$. This refers to the dynamic scaling of the gauge field propagator, which has[96] exactly twice the value of z from that defined by Eq. (8).

For our discussion below, we need the thermal entropy density, S, of this non-Fermi liquid compressible state at low temperatures. This is found to be

$$S \sim T^{1/z}. \tag{9}$$

This can be viewed as an analog of the Stefan-Boltzmann law, which states that $S \sim T^{d/z}$ for a d-dimensional quantum system with excitations which disperse as $\omega \sim |\boldsymbol{k}|^z$. In the present case, our critical fermion excitations disperse only transverse to the Fermi surface, and so they have the phase space, and corresponding entropy, of effective dimension $d_{\mathrm{eff}} = 1$. Following critical phenomena terminology, let us rewrite Eq. (9) in the form

$$S \sim T^{(d-\theta)/z}, \tag{10}$$

where θ is the violation of hyperscaling exponent, defined by $d_{\mathrm{eff}} = d - \theta$. The present non-Fermi liquid therefore has

$$\theta = d - 1. \tag{11}$$

We conclude this section by giving a few more details of the derivation of the above scaling properties of Eq. (7) for the case of a U(1) gauge group, focusing on the determination of the value of z. In the low energy limit, it has been argued[95,96] that we can focus on the gauge field fluctuations collinear to single direction $\boldsymbol{p}$, and these couple most efficiently to fermions at antipodal points on the Fermi surface where the tangent to the Fermi surface is also parallel to $\boldsymbol{p}$: see Fig. 18. From Eq. (7), it is then straightforward to derive the following low energy action for the

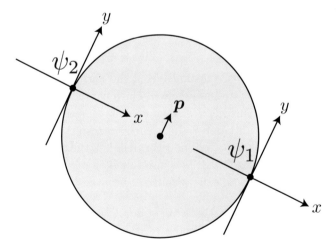

Fig. 18.   Low energy limit of Eq. (7). The A gauge-field fluctuations with momenta $\boldsymbol{p}$ couple most efficiently to fermions $\psi_{1,2}$ near Fermi surface points where the tangent is collinear to $\boldsymbol{p}$.

long-wavelength fermions, $\psi_{1,2}$ at the antipodal points, and the gauge field A

$$\mathcal{S} = \int d\tau dx dy \left[\psi_1^\dagger \left(\partial_\tau - i\partial_x - \partial_y^2 \right) \psi_1 + \psi_2^\dagger \left(\partial_\tau + i\partial_x - \partial_y^2 \right) \psi_2 \right.$$
$$\left. - g A \left(\psi_1^\dagger \psi_1 - \psi_2^\dagger \psi_2 \right) + \frac{1}{2} \left(\partial_y A \right)^2 \right]. \tag{12}$$

Here g is the gauge coupling constant, and A is the single component of the photon in $d = 2$ which is transverse to $\mathbf{q}$. This theory has been studied in great detail in recent work,[95–97] and it was found that the fermion temporal derivative terms are irrelevant in the scaling limit. Here we will assume that this is the case, and show how this fixes the value of z. It is easy to see that the spatial gradient terms in $\mathcal{S}$ are invariant under the following scaling transformations:

$$x \to x/s, \quad y \to y/s^{1/2}, \quad \tau \to \tau/s^z,$$
$$A \to A\, s^{(2z+1)/4}, \quad \psi \to \psi\, s^{(2z+1)/4}. \tag{13}$$

Then the gauge coupling constant in Eq. (12) is found to transform as

$$g \to g\, s^{(3-2z)/4}, \tag{14}$$

and we see that a fixed point theory requires $z = 3/2$ at tree level. The unusual feature of this computation is that we have used the invariance of an interaction term to fix the value of z. Usually, z is determined by demanding invariance of the temporal derivative terms which are quadratic in the fields; however, such terms are strongly irrelevant here, and so can be set to zero at the outset. Indeed the irrelevance of terms like $\psi^\dagger \partial_\tau \psi$ is an inevitable characteristic of a non-Fermi liquid, because then the dominant frequency dependence of the fermion Green's function arises from the self energy. This opens the possibility of determining z by fixing the strength of a boson-fermion interaction. In the present case, a lengthy computation[96] with $\mathcal{S}$ shows that such a tree-level value of z has no corrections to three loops.

5. Connections to String Theory

Recent years have seen a significant effort to realize strongly-coupled conformal and compressible phases of matter using the methods of gauge gravity duality.[76] Underlying this connection is the AdS/CFT correspondence which provides a duality between CFTs in $d + 1$ spacetime dimensions, and theories of gravity in $d + 2$ dimensional anti-de Sitter space (AdS$_{d+2}$).

An intuitive picture of the correspondence is provided by the picture of a d-dimensional D-brane of string theory shown in Fig. 19. The low energy limit of the string theory is a CFT-$(d + 1)$, representing the quantum matter we are interested in. The strings move in AdS$_{d+2}$, and can be seen as the source of long-range entanglement in the quantum matter. This is highlighted by the similarity between Fig. 19 and the tensor network representation of entanglement[100,101] in Fig. 20. In

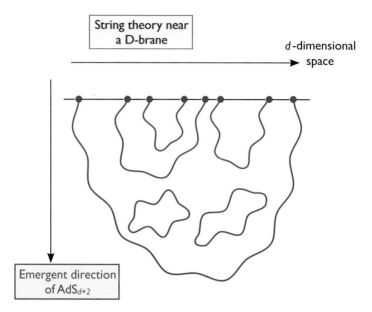

Fig. 19.   A D-brane in string theory. The strings end on a d-dimensional spatial surface. The blue circles represent the particles of quantum matter.

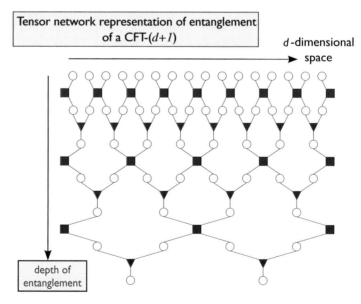

Fig. 20.   Pictorial tensor network representation (from ref. 102) of entanglement on a lattice model of quantum degrees of freedom represented by the open circles in the top row.

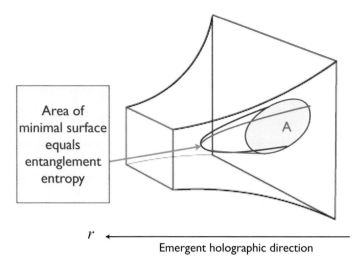

Fig. 21. Computation of the entanglement entropy defined in Fig. 7 of region A. The Ryu-Takayanagi formula equates S_E to the area of the minimal surface enclosing region A in the gravity theory.

this connection, the emergent spatial direction of AdS_{d+2} is seen to represent the depth of entanglement between the quantum matter degrees of freedom.[102] The Ryu-Takayanagi formula[103] for the entanglement entropy (Fig. 21) also emerges from this connection via a computation of the entanglement entropy from the tensor network.

The early applications of gauge-gravity duality to condensed matter physics[104] addressed issues related to the $T > 0$ quantum-critical dynamics (Fig. 15) of conformal quantum matter; these have been reviewed recently elsewhere.[76] Here, I will briefly describe recent ideas on its application to compressible quantum matter.

Let us take as our objective the determination of the gravity dual of the theory in Eq. (7) describing non-zero density fermions coupled to a gauge field. As argued by 't Hooft,[105] such duals are obtained in a suitable large N limit. In the condensed matter latter literature, the fermion ψ is endowed with N_f flavors and the large N_f has been intensively examined.[87–99] At leading order in $1/N_f$ in $d = 2$, computations from Eq. (12) show that the fermion Green's function is modified from the Fermi liquid form in Eq. (6) by a singular correction which approaches the non-Fermi liquid scaling structure in Eq. (8); schematically, this correction is of the form

$$G_\psi^{-1} \approx \omega - v_F q + i\frac{c}{N_f}\omega^{2/3},\tag{15}$$

which exhibits the $z = 3/2$ scaling discussed below Eq. (12). Note that the term of Eq. (15) which is most singular in the low energy limit has a prefactor of $1/N_f$. This is dangerous, and leads to a breakdown in the bare Feynman graph structure of the $1/N_f$ expansion;[95,96] even at first order in $1/N_f$, it is necessary to at least sum all planar graphs.

So let us consider an alternative case[77,106–108] where the gauge group is $SU(N_c)$, and we take the fermions ψ to transform under the *adjoint* representation of $SU(N_c)$. Then, an analysis of the Feynman graph expansion shows that the low loop contributions to the $N_c \to \infty$ theory have the same low frequency structure as in Eq. (15) in $d = 2$, but *without* suppression of the singular terms by powers of $1/N_c$:

$$G_\psi^{-1} \approx \omega - v_F q + i\tilde{c}\,\omega^{2/3}. \tag{16}$$

This indicates that the Feynman graph counting of powers of $1/N_c$ holds in the $N_c \to \infty$ limit, and is identical to that in the classic paper by 't Hooft.[105] Consequently, even at non-zero density and in the critical low energy theory, the $1/N_c$ expansion is an expansion in powers of the genus of the surface defined by the double-line Feynman graphs. By the arguments of 't Hooft,[105] we can reasonably hope that the $N_c \to \infty$ theory is described by a dual gravity theory. Furthermore, given the issues with the $1/N_f$ expansion noted above, the $N_c \to \infty$ limit appears to be suited to capture the physics of condensed matter systems.

Now, we will constrain the background metric of this hypothetical gravity theory by general scaling arguments.[107] We represent the d-dimensional spatial displacement by $\boldsymbol{dx}$, time displacement by dt, the emergent direction by dr, and proper distance on the holographic space by ds. We are interested in states with a low energy scaling symmetry, and so we demand that the low energy metric obey

$$\boldsymbol{x} \to \zeta\,\boldsymbol{x}$$
$$t \to \zeta^z\,t$$
$$ds \to \zeta^{\theta/d}\,ds. \tag{17}$$

under rescaling by a factor ζ. This defines z as the dynamic critical exponent, and we now argue that θ is the violation of hyperscaling exponent which was defined earlier by Eq. (10). Using translation and rotational invariance in space, and translational invariance in time, we deduce the metric

$$ds^2 = \frac{1}{r^2}\left(-\frac{dt^2}{r^{2d(z-1)/(d-\theta)}} + r^{2\theta/(d-\theta)}dr^2 + \boldsymbol{dx}^2\right), \tag{18}$$

as the most general solution to (17) modulo prefactors and reparametrization invariance in r. For our choice of the co-ordinates in (18), r transforms as

$$r \to \zeta^{(d-\theta)/d}r. \tag{19}$$

Now let us take this gravity theory to a temperature $T > 0$. This thermal state requires a horizon, and let us assume the horizon appears at $r = r_H$. The entropy density of this thermal state, S, will be proportional to the spatial area of the horizon, and so from Eq. (18) we have $S \sim r_H^{-d}$. Now T scales as $1/t$, and so from Eqs. (17) and (19) we deduce $r_H^{-d} \sim T^{(d-\theta)/z}$ and so $S \sim T^{(d-\theta)/z}$, which matches the definition in Eq. (10). This justifies our identification of the θ appearing in Eq. (18) as the violation of hyperscaling exponent.[107,109–116]

With this gravitational definition of z and θ, we can now obtain additional properties of these exponents which should also apply to the dual field theory. Remarkably, there is no known derivation of these properties directly from the field theory. We expect Eq. (18) to be a solution of the analog of Einstein's equations in some gravitational theory; so it is reasonable to impose the null energy condition,[109] and this yields the important inequality[107]

$$z \geq 1 + \frac{\theta}{d}. \tag{20}$$

As a final general property of the quantum matter which can be computed directly from the holographic metric in Eq. (18), we turn to the entanglement entropy. This can be computed, via the Ryu-Takayanagi formula,[103] by computing the minimal surface area of Fig. 21, and we find[107,109]

$$S_E \sim \begin{cases} P & , \text{ for } \theta < d - 1 \\ P \ln P & , \text{ for } \theta = d - 1 \\ P^{\theta/(d-1)} & , \text{ for } \theta > d - 1 \end{cases} \tag{21}$$

where P is the surface area (i.e. the perimeter in $d = 2$) of region A, as in Eq. (2). Note that the 'area law' of entanglement entropy is obeyed only for $\theta < d - 1$. The regime $\theta > d - 1$ has strong violations of the area law, and so this is unlikely to represent a generic local quantum field theory.

Note that here we defined z and θ as exponents which appear in the metric of the gravitation theory in Eq. (18). However, as we have shown above, they also have independent definitions in terms of the boundary quantum theory via Eq. (10). One of the important consequences of the gravitational definition is that we are now able to conclude that these exponents obey the inequality Eq. (20), and constrain the entanglement entropy as in Eq. (21). No independent field-theoretic derivation of these results is known. Indeed, Eqs. (20) and (21) may be taken as necessary conditions for the existence of a reasonable gravity dual of the field theory.

So far, our gravitational scaling analysis has been very general, and could apply to any dual critical theory. We now compare to the results of the field theoretic analysis discussed in Sec. 4 for the non-Fermi liquid state of fermions coupled to a gauge field. We use the temperature dependence of the thermal entropy to fix the value of $\theta = d - 1$ found in Eq. (11). The combination of Eqs. (18) and (11) is then the metric of the hypothetical gravitational dual[107,109] description of strongly interacting compressible quantum matter, such as that realized by the $N_c \to \infty$ limit of the theory in Eq. (7) with an $SU(N_c)$ gauge field and fermions in the adjoint representation of $SU(N_c)$.

This proposal, and in particular the value of θ in Eq. (11), can now be subjected to a number of tests:

- In $d = 2$, we have $\theta = 1$ from Eq. (11), and so from Eq. (20) the gravity dual theory requires $z \geq 3/2$. Remarkably, the lower bound, $z = 3/2$, is the value obtained from the weak-coupling field theory analysis extended to three loops,[96] as we discussed below Eq. (12).

- We see from Eq. (21) that for the value of θ in Eq. (11), there is logarithmic violation of the area law,[109] as expected for a system with a Fermi surface.[32,117]
- The metric in Eq. (18) appears as a solution[118–120] of a class of Einstein-Maxwell-dilaton theories. In this realization, there is a non-zero charge density $\langle \mathcal{Q} \rangle$ on the d-dimensional boundary, and the compressibility $d\langle \mathcal{Q} \rangle / d\mu$ is non-zero.
- A complete computation of the entanglement entropy in the Einstein-Maxwell-dilaton theory yields the following expression for the entanglement entropy[107]

$$S_E = \lambda \langle \mathcal{Q} \rangle^{(d-1)/d} P \ln P, \qquad (22)$$

A key feature[107] is that the dependence upon the shape of region A is only through the value of P, and the prefactor λ is *independent* of the shape or any other geometric of property of region A: this matches the characteristics of the entanglement entropy of a spherical Fermi surface.[117,121]

- The value of λ in Eq. (22), by a variant of the attractor mechanism,[122–124] is independent of all ultraviolet details of the gravity theory,[107] and S_E depends only upon the value of $\langle \mathcal{Q} \rangle$ as shown. This supports the conclusion that the prefactor of the entanglement entropy of a non-Fermi liquid is universal, as it is believed to be for an interacting Fermi liquid.[117]
- We expect a Luttinger relation for the volume of the 'hidden' Fermi surface, with $\langle \mathcal{Q} \rangle \sim k_F^d$. Then the k_F dependence of Eq. (22) is that expected for a Fermi surface: this can be viewed as indirect evident for the Luttinger relation.[107] In this manner, the Luttinger relation, which is one of the deepest results of condensed matter physics, is surprisingly connected to two fundamental features of the holographic theory: Gauss' Law and the attractor mechanism.[107,125]
- Refs. 107, 126 also studied the transition of the Einstein-Maxwell-dilaton theory to a state with *partial* confinement, in which there were additional Fermi surfaces of gauge-neutral particles. The resulting state is analogous to the 'fractionalized Fermi liquid' of Kondo and Hubbard models.[79,127] It was found[107] that the holographic entanglement entropy of this partially confined state was given by Eq. (22) but with $\langle \mathcal{Q} \rangle \to \langle \mathcal{Q} - \mathcal{Q}_{\text{conf}} \rangle$, where $\langle \mathcal{Q}_{\text{conf}} \rangle$ is the density associated with the Fermi surfaces of gauge-neutral particles. If we now use Eq. (22) to fix the k_F of the hidden Fermi surfaces of gauge-dependent particles just as above, then we see that the Luttinger relation for $\langle \mathcal{Q} \rangle$ equates it to the sum of the gauge-neutral and gauge-charged Fermi surfaces, again as expected from the gauge theory analysis.[127]

Clearly, it would be useful to ultimately obtain evidence for the wavevector k_F of the hidden Fermi surfaces by the spatial modulation of some response function. Short of such confirmation, the above tests do provide strong evidence for the presence of a hidden Fermi surface in such gravity theories of compressible quantum matter. These gravity theories appear as solutions of the Einstein-Maxwell-dilaton theories[118–120] which contain only bosonic degrees of freedom; so they may be viewed as analogs of the 'bosonization' of the Fermi surface.[128–134]

Acknowledgments

I am very grateful to D. Chowdhury, P. Fendley, E. Fradkin, S. Kachru, L. Huijse, M. Metlitski, R. Nandkishore, and D. Vegh for valuable comments. I would like to especially thank S. Kivelson for detailed comments on all aspects of the manuscript. This research was supported by the National Science Foundation under grant DMR-1103860, and by a MURI grant from AFOSR.

References

1. A. Einstein, B. Podolsky, and N. Rosen, Can quantum-mechanical description of physical reality be considered Complete?, *Phys. Rev.* **47**, 777 (1935).
2. J. S. Bell, On the Einstein Podolsky Rosen paradox, *Physics* **1**, 195 (1964).
3. S. R. White and A. L. Chernyshev, Neel order in square and triangular lattice Heisenberg models, *Phys. Rev. Lett.* **99**, 127004 (2007). [arXiv:0705.2746]
4. P. Fazekas and P. W. Anderson, On the ground state properties of the anisotropic triangular antiferromagnet, *Philos. Mag.* **30**, 423 (1974).
5. L. Pauling, A resonating-valence-bond theory of metals and intermetallic compounds, *Proc. Roy. Soc. London A* **196**, 343 (1949).
6. P. W. Anderson, The resonating valence bond state in La_2CuO_4 and superconductivity, *Science* **235**, 1196 (1987).
7. S. A. Kivelson, D. S. Rokhsar, and J. P. Sethna, Topology of the resonating valence-bond state: Solitons and high-T_c superconductivity, *Phys. Rev. B* **35**, 8865 (1987).
8. G. Baskaran and P. W. Anderson, Gauge theory of high-temperature superconductors and strongly correlated Fermi systems, *Phys. Rev. B* **37**, 580 (1988).
9. D. Rokhsar and S. A. Kivelson, Superconductivity and the quantum hard-core dimer gas, *Phys. Rev. Lett.* **61**, 2376 (1988).
10. E. Fradkin, The spectrum of short-range resonating valence bond theories, in *Field Theories in Condensed Matter Physics: A Workshop*, ed. Z. Tesanovic, Addison-Wesley (1990); E. Fradkin and S. A. Kivelson, Short range resonating valence bond theories and superconductivity, *Mod. Phys. Lett. B* **4**, 225 (1990).
11. N. Read and S. Sachdev, Spin-Peierls, valence bond solid, and Neel ground states of low dimensional quantum antiferromagnets, *Phys. Rev. B* **42**, 4568 (1990).
12. V. Kalmeyer and R. B. Laughlin, Equivalence of the resonating-valence-bond and fractional quantum Hall states, *Phys. Rev. Lett.* **59**, 2095 (1987).
13. X.-G. Wen, F. Wilczek, and A. Zee, Chiral spin states and superconductivity, *Phys. Rev. B* **39**, 11413 (1989).
14. X.-G. Wen, Mean-field theory of spin-liquid states with finite energy gap and topological orders, *Phys. Rev. B* **44**, 2664 (1991).
15. N. Read and S. Sachdev, Large-N expansion for frustrated quantum antiferromagnets, *Phys. Rev. Lett.* **66**, 1773 (1991).
16. S. Sachdev and N. Read, Large N expansion for frustrated and doped quantum antiferromagnets, *Int. J. Mod. Phys. B* **5**, 219 (1991) [arXiv:cond-mat/0402109].

17. R. Jalabert and S. Sachdev, Spontaneous alignment of frustrated bonds in an anisotropic, three dimensional Ising model, *Phys. Rev. B* **44**, 686 (1991). This paper notes that the frustrated classical Ising model studied is the the dual form of the gauge theory of the $\mathbb{Z}_2$ RVB phase. Notes from July 30, 1990 containing the derivation are at http://qpt.physics.harvard.edu/qdnotes.pdf, and were published in S. Sachdev and M. Vojta, Translational symmetry breaking in two-dimensional antiferromagnets and superconductors, *J. Phys. Soc. Japan* **69**, Suppl. B, 1 (2000) [arXiv:cond-mat/9910231].

18. T. Senthil and M. P. A. Fisher, $\mathbb{Z}_2$ gauge theory of electron fractionalization in strongly correlated systems, *Phys. Rev. B* **62**, 7850 (2000). [arXiv:cond-mat/9910224]

19. R. Moessner, S. L. Sondhi, and E. Fradkin, Short-ranged RVB physics, quantum dimer models and Ising gauge theories, *Phys. Rev. B* **65**, 024504 (2001). [arXiv:cond-mat/0103396]

20. E. Ardonne, P. Fendley, and E. Fradkin, Topological order and conformal quantum critical points, *Ann. Phys.* **310**, 493 (2004). [arXiv:cond-mat/0311466]

21. S. Sachdev, Kagome and triangular lattice Heisenberg antiferromagnets: ordering from quantum fluctuations and quantum-disordered ground states with deconfined bosonic spinons, *Phys. Rev. B* **45**, 12377 (1992). This paper discusses $\mathbb{Z}_2$ RVB phases on the kagome and triangular lattices in general, and notes that these should appear in the quantum dimer model on these lattices (page 12393).

22. R. Moessner and S. L. Sondhi, Resonating valence bond phase in the triangular lattice quantum dimer model, *Phys. Rev. Lett.* **86**, 1881 (2001). [arXiv:cond-mat/0007378]

23. G. Misguich, D. Serban, and V. Pasquier, Quantum dimer model on the kagome lattice: solvable dimer liquid and Ising gauge theory, *Phys. Rev. Lett.* **89**, 137202 (2002). [arXiv:cond-mat/0204428]

24. H. Yao and S. A. Kivelson, Exact spin liquid ground states of the quantum dimer model on the square and honeycomb lattices, arXiv:1112.1702.

25. D. J. Thouless, Fluxoid quantization in the resonating-valence-bond model, *Phys. Rev. B* **36**, 7187 (1987).

26. N. Read and B. Chakraborty, Statistics of the excitations of the resonating-valence-bond state, *Phys. Rev. B* **40**, 7133 (1989).

27. A. Y. Kitaev, Fault-tolerant quantum computation by anyons, *Ann. Phys.* **303**, 2 (2003). [arXiv:quant-ph/9707021]

28. A. Y. Kitaev, Anyons in an exactly solved model and beyond, *Ann. Phys.* **321**, 2 (2006). [arXiv:cond-mat/0506438]

29. T. Grover, A. M. Turner, and A. Vishwanath, Entanglement entropy of gapped phases and topological order in three dimensions, *Phys. Rev. B* **84**, 195120 (2011). [arXiv:1108.4038]

30. A. Kitaev and J. Preskill, Topological entanglement entropy, *Phys. Rev. Lett.* **96**, 110404 (2006). [arXiv:hep-th/0510092]

31. M. Levin and X.-G. Wen, Detecting topological order in a ground state wave function, *Phys. Rev. Lett.* **96**, 110405 (2006). [arXiv:cond-mat/0510613]

32. Y. Zhang, T. Grover, and A. Vishwanath, Topological entanglement entropy of $\mathbb{Z}_2$ spin liquids and lattice Laughlin states, *Phys. Rev. B* **84**, 075128 (2011). [arXiv:1106.0015]

33. F. A. Bais, Flux metamorphosis, *Nucl. Phys. B* **170**, 3243 (1980).

34. F. A. Bais, P. van Driel, and M. de Wild Propitius, Quantum symmetries in discrete gauge theories, *Phys. Lett. B* **280**, 63 (1992).

35. J. M. Maldacena, G. W. Moore and N. Seiberg, D-brane charges in five-brane backgrounds, *JHEP* **0110**, 005 (2001). [arXiv:hep-th/0108152]

36. M. Freedman, C. Nayak, K. Shtengel, K. Walker, and Z. Wang, A class of P,T-invariant topological phases of interacting electrons, *Ann. Phys.* **310**, 428 (2004). [arXiv:cond-mat/0307511]

37. C. Xu and S. Sachdev, Global phase diagrams of frustrated quantum antiferromagnets in two dimensions: doubled Chern-Simons theory, *Phys. Rev. B* **79**, 064405 (2009). [arXiv:0811.1220]

38. S. T. Flammia, A. Hamma, T. L. Hughes, and X.-G. Wen, Topological entanglement Renyi entropy and reduced density matrix structure, *Phys. Rev. Lett.* **103**, 261601 (2009). [arXiv:0909.3305]

39. I. R. Klebanov, S. S. Pufu, S. Sachdev and B. R. Safdi, Renyi entropies for free field theories, arXiv:1111.6290.

40. S. Sachdev, Kagome and triangular lattice Heisenberg antiferromagnets: ordering from quantum fluctuations and quantum-disordered ground states with deconfined bosonic spinons, *Phys. Rev. B* **45**, 12377 (1992).

41. S. Yan, D. A. Huse, and S. R. White, Spin liquid ground state of the $S = 1/2$ Kagome Heisenberg model, *Science* **332**, 1173 (2011). [arXiv:1011.6114]

42. Y. Huh, M. Punk and S. Sachdev, Vison states and confinement transitions of $\mathbb{Z}_2$ spin liquids on the kagome lattice, *Phys. Rev. B* **84**, 094419 (2011). [arXiv:1106.3330]

43. L. Messio, B. Bernu, and C. Lhuillier, The Kagome antiferromagnet: a chiral topological spin liquid? arXiv:1110.5440.

44. H.-C. Jiang, Z. Wang, and L. Balents, Identifying topological order by entanglement entropy, arXiv:1205.4289.

45. S. Depenbrock, I. P. McCulloch, and U. Schollwöck Nature of the spin liquid ground state of the $S = 1/2$ Kagome Heisenberg model, arXiv:1205.4858.

46. Y. Lee, Experimental signatures of spin liquid physics on the $S = 1/2$ Kagome lattice, *Bull. Am. Phys. Soc.* **57**, 1, H8.00005 (2012).

47. L. Balents, M. P. A. Fisher, and S. M. Girvin, Fractionalization in an easy-axis Kagome antiferromagnet, *Phys. Rev. B* **65**, 224412 (2002). [arXiv:cond-mat/0110005]

48. S. V. Isakov, Y.-B. Kim, and A. Paramekanti, Spin liquid phase in a $S = 1/2$ quantum magnet on the kagome lattice, *Phys. Rev. Lett.* **97**, 207204 (2006). [arXiv:cond-mat/0607778]

49. S. V. Isakov, M. B. Hastings, and R. G. Melko, Topological entanglement entropy of a Bose-Hubbard spin liquid, *Nature Physics* **7**, 772 (2011). [arXiv:1102.1721]

50. F. Figueirido, A. Karlhede, S. Kivelson, S. Sondhi, M. Rocek, and D. S. Rokhsar, Exact diagonalization of finite frustrated spin-1/2 Heisenberg models, *Phys. Rev. B* **41**, 4619 (1990).

51. H.-C. Jiang, H. Yao, and L. Balents, Spin liquid ground state of the spin-1/2 square J_1-J_2 Heisenberg model, arXiv:1112.2241.

52. L. Wang, Z.-C. Gu, X.-G. Wen, and F. Verstraete, Possible spin liquid state in the spin 1/2 J_1-J_2 antiferromagnetic Heisenberg model on square lattice: A tensor product state approach, arXiv:1112.3331.

53. T. Li, F. Becca, W. Hu, and S. Sorella, Gapped spin liquid phase in the J_1-J_2 Heisenberg model by a Bosonic resonating valence-bond ansatz, arXiv:1205.3838.

54. S. Wenzel and W. Janke, Comprehensive quantum Monte Carlo study of the quantum critical points in planar dimerized/quadrumerized Heisenberg models, *Phys. Rev. B* **79**, 014410 (2009). [arXiv:0808.1418]

55. M. A. Metlitski, C. A. Fuertes, and S. Sachdev, Entanglement entropy in the O(N) model, *Phys. Rev. B* **80**, 115122 (2009). [arXiv:0904.4477].

56. D. S. Chow, P. Wzietek, D. Fogliatti, B. Alavi, D. J. Tantillo, C. A. Merlic, and S. E. Brown, Singular behavior in the pressure-tuned competition between spin-

Peierls and antiferromagnetic ground states of (TMTTF)$_2$PF$_6$, *Phys. Rev. Lett.* **81**, 3984 (1998).

57. Ch. Ruegg, B. Normand, M. Matsumoto, A. Furrer, D. F. McMorrow, K. W. Krämer, H. -U. Güdel, S. N. Gvasaliya, H. Mutka, and M. Boehm, Quantum magnets under pressure: Controlling elementary excitations in TlCuCl$_3$, *Phys. Rev. Lett.* **100**, 205701 (2008). [arXiv:0803.3720]

58. B. Normand and T. M. Rice, Dynamical properties of an antiferromagnet near the quantum critical point: Application to LaCuO$_{2.5}$, *Phys. Rev. B* **56**, 8760 (1997). [arXiv:cond-mat/9701202]

59. S. Sachdev, Theory of finite temperature crossovers near quantum critical points close to, or above, their upper-critical dimension, *Phys. Rev. B* **55**, 142 (1997). [arXiv:cond-mat/9606083]

60. S. Sachdev, Exotic phases and quantum phase transitions: model systems and experiments, Rapporteur presentation at the 24th Solvay Conference on Physics, *Quantum Theory of Condensed Matter*, Brussels, Oct 11-13, 2008, arXiv:0901.4103.

61. N. Read and S. Sachdev, Valence bond and spin-Peierls ground states of low dimensional quantum antiferromagnets, *Phys. Rev. Lett.* **62**, 1694 (1989).

62. T. Senthil, A. Vishwanath, L. Balents, S. Sachdev, and M. P. A. Fisher, Deconfined quantum critical points, *Science* **303**, 1490 (2004). [cond-mat/0311326]

63. T. Senthil, L. Balents, S. Sachdev, A. Vishwanath, and M. P. A. Fisher, Quantum criticality beyond the Landau-Ginzburg-Wilson paradigm, *Phys. Rev. B* **70**, 144407 (2004). [cond-mat/0312617]

64. A. W. Sandvik, Evidence for deconfined quantum criticality in a two-dimensional Heisenberg model with four-spin interactions, *Phys. Rev. Lett.* **98**, 227202 (2007). [arXiv:cond-mat/0611343]

65. R. K. Kaul and A. W. Sandvik, A lattice model for the SU(N) Neel-VBS quantum phase transition at large N, arXiv:1110.4130

66. A. W. Sandvik, Continuous quantum phase transition between an antiferromagnet and a valence-bond-solid in two dimensions; evidence for logarithmic corrections to scaling, *Phys. Rev. Lett.* **104**, 177201 (2010). [arXiv:1001.4296]

67. A. Banerjee, K. Damle, and F. Alet, Impurity spin texture at a deconfined quantum critical point, *Phys. Rev. B* **82**, 155139 (2010). [arXiv:1002.1375]

68. M. Greiner, O. Mandel, T. Esslinger, T. W. Hänsch, and I. Bloch, Quantum phase transition from a superfluid to a Mott insulator in a gas of ultracold atoms, *Nature* **415**, 39 (2002).

69. M. P. A. Fisher, P. B. Weichman, G. Grinstein, and D. S. Fisher, Boson localization and the superfluid-insulator transition, *Phys. Rev. B* **40**, 546 (1989).

70. X. Zhang, C.-L. Hung, S.-K. Tung, and C. Chin, Quantum critical behavior of ultracold atoms in two-dimensional optical lattices, *Science* **335**, 1070 (2012). [arXiv:1109.0344]

71. J. Simon, W. S. Bakr, Ruichao Ma, M. E. Tai, P. M. Preiss, and M. Greiner, Quantum simulation of antiferromagnetic spin chains in an optical lattice, *Nature* **472**, 307 (2011). [arXiv:1103.1372]

72. C. Weitenberg, M. Endres, J. F. Sherson, M. Cheneau, P. Schauß, T. Fukuhara, I. Bloch, and S. Kuhr, Single-spin addressing in an atomic Mott insulator, *Nature* **471**, 319 (2011). [arXiv:1101.2076]

73. S. Chakravarty, B.I. Halperin, and D.R. Nelson, Low-temperature behavior of two-dimensional quantum antiferromagnets, *Phys. Rev. Lett.* **60**, 1057 (1988).

74. S. Sachdev and J. Ye, Universal quantum critical dynamics of two-dimensional antiferromagnets, *Phys. Rev. Lett.* **69**, 2411 (1992). [cond-mat/9204001]

75. A. V. Chubukov, S. Sachdev, and J. Ye, Theory of two-dimensional quantum Heisenberg antiferromagnets with a nearly critical ground state, *Phys. Rev. B* **49**, 11919 (1994). [cond-mat/9304046]

76. S. Sachdev, What can gauge-gravity duality teach us about condensed matter physics? *Ann. Rev. Condensed Matter Physics* **3**, 9 (2012). [arXiv:1108.1197]

77. L. Huijse and S. Sachdev, Fermi surfaces and gauge-gravity duality, *Phys. Rev. D* **84**, 026001 (2011). [arXiv:1104.5022]

78. T. M. Rice, Kai-Yu Yang, and F. C. Zhang, A phenomenological theory of the anomalous pseudogap phase in underdoped cuprates, *Rep. Prog. Phys.* **75**, 016502 (2012). [arXiv:1109.0632]

79. S. Sachdev, M. A. Metlitski, and M. Punk, Antiferromagnetism in metals: from the cuprate superconductors to the heavy fermion materials, *Proc. SCES 2011*; arXiv:1202.4760.

80. M. Yamashita, N. Nakata, Y. Senshu, M. Nagata, H. M. Yamamoto, R. Kato, T. Shibauchi, and Y. Matsuda, Highly mobile gapless excitations in a two-dimensional candidate quantum spin liquid, *Science* **328**, 1246 (2010).

81. O. I. Motrunich, Variational study of triangular lattice spin-1/2 model with ring exchanges and spin liquid state in κ-$(ET)_2Cu_2(CN)_3$, *Phys. Rev. B* **72**, 045105 (2005).

82. S.-S. Lee and P. A. Lee, U(1) gauge theory of the Hubbard model: Spin liquid states and possible application to κ-$(BEDT$-$TTF)_2Cu_2(CN)_3$, *Phys. Rev. Lett.* **95**, 036403 (2005). [arXiv:cond-mat/0502139]

83. T. Grover, N. Trivedi, T. Senthil, and P. A. Lee, Weak Mott insulators on the triangular lattice: possibility of a gapless nematic quantum spin liquid, *Phys. Rev. B* **81**, 245121 (2010). [arXiv:0907.1710]

84. R. R. Biswas, Liang Fu, C. Laumann, and S. Sachdev, SU(2)-invariant spin liquids on the triangular lattice with spinful Majorana excitations, *Phys. Rev. B* **83**, 245131 (2011). [arXiv:1102.3690]

85. H.-H. Lai and O. I. Motrunich, SU(2)-invariant Majorana spin liquid with stable parton Fermi surfaces in an exactly solvable model, *Phys. Rev. B* **84**, 085141 (2011). [arXiv:1106.0028]

86. G. Chen, A. Essin, and M. Hermele, Majorana spin liquids and projective realization of SU(2) spin symmetry, *Phys. Rev. B* **85**, 094418 (2012). [arXiv:1112.0586]

87. M. Yu Reizer, Effective electron-electron interaction in metals and superconductors, *Phys. Rev. B* **39**, 1602 (1989).

88. P. A. Lee, Gauge field, Aharonov-Bohm Flux, and high-T_c superconductivity, *Phys. Rev. Lett.* **63**, 680 (1989).

89. B. Blok and H. Monien, Gauge theories of high-T_c superconductors, *Phys. Rev. B* **47**, 3454 (1993).

90. B. I. Halperin, P. A. Lee and N. Read, Theory of the half-filled Landau level, *Phys. Rev. B* **47**, 7312 (1993).

91. J. Polchinski, Low energy dynamics of the spinon-gauge system, *Nucl. Phys. B* **422**, 617 (1994). [arXiv:cond-mat/9303037]

92. C. Nayak and F. Wilczek, Non-Fermi liquid fixed point in 2+1 dimension, *Nucl. Phys. B* **417**, 359 (1994). [arXiv:cond-mat/9312086]

93. B. L. Altshuler, L. B. Ioffe and A. J. Millis, On the low energy properies of fermions with singular interactions, *Phys. Rev. B* **50**, 14048 (1994). [arXiv:cond-mat/9406024]

94. Y.-B. Kim, A. Furusaki, X.-G. Wen and P. A. Lee, Gauge-invariant response functions of fermions coupled to a gauge field, *Phys. Rev. B* **50**, 17917 (1994). [arXiv:cond-mat//9405083]

95. S.-S. Lee, Low energy effective theory of Fermi surface coupled with U(1) gauge field in 2+1 dimensions, *Phys. Rev. B* **80**, 165102 (2009). [arXiv:0905.4532]

96. M. A. Metlitski, and S. Sachdev, Quantum phase transitions of metals in two spatial dimensions: I. Ising-nematic order, *Phys. Rev. B* **82**, 075127 (2010). [arXiv:1001.1153]

97. D. F. Mross, J. McGreevy, H. Liu, and T. Senthil, A controlled expansion for certain non-Fermi liquid metals, *Phys. Rev. B* **82**, 045121 (2010). [arXiv:1003.0894].

98. S. C. Thier and W. Metzner, Singular order parameter interaction at the nematic quantum critical point in two-dimensional electron systems, *Phys. Rev. B* **84**, 155133 (2011). [arXiv:1108.1929]

99. C. Drukier, L. Bartosch, A. Isidori, and P. Kopietz, Functional renormalization group approach to the Ising-nematic quantum critical point of two-dimensional metals, arXiv:1203.2645.

100. M. Levin and C. P. Nave, Tensor renormalization group approach to 2D classical lattice models, *Phys. Rev. Lett.* **99**, 120601 (2007). [arXiv:cond-mat/0611687]

101. G. Vidal, Entanglement renormalization, *Phys. Rev. Lett.* **99**, 220405 (2007). [arXiv:cond-mat/0512165]

102. B. Swingle, Entanglement renormalization and holography, arXiv:0905.1317.

103. S. Ryu and T. Takayanagi, Holographic derivation of entanglement entropy from AdS/CFT, *Phys. Rev. Lett.* **96**, 181602 (2006). [arXiv:hep-th/0603001]

104. C. P. Herzog, P. Kovtun, S. Sachdev, and D. T. Son, Quantum critical transport, duality, and M-theory, *Phys. Rev. D* **75**, 085020 (2007) [arXiv:hep-th/0701036]

105. G. 't Hooft, A planar diagram theory for strong interactions, *Nucl. Phys. B* **72**, 461 (1974).

106. D. Yamada and L. G. Yaffe, Phase diagram of $\mathcal{N}$=4 super-Yang-Mills theory with R-symmetry chemical potentials, *JHEP* **0609**, 027 (2006). [hep-th/0602074]

107. L. Huijse, S. Sachdev, and B. Swingle, Hidden Fermi surfaces in compressible states of gauge-gravity duality, *Phys. Rev. B* **85**, 035121 (2012). [arXiv:1112.0573]

108. J. Bhattacharya, N. Ogawa, T. Takayanagi and T. Ugajin, Soliton stars as holographic confined Fermi liquids, *JHEP* **1202**, 137 (2012). [arXiv:1201.0764]

109. N. Ogawa, T. Takayanagi and T. Ugajin, Holographic Fermi surfaces and entanglement entropy, *JHEP* **1201**, 125 (2012). [arXiv:1111.1023]

110. E. Shaghoulian, Holographic entanglement entropy and Fermi surfaces, arXiv:1112.2702.

111. X. Dong, S. Harrison, S. Kachru, G. Torroba and H. Wang, Aspects of holography for theories with hyperscaling violation, arXiv:1201.1905.

112. K. Narayan, On Lifshitz scaling and hyperscaling violation in string theory, arXiv:1202.5935.

113. B. S. Kim, Schrödinger holography with and without hyperscaling violation, arXiv:1202.6062.

114. H. Singh, Lifshitz/Schrödinger Dp-branes and dynamical exponents, arXiv:1202.6533.

115. P. Dey and S. Roy, Lifshitz-like space-time from intersecting branes in string/M theory, arXiv:1203.5381.

116. P. Dey and S. Roy, Intersecting D-branes and Lifshitz-like space-time, arXiv:1204.4858.

117. B. Swingle, Entanglement entropy and the Fermi surface, *Phys. Rev. Lett.* **105**, 050502 (2010). [arXiv:0908.1724]

118. C. Charmousis, B. Gouteraux, B. S. Kim, E. Kiritsis and R. Meyer, Effective holographic theories for low-temperature condensed matter systems, *JHEP* **1011**, 151 (2010). [arXiv:1005.4690]

119. N. Iizuka, N. Kundu, P. Narayan and S. P. Trivedi, Holographic Fermi and non-Fermi liquids with transitions in dilaton gravity, *JHEP* **1201**, 094 (2012). [arXiv:1105.1162]

120. B. Gouteraux and E. Kiritsis, Generalized holographic quantum criticality at finite density, *JHEP* **1112**, 036 (2011). [arXiv:1107.2116]

121. D. Gioev and I. Klich, Entanglement entropy of fermions in any dimension and the Widom conjecture, *Phys. Rev. Lett.* **96**, 100503 (2006). [arXiv:quant-ph/0504151]

122. S. Ferrara, R. Kallosh and A. Strominger, N=2 extremal black holes, *Phys. Rev. D* **52**, 5412 (1995). [hep-th/9508072]

123. A. Sen, Black hole entropy function and the attractor mechanism in higher derivative gravity, *JHEP* **0509**, 038 (2005). [hep-th/0506177]

124. K. Goldstein, S. Kachru, S. Prakash and S. P. Trivedi, Holography of charged dilaton black holes, *JHEP* **1008**, 078 (2010). [arXiv:0911.3586]

125. S. Sachdev, A model of a Fermi liquid using gauge-gravity duality, *Phys. Rev. D* **84**, 066009 (2011). [arXiv:1107.5321]

126. S. A. Hartnoll and L. Huijse, Fractionalization of holographic Fermi surfaces, arXiv:1111.2606.

127. T. Senthil, S. Sachdev, and M. Vojta, Fractionalized Fermi liquids, *Phys. Rev. Lett.* **90**, 216403 (2003). [arXiv:cond-mat/0209144]

128. A. H. Castro Neto and E. Fradkin, Bosonization of the low energy excitations of Fermi liquids, *Phys. Rev. Lett.* **72**, 1393 (1994). [arXiv:cond-mat/9304014]

129. A. H. Castro Neto and E. Fradkin, Exact solution of the Landau fixed point via bosonization, *Phys. Rev. B* **51**, 4084 (1995). [arXiv:cond-mat/9310046]

130. F. D. M. Haldane, Luttinger's theorem and bosonization of the Fermi surface, *Proc. Int. School of Physics "Enrico Fermi"*, Course CXXI "Perspectives in Many-Particle Physics," eds. R. A. Broglia and J. R. Schrieffer (North-Holland, Amsterdam 1994). [arXiv:cond-mat/0505529]

131. A. Houghton, H.-J. Kwon, and J. B. Marston, Multidimensional Bosonization, *Adv. Phys.* **49**, 141 (2000). [arXiv:cond-mat/9810388]

132. M. J. Lawler, V. Fernandez, D. G. Barci, E. Fradkin, and L. Oxman, Non-perturbative behavior of the quantum phase transition to a nematic Fermi fluid, *Phys. Rev. B* **73**, 085101 (2006). [arXiv:cond-mat/0508747]

133. M. J. Lawler and E. Fradkin, 'Local' quantum criticality at the nematic quantum phase transition, *Phys. Rev. B* **75**, 033304 (2007). [arXiv:cond-mat/0605203]

134. D. F. Mross and T. Senthil, Decohering the Fermi liquid: A dual approach to the Mott transition, *Phys. Rev. B* **84**, 165126 (2011). [arXiv:1107.4125]

Discussion

B. Halperin Thank you very much Subir, particularly for sticking to the time so closely, and this indeed gives us time for discussion. So I will entertain questions, or comments. Tony (Leggett).

T. Leggett You discussed the question of the phase diagram of the cuprates and quite a lot of the important evidence, if I understand it correctly, came from quantum oscillation experiments. I know there is a very well developed theory of the interpretation of the quantum oscillation data in a Fermi liquid. How far do we know the standard interpretation is valid beyond that?

S. Sachdev The data that I showed you was on the electron doped cuprates and there is also similar data on the hole-doped cuprates that I did not show. All of this data is at relatively high fields, 40-50 Tesla. And all indications are, from measurements at those high fields (which are somewhat incomplete, admittedly), that the system is a Fermi liquid. Particularly the temperature dependence of the amplitude suggests it is an ordinary Fermi liquid. So even if there is something more exotic at low fields I think most people expect at these high fields for these exotic phases to disappear. Now the second part of your question, what do we know about quantum oscillations in exotic compressible phases? I think that is still an open subject. Few people have studied it, and you find that the quantum oscillations on the whole do survive if you have a Fermi surface although somehow the temperature dependence in this famous Lifshitz-Kosevich prefactor will change. But our best understanding is that you will still have quantum oscillations also in these more exotic non-Fermi liquid states.

T. Leggett And there would still be an accurate measurement of the area of the Fermi surface?

S. Sachdev Correct, yes.

B. Altschuler Subir (Sachdev), I just want to understand, it seems that you assume that these transitions between two phases are always second order. Is it an assumption or are there some theoretical reasons to believe that first order transitions between these two states are impossible?

S. Sachdev There certainly can be first order transitions, but, in the first two parts of my talk, when I talked about the gapped quantum antiferromagnet and also the superfluid-insulator transition I think there is overwhelming evidence from numerical studies that it is second order, in fact the conformal field theory...

B. Altschuler So the evidence is based on numerical studies and there is no...

S. Sachdev Well, no, so you can assume it is a second order transition and work out a field theory and then see if in the field theory there is an internal inconsistency. That certainly has been done and there is not any. But that does not tell you whether a specific model like the Hubbard model or your antiferromagnet has a second order transition. That is a separate question, for your specific microscopic model, and for that you do need numerical studies to settle whether it has a first or second order transition. Now for the metallic systems, the spin-density wave transition involving change in shape of Fermi surface, there we are working with the assumption of a second order critical point, but as I also mentioned, we know for sure that that critical point, at least at zero field, is masked by the appearance of superconductivity. So the transition ultimately disappears and you get a superconducting phase, and again, in a superconductor, you have a reason to believe you have a continuous magnetic ordering transition.

B. Altschuler But for instance, the Clogston-Chandrasekhar transition is first order.

S. Sachdev Sure, well that is a different transition. I was referring to the transitions that are present in this class of materials at zero magnetic field. Sure.

R. Dijkgraaf I have a question about this holography using these tensor networks. So, what kind of sets the length scale or even the metric on that extra spacetime dimension?

S. Sachdev I think that Xiao-Gong Wen is the expert on this and probably will say something, but I showed you a picture for a quantum critical state of a conformal field theory. In that case it is continuous forever. If you have a gapped state, I think it does terminate at some point and that termination, exactly how it terminates tells you something about the topological order in the state you began with. Was that correct, Xiao-Gong?

X-G. Wen You know, for the critical states, the (MERA) tensor network somehow correctly captures the whole entanglement strength at a different length-scale. (Since MERA tensor network is simliar to anti-de Sitter space), this is why I have this minimal surface in the anti-de Sitter space that happens to give you a correct result. For the gapped states, I think this is a totally different story.

J. Maldacena Is this tensor network not another way of talking about the Wilsonian renormalization group?

S. Sachdev Sure, sure, it is. It is formulated somewhat differently and, you know, the advantage is that it can be numerically implemented on systems that would otherwise have a sign problem. But it is a different way of doing the numerical implementation of the Wilsonian renormalization group, correct.

B. Halperin OK, we will take one more question and then we will go on with the next...

A. Polyakov I think one should be very careful with applying gauge-string duality to such systems. There are two conditions for the applicability of this. One is that you need to have something similar to a large number of colors which we have in field theory to select planar diagrams on the gauge theory side. The other thing is that without supersymmetry, generally you have to compare gauge symmetry in D dimensions not with some Lagrangian theory in D+1 dimensions but with a full-fledged string theory on a curved background which we generally do not know how to solve. That is the reason, for example, why we still do not have gauge-string duality in ordinary QCD. It is just a string theory which is very difficult to solve, nonlinear sigma models and so on. So, that is basically the point. You see, it is very important to distinguish gauge-string duality from the trivial symmetry statement that you have: in the bulk, you have conformal symmetry, on the boundary you also have conformal symmetry, they are the same, so the theories are the same. That is certainly not true.

S. Sachdev If you give me a few minutes to respond, so they are different types of questions. I would still be happy with interesting information on supersymmetric gauge theory at finite chemical potential because the examples we have of compressible phases are so small that it would be nice to have any theory under control, even with your favorite symmetries, for which you have a new description of a compressible phase. Of course we are not claiming that this compressible phase is realized in the high-temperature superconductor. But it will then help our understanding of different types of exotic phases that quantum matter can exhibit. The other level of the answer, on conformal field theories itself. So, yes it is not under control, you know it is only valid in certain large N limits with additional supersymmetry, so we are admittedly doing something uncontrolled. Now for something like critical exponents certainly we do not need gauge-gravity duality. We have other methods like the Wilsonian method of getting these results very accurately, but for questions at finite temperatures involving an eventual crossover to hydrodynamic behavior, all those methods fail. So, again we are looking for examples which are controllable in some limits where we can say something systematic about this finite temperature behavior. And so it does not have a formal expansion parameter but you could also view it as some effective field theory in powers of gradients of these dual fields and it does seem that if you terminate that expansion at low order, you get rather sensible results. So, it remains to be seen whether it is quantitatively accurate, but it is better than anything else for these questions.

A. Polyakov OK, I will add something too, which will be in contradiction to what I said before. Namely I think there is also another possible – maybe in the far future possible – application of gauge-string correspondence, and this is the three dimensional Ising model, in which we have a conformal field theory, a three dimensional conformal field theory at the critical point. We do not have large N, but still by some change of variables it tantalizingly reduces to free fermionic string theory, but it is a terribly difficult problem for the future I guess.

B. Halperin Maybe we should go on to the first contributed talk, by Leon Balents, recent developments in topological aspects of the electronic states. Leon.

Prepared comments

L. Balents: Recent Developments in Topological Aspects of Electronic States

I would like to enlarge upon something Subir mentioned very briefly, but which is really a large and growing enterprise in condensed matter physics. This subject grew out of the Bloch theory of electrons in solids, and this band theory is still a workhorse today. Indeed in the form of local density approximation and its variants, it is used intensively by a large community of computational "ab initio" physicists, materials scientists, and engineers. But today there is an ever growing appreciation that the band theory of solids has a strong topological component. That is, electronic structures of systems within a given symmetry class can be distinguished by topology of the bands. We can think of the IQHE as the first example of this, where the topology is classified by first Chern number. Relatively recently, work mostly by Charlie Kane showed that time reversal invariant systems in 2 and 3 dimensions have a Z_2 topological classification. In principle, this type of understanding can extend to consideration of more complex symmetries than just time reversal, even to the full venue of space groups of periodic solids.

Topological distinctions are interesting because they bring robustness to small perturbations. In band topology we've seen also that they generally involve some bulk-boundary correspondence. This links the bulk topology to robust electronic states at the surface of the solid. Also the global nature of the topological structure leads to fermionic excitations that, when looked at locally (either in real space or k space) are "anomalous". By this I mean that they could not exist in isolation, but only as the "boundary" of some larger space.

Where this is best understood is for gapped insulators, for the case of "generic symmetries": particle conservation, time-reversal, and to some extent for charge conjugation and parity. This is what we call topological insulators (or topological superconductors). It is in a way the most interesting case because only time reversal and particle conservation remain unbroken in imperfect samples with disorder and/or defects. Two examples of this are the integer quantum Hall effect, where the boundary states are chiral Fermion edge states, and three-dimensional topological insulators, where the surface states are chiral Dirac fermions (1/4 of graphene). These surface states could not exist in isolated one-dimensional or two-dimensional systems, respectively.

Topological classification can also extend to gapless situations (compressible in Subir's language). The reason is that usually over most of reciprocal space there is still a "direct gap" between states of a fixed momentum. So over this part of phase space one can still consider adiabatic deformations that preserve the gap, and there can be topological distinctions.

A example of this which has recently arisen is what we call a "Weyl semimetal", where a three-dimensional system is gapless at points, and the spectrum near these points is that of massless Weyl fermions. These points act as monopoles of Berry curvature, which confers on them topological stability. There is also a bulk-boundary correspondence. The Weyl points lead to gapless chiral edge states, similar to in the integer quantum Hall effect, and what you might call a "semi-quantized" anomalous Hall effect. Experimentally this should also manifest itself as "Fermi arcs" in photoemission. One reason this is interesting is that it allows formation of quantum Hall-like physics without requiring strong time reversal symmetry breaking, and may be a more feasible way to obtain "dissipationless" edge physics in a real material.

I think this is a trend which is not going to stop any time soon. Topology has a rich home in the diverse band structures of solids. And there are many interesting theoretical problems in understanding transport in these types of states, and especially in the role of strong electron-electron interactions.

M. Fisher: Bose and Non-Fermi Liquid Metals

Underlying the quantum theory of many-particle non-relativistic systems are the canonical quantum fluids – superfluid phases of Bosons and Fermi-liquid phases of Fermions. These two quantum phases of matter are extremely well understood, and are both accessible from models of weakly interacting particles. The only low energy excitation in a superfluid is the Goldstone mode of the broken particle number conservation, while the dominant excitations in a Fermi-liquid are Fermionic quasiparticle excitations which are adiabatically connected to the single particle excitations in the free Fermi gas.

But quantum many-particle systems are potentially *much* richer than this. One of the frontiers of quantum condensed matter physics seeks to access and classify quantum phases of Bosons/Fermions in two-dimensions which are *not* superfluids/Fermi-liquids. Most challenging are gapless, compressible phases which exhibit correlations which are singular along lines in the two-dimensional momentum space.

Significant recent progress has been made in accessing and analyzing so-called "Bose-Metal" phases. A Bose-Metal is a stable zero temperature quantum phase of Bosons which breaks no symmetries whatsoever, neither the global $U(1)$ symmetry as in the superfluid nor translational symmetry as in a crystalline phase of Bosons. Moreover, the Bosons in a two-dimensional Bose-Metal possess a momentum distribution function which is singular along closed curves in momentum space. This is reminiscent of the Fermi surface in a Fermi-liquid, but a Bose-Metal phase affords no weakly interacting quasiparticle description.

The most successful theoretical approach to access Bose-Metals has been the parton construction, wherein the Bose operator is written as a bi-linear of Fermionic "partons", $b^\dagger = d_1^\dagger d_2^\dagger$. The Fermionic partons $d_1^\dagger, d_2^\dagger$ are taken to fill (generally different) Fermi seas - a Bosonic wavefunction is obtained by an appropriate Gutzwiller projection. Recent progress has identified a rather simple Hamiltonian of hard core Bosons hopping on (say) a two-dimensional square lattice, augmented by a four-site "ring-exchange" term. This model can be successfully attacked using a combination of density-matrix-renormalization-group on quasi-1d systems, Gutzwiller variational wavefunctions as well as a bosonized description of a parton gauge theory. Remarkably, the resulting Bose-Metal phase retains signatures of the Fermi-surfaces of the Fermionic partons, despite the fact that these partons are mere theoretical constructs.

A generalization of this parton approach to two-dimensional strongly interacting Fermionic systems is also very promising. In this case the electron operator is decomposed in terms of a product of three Fermionic partons, $c_\sigma^\dagger = d_1^\dagger d_2^\dagger f_\sigma^\dagger$. For spinful electrons the Gutzwiller wavefunction obtained from this parton construction is expressed as a product of four determinants. By construction, this wavefunction is a non-Fermi liquid state which violates Luttinger's theorem which states that in *any* Fermi-liquid the Fermi surface will enclose a volume equal (in appropriate units) to the total number of Fermions.

In very recent unpublished work, a model Hamiltonian of electrons hopping on a two-dimensional square lattice interacting via a four-site singlet rotation term, has shown compelling evidence for a non-Fermi liquid ground state. Incredibly, the properties of this non-Fermi liquid phase are consistent with the parton construction and the Gutzwiller wavefunction involving a product of four fermion determinants.

The field of strong interacting non-Fermi liquid phases in two and three dimensions is in its infancy. While there is a pressing need for theories of such states coming from experiments on a number of strongly correlated electronic materials, the theoretical challenges are quite daunting. Numerical approaches are strongly limited by the so-called Fermion sign problem. Nevertheless, a combination of various analytic lines of attack, such as the parton construction, in tandem with increasingly sophisticated and powerful numerical simulations points towards an encouraging and exciting future.

Discussion

S. Kachru Yes, this is a question directed at Matthew (Fisher). My question is: we have heard a lot of discussion of entanglement and in the context of strongly correlated systems you might have then thought that candidate ground states that have non-Fermi liquid behavior would show different signatures in their ground state entanglement than more conventional phases like Fermi liquids. So given that you have candidate wave functions, can you distinguish them just through their entanglement behavior from more conventional ground states?

M. Fisher I think the qualitative scaling of the entanglement entropy in eg. the Bose metal state that I was discussing is unfortunately going to be the same as in a Fermi liquid, that is $L \log(L)$ in two dimensions for a size L region. My understanding is that in a Fermi liquid there are actually explicit expressions relating the coefficient of the $L \log(L)$ in real space to the shape in real space of the region that one is taking and there may be differences there between a Bose metal and a Fermi liquid metal. So I that is something that actually is worth exploring further.

T. Leggett Question to Matthew (Fisher). You said that the Bose metal you described was not a superfluid but your justification was that is does not break U(1) symmetry. Does it behave phenomenologically like a superfluid? That is, does it for example show Meissner diamagnetism?

M. Fisher No it would not behave phenomenologically like a superfluid, but the main difference in two dimensions would be the absence of off-diagonal long range order and an absence of a superfluid density. So it would not show a Meissner effect, with a vanishing superfluid density. Moreover there would be low-energy excitations along lines in momentum space in a Bose metal in contrast to a superfluid where the only low energy excitation occurs at zero momentum. I mean, one might have a roton minimum but it does not come down to zero energy so all the spectral weight at low energy is at zero momentum in the superfluid and that is not the case in this Bose metal state.

W. Phillips Again for Matthew (Fisher), having to do with the Bose metal. What keeps this Bose system from Bose condensing at T=0 ?

M. Fisher The Hamiltonian. [laughter] I am not joking.

W. Phillips OK, what about the Hamiltonian?

M. Fisher Well, no, but quite seriously, for the bosons in the continuum, like in an atom trap with no optical lattice, the kinetic energy is $p^2/(2m)$. That likes to minimize the energy at zero momentum, i.e. it wants to condense the bosons at zero momentum. Now in the Hamiltonian that I have been looking at, which exhibits the Bose metal, the bosons are hopping on a lattice and indeed the nearest-neighbor hopping term is mimicking the $p^2/(2m)$. In fact if we just have a Hamiltonian which is entirely that $p^2/(2m)$ term on the lattice, maybe with a short range repulsion, we will get a superfluid. But we add another term in, which essentially is a four-body term, and what is most important, it is a four-body term that destroys the sign structure of the ground state wave function. I mean, you can see explicitly that the ground state wave function is not going to be non-negative. The Marshall sign rule, if you will, is not satisfied. And so the ground state wave function in the Bose metal is riddled with minus signs in much the same way that a ground state for a free Fermi gas is. So that is why I think having the lattice is important, or having some way to scramble the sign of the wave function, with a gauge potential perhaps or with a magnetic field perhaps or with frustration in a magnet. I think that is the hint we have been trying to use to find such phases.

B. Halperin Sasha (Polyakov), you had a question.

A. Polyakov I almost decided not to ask, because it may be a stupid question. You see, there is a phenomenon in field theory that if you have for example a Fermi field, a charged Fermi field, and it is charged to the Chern-Simons gauge field, then there is a Fermi-Bose transmutation. The fermion becomes a boson. And I was just wondering, without any reason actually, whether it has any relation to what you have discussed.

M. Fisher In fact, the way that we stumbled into the particular Bose metal construction that I was talking about was by trying to do a time-reversal analogue of statistical transmutation in two dimensions. The challenge in the normal Chern-Simons theory, is, even if you attach say 2π flux, you did it rigorously on the lattice, you would not have broken time-reversal invariance. As soon as you make any approximation whatsoever, you do. So it is not really, as far as I know, extremely useful in describing quantum states of matter in zero magnetic field.

Prepared comments

X. G. Wen: Topological Order: a New Order beyond Symmetry Breaking

I would like to give a brief overview on some new understanding of quantum matter.

For a long time, we thought that all possible phases and phase transitions can be understood through Landau's symmetry breaking theory. However, in the study of chiral spin states and the fractional quantum Hall states, we realized that those states contain a totally new kind of orders which cannot be characterized by any conventional means. We have to introduced topological probes, such as the ground state degeneracy on spaces with various topologies and the non-Abelian geometric phases of the ground states from deforming the spaces, to characterize/define the new orders (which is named as topological order).

Recently, we realized that topological orders are nothing but the patterns of long-range entanglements defined via local unitary transformations. A topologically ordered state is a state that cannot be transformed into a product state.

One way to make a long-range entangled state (or topologically ordered state) in a spin system is to first join the up-spins into strings, then make a superposition of all the loop states of strings. This leads to a string-net condensed state – a long-range entangled state.

The long-range entangled states can produce some amazing phenomena. String-net liquid can produce gauge fields, in the sense that string density waves in string-net liquid can satisfy Maxwell-Yang-Mills equation and behave like gauge fields.

String-net condensation can also produces Fermi statistics from qubits: The ends of string behave like point particles which can carry Fermi statistics. So string-net condensation in qubit system provides a way to unify gauge interactions and Fermi statistics in any dimensions.

Such a notion of viewing quantum states as patterns of entanglements leads to a much deeper understanding of quantum matter. In addition to symmetry breaking states that are described by group theory, we find that, in 2+1D, non-chiral long-range entangled states are classified by string-net theory (or a tensor category theory). We also find that short-range entangled states with symmetries are classified by group cohomology theory of the symmetry group. Such a theory generalizes the free fermion K-theory for the topological insulators and superconductors to interacting boson/fermion systems.

We see that quantum entanglements present to us a rich and new quantum world. Topological order and quantum entanglements open up a new chapter in condensed matter physics.

D. Haldane: Geometry of FQHE Phases

I will discuss the geometry of the FQHE (2D fractional quantum Hall effect), exhibited by an incompressible fluid state of quantum matter. Unlike the integer QHE, the fractional effect in a partially-filled Landau level is not explained by the Pauli principle. Electron dynamics is then characterized by the non-commutative geometry of their "guiding centers" with metric-independent commutation relations $[R^x, R^y] = -i\ell_B^2$, where $2\pi\ell_B^2$ is the area per magnetic flux quantum h/e, (the area per independent orbital in the Landau level). This "quantum area" characterizes an uncertainty principle that makes the guiding-center geometry "fuzzy" on smaller area scales, and (when the Landau orbits are quantized) provides a short-distance regularization on a much larger scale than the usual cutoff at atomic dimensions. In the FQHE at Landau level filling $\nu = p/q$, the particles condense into a fluid with an "elementary droplet" or "composite boson" of p particles bound in q orbitals (q "attached flux quanta"). The incompressibility is due to short-range Coulomb repulsion, which gives rise to an energy gap (like the Hubbard repulsive energy U in Mott insulators) that prevents an electron from visiting a region that has been occupied by another electron, or (more generally) by a fluid droplet. However, unlike the Mott-Hubbard case, there is no underlying geometry of a lattice to define the orbitals that resist multiple occupation, so what defines the geometry of the FQHE fluid? There are two physical sources of geometry (see Fig.(1)): the shape of the Landau orbits (which controls the integer QHE) and the shape of the Coulomb equipotentials near a point charge on the 2D surface (which, together with a Landau-orbit form-factor, controls the FQHE).

Incompressibility is clearly seen using numerical finite-size exact diagonalization, which reveals the nature of the excitations as a dipolar pair of a (fractional charge $\pm e* = \pm e/q$) quasiparticle and quasihole that carries a net electric dipole, and moves in a direction transverse to the dipole with a momentum $P_a = B\epsilon_{ab}(e^*d^b)$ where d^a is the relative displacement from quasihole to quasiparticle.

At small momentum or dipole moment, this collective mode is seen in Fig.(1) to disappear into the continuum of two quasiparticle + two quasihole states. The FQHE state (in this case the Laughlin $\nu = \frac{1}{3}$ state) is separated from other states by a gap. There have been various narrative explanations of FQHE incompressibility (composite-boson Ginzburg-Landau theory, composite fermions filling "composite-fermion Landau levels", *etc.*, but none of these account for a fundamental property of FQHE states found in 1985 by Girvin, MacDonald and Platzman (GMP): the "guiding-center structure factor" $S_{\text{gc}}(\mathbf{Q})$ vanishes as $|Q|^4$ in the limit of small wavenumber Q. A key idea in the FQHE is the description of its topological order by a Chern-Simons topological field theory, but this contains nothing about length or energy scales (the Hamiltonian vanishes!) and *assumes* incom-

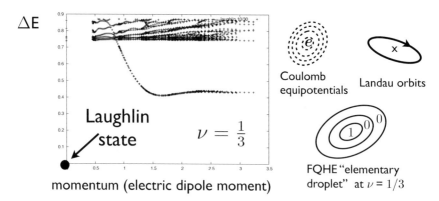

Fig. 1. (Left): Excitation spectrum of the incompressible Laughlin $\nu = \frac{1}{3}$ state (short-range V_1 pseudopotential only), from exact finite-size diagonalization. The collective mode becomes a quadrupolar spin-2 "graviton analog" at small momentum, but is hidden in the many-quasiparticle-quasihole continuum. (Right): The in-principle-distinct shapes of Landau orbits and point-charge Coulomb equipotentials are the only sources of QHE geometry. Occupations of orbitals (relative to uniform occupation $\nu = \frac{1}{3}$) in the elementary droplet determines its geometric spin as $s = -1$.

pressibility rather than describing it.

The new idea[a] is that quantum geometry of the emergent *shape* of the orbitals (or "flux attachment") defining the FQHE elementary droplet describes incompressibility: previous ideas about "flux attachment" described it abstractly, or implicitly assumed its shape was that of the Landau orbits. But the shape determines the correlation energy and there is a preferred shape that minimizes this. The fluid then has an analog of a "shear modulus" describing the quadratic energy cost of small area-preserving local deformations of the shape.

The shape of the elementary droplet defines the natural "quantum metric" of the FQHE state. This (spatial, not space-time) metric can locally fluctuate in space and time, giving rise to flows of locally-conserved Gaussian curvature. Around 1990, Wen and Zee, and Fröhlich and coworkers, showed how to add curvature to Chern-Simons theory through the gauge field of curvature (spin connection), but were discussing the FQHE fluid on a static curved 2D surface embedded in 3D Euclidean space. Now the curvature is that of an intrinsic and dynamic emergent metric associated with incompressibility, but Gaussian curvature formulas are valid independent of the source of the curvature, and it turns out that that the electron density in not just fixed by the magnetic flux density but also couples to the curvature gauge field with a topologically-quantized "geometric spin" s that depends on the arrangement of the p electrons in the q orbitals of the $\nu = p/q$ FQHE droplet: if the total curvature increases by 2π, the num-

---

[a]F. D. M. Haldane, *Phys. Rev. Lett.* **107**, 116801 (2011).

ber of linearly-independent states of the elementary droplet on the surface (a topological number, provided the state is gapped) increases by $2s$. The local charge density can change (and adjust to non-uniform potentials) at the expense of local deformations of the metric that cost correlation energy, and in particular, the fluctuation of guiding-center density is proportional to Gaussian curvature, which is essentially the second derivative of the metric, explaining GMP's result $S_{gc}(\mathbf{Q}) \propto Q^4$.

The metric is a quasi-local quantum object with components that do not commute with each other (but commute with the determinant), and $S_{gc}(\mathbf{Q})$ has a direct interpretation in terms of the zero-point fluctuations of the metric. The collective mode at long wavelengths becomes a spin-2 analog of the graviton, unfortunately hidden in Fig.(1). When quantum-geometry is added to the topological description of the FQHE, it provides a unified description that now includes both energy and length scales, and is a condensed-matter example of quantum geometry which may have lessons for quantum geometry of gravity.

Discussion

B. Halperin The floor is open for questioning.

J. Maldacena Yes, this is for Wen. So when we talk about particles that condense, we have some particle-like excitation which can carry energy, momentum, etc. So when you are talking about these strings that condense, do these strings exist in any way or how should I think of these strings as physical objects?

X-G. Wen Actually strings are not physical objects. "String condensation" is a word for a particular order in the ground state. So the physical objects are qubits. Actually the ground state of qubits may have many different entanglement patterns, and one of the entanglement patterns is called string condensation. So the emergent gauge fields from this ground state actually are just collective modes of this spin fluctuation.

S. Das Sarma Yes my question is also for Xiao-Gong (Wen). I just want to make sure that you do not disagree with the statement that if we have a Hamiltonian –and I am not talking about an effective Hamiltonian, all spin Hamiltonians are effective Hamiltonians, but I am talking about Hamiltonians of real electrons or if it is a bosonic system, bosons– if you or I or someone else is given a Hamiltonian like that, we have no way of stating (this goes back to Juan Maldacena's question) whether it is topological or not, because you have to solve the problem. I mean, you have a construct of how to create a topological phase using this string condensation which one may even be able to map to a spin model, but if I have a real Hamiltonian, short of solving the problem I know not if it is topological or not because I do not know which TQFT describes it. I hope we agree, that is all I want to make sure.

X-G. Wen So I feel this is a very important question. Basically, given a Hamiltonian, how do you find what is the ground state. Let me try to put this from another angle. You know we have a lot of mean field theory approaches. But using the mean field theory approaches, we already assume that the ground state is short-range entangled. So we never get interesting phases using a mean field theory approach. Thus the difficulty to obtain these highly entangled ground states is because we do not have anything in our toolbox to really do it. So this is really a challenge for theorists to come up with something which is beyond the standard mean field theory approach, so that we can access these new highly entangled states. Actually the tensor network approach which developed in quantum information science could be one of the tools to achieve this. But once we obtain the ground state described by a tensor in tensor network, there is yet another question: how do you know which tensor describe which phase? Again these are all new theoretical challenges, so there are a lot of things to be done. Actually symmetry breaking theory is like an empire but for long-range entangled states we almost have nothing. We need to build another empire, and there is a totally new theoretical structure to go along with this.

M. Shifman I have a question, and my question is to professor Sachdev and maybe to some other speakers who discuss this issue of the holographic description of systems in condensed matter and in some liquids. And my question is: today, what is the success story? I mean, is there a physical system which is really realized in nature and is nicely described in this holographic way, and to which degree is it described, is it just some general characteristics or fine details of this system?

S. Sachdev OK, let us see, I can answer that at two levels. So, personally I think the most promising avenue for having really close contact between the theory and experiment and perhaps even making it quantitative is probably something like the 2D superfluid-insulator transition of bosons. So one could model that by some dual theory of a gauge field which is dual to the charge density of the Bose system and do some kind of generalized expansion in gradients of the various fields in the dual description. So, in collaboration with Rob Myers, we have done that to fourth order in gradients and deduced a frequency dependent conductivity which turns out to have a very physical looking form, and the cost of this expansion is that there is one free parameter to every order that you go out and you could try to now match these parameters to zero temperature correlation functions that you could get from the CFT by other methods. So we are carrying out that programme, we will see how well that works. We would have to compare it with experiments. There are not any clear-cut experiments of this quantity either, but I think that in cold atom systems it is possibly achievable.

M. Shifman May I say that you hope that in the near future there will be a success story? Right? Of this approach?

S. Sachdev It depends what your definition of success is, so if you want a quantitative comparison between one experiment and one theory we are not there yet. If you want some general new type of understanding of properties of strongly interacting phases of matter, I think we already have that. The other system to which this has been applied a lot is this strange metal that I mentioned. Now it is I think quite remarkable that the very simplest theory starting from the dual gravity gives you a strange metal which has some similarities to the experiments but it also has a lot of unsatisfactory features like finite ground state entropy in the very simplest description. And now it remains to be seen whether if we try to fix these features, do we also lose the interesting part of the physics? There is a lot of work on that question. But it is encouraging. I mean it is giving us a new way of looking at the strange metal and I would say, that is my personal opinion, unlike previous approaches in condensed matter we also have a route to improving it. Many of the older methods were some kind of uncontrolled mean field theory which we did not know how to improve. And that is why there is a lot of excitement and discussion on exactly how we are going to improve these descriptions.

B. Halperin Let me just ask one question. Just a clarification. It also came up, someone asked me during the break, to make sure. When you talk about the transition between the insulator and the superfluid, if we are applying it say to cold atoms we want to be on a lattice where you are going from a state which is commensurate with a Mott insulator with N particles per unit cell, into one in which it is not quantized. Is that correct?

S. Sachdev That is certainly the simplest case. Yes. There are more exotic critical points with rational fraction numbers, but OK. Experimentally...

B. Halperin I just wanted to make it concrete, because you did refer to possible experiments. That would be the experiment one might want to do.

S. Sachdev All right, so there is an experiment in the university of Chicago in a 2D Bose gas at integer filling by Cheng Chin. The quantities they measure are not the quantities we find easy to calculate in condensed matter and hopefully that gap will be bridged in the future.

I. Klebanov I have a comment and a question related to Misha's (Shifman) question and Subir's (Sachdev) talk actually, concerning the status of predictive power of the dualities. So one thing that we hear a lot about in this session is entanglement entropy. And at least one sort of qualitative thing I learned from applications of gauge gravity techniques is actually, so if you have a confining theory with some confining scale and you compute the entanglement entropy, say between two parallel plates, and study it as a function of the distance between the plates, it turns out that there is large entanglement when the plates are closer than the confinement scale. It is of order N^2 say at large N. And then, when you move the plates beyond a certain distance there is a phase transition to basically having the entanglement only of order of N^0. So this is something that you can see from the application of these types of techniques. So that story of course is not literally for QCD but apparently people have studied on a lattice similar things and perhaps this is what also happens in QCD. So that is not a big deal, so I do not know if you would call it a big success, but it kind of clues you in to some phenomena that may be going on in the real world. My impression is that these compressible phases that Subir (Sachdev) talked about had a certain scale in two spatial dimensions, and I was wondering if one expects also some kind of qualitative transition in the behavior of entanglement entropy as a function of distance scale.

S. Sachdev You do expect some crossover. So now you could imagine starting from some conformal field theory which is very strongly coupled in the UV. You turn on a chemical potential. Then there is scale set by the chemical potential where you cross over from the conformal behavior to some new behavior characteristic of the compressible phase. And so there presumably would be a similar crossover in the entanglement but that is something we do not understand at all, how to do that in this AdS description. There are some ideas but there is no complete theory yet.

D. Gross I think it is also worth mentioning, Subir (Sachdev), that the quark-gluon plasma which in fact historically was the first of these applications to real systems of the duality, and at least qualitatively successful in explaining or suggesting that the RHIC experiments were seeing a perfect fluid with very small viscosity. That I regard as certainly a qualitative success in the real world.

S. Das Sarma I also have a comment on professor Shifman's question, and then a question for Subir (Sachdev). So there is no question that AdS-CFT is a deep and very interesting way of approaching the question of quantum phases of condensed matter as Subir (Sachdev) explained very beautifully. Interesting, deep, yes, but it remains to be seen whether it is also useful. That is not clear at all. Meaning, in condensed matter we, at least I and I would say much of the community, define usefulness in the context of answering a question that is out there which we have been trying to answer and which we have not been able to answer. That I would say has not happened. Many questions have been answered very beautifully, but we do not know what these questions are, because these are quantum critical... [laughter] No I am not being facetious, because those quantum criticality, we do not know which physical system they occur, maybe tomorrow Seamus (Davis) will discover them and they will be very, very useful. I mean, the moment that happens, a huge number of condensed matter theorists will start using these things. That has not happened. But it is certainly a very deep way of looking at things and I will be very shocked if it does not develop usefulness as Subir says in the near future because this sort of theory building usually has applications. But, Subir (Sachdev), I had a question, I had a question on just a garden variety high-temperature superconductivity, nothing to do with AdS-CFT.

S. Sachdev Before you go to that question, my answer would be similar to the previous answer but I think this is a condensed matter question and there will be another discussion of AdS-CFT on Friday I believe, so we just stick to condensed matter.

S. Das Sarma Right, so my question to you is that about the high-T_c phase diagram there is a recent from Rick Greene, my collaborator, for electron doped materials where they find something that I found astonishing. They measure the coupling constant from just looking at T_c and then they extract coupling constants from transport measurements. It is a classic technique that was developed for regular superconductors, and for phonon-induced superconductivity, these two coupling constants, the λ, renormalization constant, are the same. And here they find something very identical, I mean, phenomenology, which kind of is indicating that it is just a garden variety superconductor where magnon fluctuation is permeating things. Do you have a comment on that? I mean, I found this astonishing, I find this hard to believe, but the numbers are very compelling. I assume you know the experiment.

S. Sachdev Yes, so I am not aware of the very latest analysis you present, but I did discuss in my talk the evolution from the small Fermi pockets in the presence of the antiferromagnetism to the large Fermi surface. And the simplest theory of fluctuations near that critical point – I guess going back to work of Scalapino before the discovery of high T_c – did argue that you have d-wave pairing near such a critical point. So I think the analysis Rick is doing is probably some dressed up version of that using Eliashberg generalization. So that is a question that has been studied in great detail. Does that theory really work near this 2D quantum critical point. And especially my student Max Metlitski and I have looked at that and our conclusion is that, yes, the answer is correct if there is high-temperature superconductivity but it is a strongly coupled problem where you can not apply Eliashberg. It is not under any control. So I think any quantitative understanding at this point is probably fortuitous, or something we do not understand.

S. Das Sarma Because they use the BCS model, they do not use any strong coupling.

S. Sachdev I think we do not have a complete theory of pairing in two dimensions near such a critical point and that is certainly one of the most interesting open problems.

G. Gibbons I would like to ask Subir, without wishing to pre-empt Gary Horowitz's talk, your remark about the difficulty with entropy at zero temperature, has that to do with the entropy of zero temperature black holes?

S. Sachdev That is correct. In the simplest solutions you get an $AdS_2 \times R^2$ near-horizon entropy and it comes from that. And that factorization into AdS_2 seems rather unphysical from a condensed matter perspective.

G. Gibbons I mean I just want to remark that in the past the entropy of the extreme black holes has been slightly controversial but I mean the consensus view is that they have entropy but the real situation is that it is somewhat directional dependent in phase space how you get that entropy. There might actually be a way of resolving this issue somehow within condensed matter, or within this model.

A. Zeilinger Thank you. For whatever it may be worth, I just would like to remark that, using this kind of measurement based optical quantum computation scheme, with my young colleague Philip Walther we recently did this simulation of a Heisenberg spin system, a small system, only four qubits, as this is what we can do now easily. But the interesting point there is that we can – this is measurement based – actually tune the interaction between the systems and see things like shifting entanglement around between the four spins, seeing frustration and so on. So this is just the start but I want people to know so that maybe they make suggestions what else we can do. I have a question, a specific question to professor Wen, when we talk in these systems about using entanglement I have a very naive question, that is there are very different classes of entanglement. There are classes where, when you take one of the entangled systems out, the entanglement breaks completely down. There is no entanglement left. There are other classes like cluster states or some Dicke states when you take part of the system out there is lot of entanglement left and so on. This is related to the notion of monogamy of entanglement. Does this play any role in applying entanglement to complex systems, large systems, because there I would naively expect that you do not want entanglement to disappear when you take just one of the participants out of the game.

X-G. Wen Yes, this notion of long-range entanglement actually is some residual structure of entanglement which is robust against any local modification. If you do unitary transformations you can change entangled states to unentangled states. But however, when we have a large system, you only do the local unitary transformation among nearby qubits. Then you have limited power to remove all the entanglement. If you do (generic) unitary transformations, you can change entangled states to unentangled states. But however, when we have a large system and you only do the local unitary transformation among nearby qubits, then you have limited power to remove all the entanglement. And the part which you cannot remove is what we call long-range entanglement. And such part, what I try to emphasize, is related to the quantum phase of the condensed matter system. So using this residual structure for entanglement which cannot be removed by a local modification actually does give us another prospect to understand quantum phases and then from there we can even get the classification theory for all different quantum phases.

E. Verlinde Yes, I have a question about emergent gravity in condensed matter systems. I heard this idea being mentioned a number of times, and I am always confused about it. I usually understand this as that one finds some spin-2 excitation. But we know in field theory that the only way to have a spin-2 excitation, if it is interacting, is that it becomes a perturbation of the metric. Hence, it is really a deformation of space-time. For instance, if we want to detect a gravity wave, we are looking for deformations in space-time. What I am asking is that, if you say that gravity is emergent in a condensed matter system, do you mean that this spin-2 excitation has really something to do with deformations of space-time? For me this sounds like a very strong statement.

X-G. Wen So, thank you for the question, and because of the limit in the slides, I only have one line for that, so there is a lot of qualifier which is dropped. And so, right now, what we can do is that we only can have this so-called linearized gravity, exactly what you mentioned. If we expand around flat space and then there is a...

E. Verlinde A gravity wave is also linear and it is really a distortion of space.

X-G. Wen Maybe let me put it in this way, we have a condensed matter system whose low energy excitations are helicity ± 2 modes, and only helicity ± 2 modes, and no other mode at the low energy. And if you write down the low energy effective theory for such helicity ± 2 mode as an only low lying excitation, this effective theory turns out to be the linearized, not Einstein gravity, but this Horava gravity with a $\omega \propto k^3$ dispersion relation. It is a linearized version of that.

E. Witten I would, if I may, add a very brief remark to that. I feel it should be extremely difficult, or impossible, to get Einstein gravity starting with a definite lattice theory because lattice theory has local observables, which a gravity theory should not have.

X-G. Wen So, I do not know, maybe let me ...

B. Halperin I do not think this was a question, this was a comment.

Prepared comments

J. C. S. Davis: Electron Correlations, Quantum Magnetism and High Temperature Superconductivity

To understand, control and apply correlated high-temperature superconductivity are among the most fundamental challenges in condensed matter physics at present.

Copper-based high temperature superconductivity was discovered by Bednortz and Muller in 1986. Its maximum superconducting critical temperature T_c is near 150 K. The CuO_2 plane electronic structure is dominated by Cu $3d$ and O $2p$ orbitals, with copper in a Cu^{2+} $3d^9$ state and oxygen in O^{2-} $2p^6$. Each Cu $d_{x^2-y^2}$ orbital containing one electron, is split energetically into singly and doubly occupied configurations by on-site Coulomb interactions so that the system is in a correlation dominated Mott insulator state that is also strongly antiferromagnetic due to inter-copper superexchange. So called "hole-doping", a process distinct from the eponymous one in semiconductors and which achieves the highest T_c superconductivity known, is carried out by removing electrons from the O atoms. At low hole-density, an unusual high energy electronic excitation which is anisotropic in k-space appears at $T^* > T_c$; this "pseudogap" phase is so called because the excitation might represent the energy gap of a distinct electronic phase. The overall phase diagram as a function of the number of holes per CuO_2, p, consists of antiferromagnetism for $p < 2 - 5\%$, the pseudogap phase $5\% < p < 20\%$, superconductivity in the range $5 - 10\% < p < 25 - 30\%$, and a likely Fermi liquid state for $p > 20 - 30\%$. T^* diminishes with increasing hole doping and the superconductivity occurs within a "dome" which appears to occur surrounding the $T^* = 0$ point. The band structure of hole-doped CuO_2 is equally mysterious because, as the electron density diminishes from the Mott insulator state at half filling, the delocalized states first appear on an incomplete Fermi surface (Fermi arc) within the Brillouin zone, and are only converted to a conventional hole-like band surrounding the (π, π) point at a much higher hole density.

Iron-based high temperature superconductivity was discovered in 2008 by Hosono and, at present, the maximum T_c exceeds 55 K. The FeAs plane electronic structure is dominated by Fe $3d$ and As $4p$ orbitals, with iron atoms in a Fe^{2+} $3d^6$ state while the arsenic atoms are As^{2-} $4p^6$. Every Fe atom has one electron each in the d_{xz}, d_{yz}, d_{xy} and $d_{3z^2-r^2}$ orbitals, all being in the same spin state due to Hund's coupling; the materials are all strongly correlated metals. Doping is achieved either by cation substitution outside the FeAs layer or by transition-metal-atom substitution within this layer. The undoped compounds are orthorhombic and antiferromagnetic correlated metals with a phase transition temperature T_{AF} near 220 K - 134 K. T_{AF} diminishes with both electron- and hole-doping, and the superconductivity occurs within a "dome" surrounding the point where the magnetic/structural transition temperatures T_{AF} are suppressed towards zero temperature. The Brillouin zone contains five electronic bands; the hole-like α_1, α_2 and γ bands surround the Γ point, and the electron-like β_1 and β_2 bands surrounding the $\tilde{M}$ point.

The discovery of iron-based superconductors provided us with an exciting new opportunity to distill the essence of high temperature superconductivity by comparing and contrasting its phenomenology with that of the apparently quite distinct copper-based materials. At an empirical level, one can distinguish several common characteristics including: (i) a predominantly layered or planar structure (CuO_2/FeAs); (ii) essential transition metal atoms (Cu/Fe) with strong onsite Hubbard, inter-orbital, and Hund's interactions; (iii) a strongly correlated antiferromagnetic and metallic phase that is suppressed efficiently leaving robust antiferromagnetic spin fluctuations; (iv) a "dome" of superconductivity surrounding the imperfectly understood critical point where the antiferromagnetic order and/or other associated orders disappear; and (v) a highly doped and weakly correlated Fermi liquid state far from the "parent" antiferromagnet that nonetheless exhibits strong superconductivity. On the theoretical side, few approaches have so far been used to compare the copper-based and iron-based superconductivity on a similar basis. One exception is functional renormalization group (FRG). In this technique, one follows the flow of the four-point vertex function for scattering between states on the Fermi surface $\Gamma(k_1, k_2, k_3, k_4)$ as the states outside an energy shell $|E| = E_F + \delta E$ are integrated out. Momentum space is discretized into a finite number of patches to determine the tangential momentum dependence of the effective interactions. The renormalization group equations are carried out at the one-loop level and numerically integrated to determine the functional renormalization-group flow of Γ as the energy cutoff δE or temperature is reduced. Although the one-loop approximation requires an initialization with appropriate band structure and bare interactions, FRG can nevertheless provide an unbiased treatment of competing instabilities (at one loop level) and indicate which instabilities are important. For copper-based superconductors FRG finds a robust $d_{x^2-y^2}$ superconducting order parameter on the single band,[a] while, for iron-based materials, the FRG prediction is that distinct superconducting energy gap functions $\Delta_i(\vec{k})$ on different bands i are anisotropic in $\vec{k}$-space, with each exhibiting distinct 90°-rotational symmetry and a specific relationship of gap minima/maxima.[a] The former is a well-known characteristic of copper-based high-T_c superconductivity, while the latter was recently discovered for iron-based high-T_c superconductivity by this author.[b]

[a]F. Wang *et al.*, *Euro. Phys. Lett.* **85**, 37005 (2009).
[b]M.P Allan *et al.*, *Science* **336**, 563 (2012).

Given the common empirical characteristics plus the FRG analysis of the superconducting energy gaps in these two types of superconductors, one can infer that quantum antiferromagnetism from strong local correlations (probably requiring transitional metal ions) which is suppressed to zero temperature by an external influence, leads to the high temperature superconductivity.

S. Das Sarma: Graphene

I will discuss physics of graphene here. Graphene is a purely two-dimensional (2D) sheet of carbon atoms arranged in a honeycomb lattice, with electrons hopping on two inequivalent sublattices as appropriate for a honeycomb arrangement. As such, graphene can be thought of as a single layer of graphite or a 1D cylindrical carbon nanotube rolled out as a 2D system or a buckeyball-shaped fullerene molecule folded out into a 2D layer. For a long time (dating back at least to 1948) theorists have been working on graphene as a hypothetical material (which should, but did not seem to, exist in nature —after all graphite which is nothing other than many layers of graphene coupled together in a 3D layered form is ubiquitous), finding that the simple single-particle band dispersion of graphene obeys at low energy precisely the chiral Dirac-Weyl (i.e. massless Dirac) equation with a velocity (often called the graphene Fermi velocity) of $v = 10^8 \mathrm{cm\,s}^{-1}$ (which is a factor of 300 smaller than the velocity of light c). Graphene is thus a unique 2D gapless semiconductor whose linear conduction (electron) and valence (hole) bands crossing at one singular point called the Dirac point. The chirality here arises not from a coupling between spin and momentum as in the Dirac theory, but from the coupling between the momentum and a pseudospin index which (acts just like spin and) arises from the underlying band structure and is associated with the A/B sublattices of graphene honeycomb structure. In addition to this chiral pseudospin index, graphene also carries the regular electron spin (two degenerate flavors up and down) and two valley flavors (arising again from the band structure). The ground state of graphene thus obeys the chiral massless Dirac equation with a degeneracy factor of 4 arising from spin and valley. The linearity of graphene band dispersion is essentially exact to a rather high momentum (and thus energy $\sim$ a few eV) value, breaking down only at very high momentum corresponding to the inverse of the lattice constant ($\sim 10^8 \mathrm{cm}^{-1}$).

The recent great excitement in graphene leading to more than 15,000 publications in the last five years (2007-2011) arises from the actual experimental development in 2004-05 (Nobel Prize in physics in 2010) of the laboratory fabrication of 'doped' graphene, so that now thousands of laboratories all over the world (essentially anybody, because all it takes is a bulk crystal of graphite and some simple adhesive tapes to peel away layers of graphite which can then be deposited as graphene layers on some other substrates!) can make 2D graphene with a finite density of electrons and holes (i.e. dope it by simply applying an external voltage through a gate). This breakthrough in not only making 2D graphene in the laboratory, but also changing the chemical potential or the Fermi level in the system at will with an external gate voltage, led to the explosive growth of the subject, making it a highly interdisciplinary field of research involving physics, chemistry, materials science, and engineering. The engineering interest arises from many unique gate-controlled opto-electronic properties of graphene involving high electrical and optical conductivity, and in addition, graphene is also the thinnest possible material which may also be the strongest possible material, making it a highly desirable nano-material for technological applications.

Fundamental interest in graphene of course arises from its unique low-energy linear, chiral, massless, and multiflavor (2 spins and 2 valleys) Dirac dispersion, which is qualitatively different from the usual parabolic low-energy band dispersion in most solid state materials. The fine-structure constant in graphene is around 0.5-2 (depending on the substrate material which provides a background polarization reducing the free space coupling), not $1/137$ as in normal QED. Graphene thus allows one to study some aspects of chiral QED in a solid state system at a very different (eV) energy scale. In particular, some of the fundamental issues which have been (and are being) actively studied both theoretically and experimentally are Klein tunneling, critical hyper-charge (which, instead of being 137 could be as low as 1-2 because of the much enhanced coupling constant), spontaneous chiral symmetry breaking (leading to massive particles, which in the context of graphene imply the spontaneous generation of a gap at the Dirac point), the interaction-induced velocity renormalization due to the ultraviolet momentum cut-off inherent in a linear dispersion (which does not arise in a parabolic band structure), various quantum Hall and fractional quantum Hall effects where interaction-induced spontaneous breaking of valley and spin symmetries may be operational, and many other phenomena (including possible superconductivity, magnetic instability, etc.). There are, however, important aspects of graphene which make it very different from the usual theoretical questions arising in nuclear physics or chiral QED. First, the interaction between the electrons is the usual nonrelativistic $1/r$ Coulomb interaction. Second, graphene, being a solid state material fabricated in the laboratory, invariably has a lot of disorder in the environment, which makes it very difficult, if not impossible, to observe true interaction-induced renormalization phenomena. In particular, the Dirac point physics for undoped graphene (i.e. no carriers or particles induced by the external gate) is very difficult to access since any remnant disorder (or finite temperature) induces some electrons and holes even if the overall system is nominally charge neutral. Thus, extrinsic thermal- or disorder-induced fluctuations turn out to be very important in graphene, particularly at the Dirac point, which is an unstable fixed point, susceptible to extrinsic density fluctuations. It is easy to see, however, that the Dirac point itself (where the valence band is completely full and the conduction band completely empty) is a non-Fermi liquid fixed point of the system (in fact, it is a marginal Fermi liquid with a logarithmic divergence in the Fermi velocity flowing possibly to an unknown strong-coupling fixed point at extreme low energy). The question of whether the interesting strong-coupling interaction physics of the Dirac point is accessible experimentally (at very low temperatures and very low particle density in very pure graphene with no disorder) or not remains open and active at this stage.

Discussion

X-G. Wen I have a comment about this high-T_c. I think it is Frank Wilczek who asked the question how to declare victory in high-T_c superconductors? And I think everybody has a different point of view on what is the important question. So the issue is that whether finding the pairing mechanism for superconductivity is the most important question in high-T_c or not. And certainly personally I think that for overdoped samples, yes, that is the important question. But I feel that for the underdoped sample there is an even more interesting question which is: what is the normal metallic state, in which experiments show there is a pseudogap. So I feel that maybe one needs to understand the pseudogap phase first, then try to understand what is the instability of the pseudogap phase. Then that is maybe a way to understand underdoped superconductivity. But this functional renormalization has an assumption which is: assume the normal state is a Fermi liquid then study the instability of the Fermi liquid, which is OK for optimal doping and overdoping, but for underdoped maybe we are facing a different challenge.

S. Davis So the only problem with the logic of that argument, which I would have agreed with completely four years ago, is that the iron superconductors have no pseudogap. So it cannot be fundamental to the mechanism.

F. Wilczek I am wondering if these new insights about high-temperature superconductors indicate that there is a limit to the temperature, or suggest ways of getting to higher temperatures?

B. Halperin Subir (Sachdev), do you want to answer that?

S. Sachdev Well, I mean, one comment is that the new superconductors have lower T_c. The highest T_c superconductors are the ones for which the problems Wen mentioned are very much present: the pseudogap phase and perhaps more exotic behavior. Stronger exotic behavior than in the pnictides and the electron...

F. Wilczek I was mainly directing my question to Davis.

S. Sachdev I can try to, I mean there does not seam to be any fundamental limit. It seems that you need to have two-dimensionality and strong spin fluctuation and a Fermi surface, but the question in my mind would be: every time you do that, will you always get this pseudogap regime, and I think probably and they seem to go hand in hand.

F. Wilczek Can you tell a chemist what they have to do in order to make a room temperature superconductor? [laughter]

S. Sachdev If I knew, I would not tell you! [laughter]

B. Halperin Seamus (Davis), did you want to comment on that as well?

S. Davis Well, yes, built into the structure of my presentation was the idea that if the functional renormalization group colleagues were able to predict another high-temperature superconductor using their scheme, that they believe they have control over, then we would definitely have a way forward in this problem. So I think there is a good deal of work going on in that channel right now.

E. Silverstein I have a kind of naive question but it is about stimulating superconductivity through time dependent effects. I know this was a development that applied to BCS superconductors in the old days successfully, and in quantum field theory it is very easy to see how time dependent effects can enhance instabilities. And my question is: is this idea being applied to the modern materials or is it just too messy? Do these time dependent effects tend to heat up the system instead of enhancing the superconductivity?

B. Halperin Does anyone want to answer that? Boris (Altshuler).

B. Altshuler Just a small comment, that stimulated superconductivity which was studied for I think at least forty years from now, is actually a very weak effect in conventional superconductors. I do not know if it was even observed in high-T_c but in conventional superconductors it is just a fraction of a percent or something like that. So I do not think that there is a hope to have huge enhancements.

T. Leggett Going back a few talks, and a question to Leon Balents. Could you say a little more about the bulk, the difference in bulk between a topological insulator and a band insulator. In particular, are there specific smoking gun experiments which at least in principle could be done on the bulk to distinguish them?

L. Balents There are impossible thought experiments that could be done in the bulk. [laughter] I would say in principle certain detailed measurements of the bulk electronic structure, if you could actually measure the electronic spectrum of the states, you could determine something. But beyond that, so I am not aware of any way to do that by a bulk measurement.

S. Sachdev Even including disorder effects? They are the same?

L. Balents Including disorder effects...

S. Sachdev Well, localization transitions in the presence of disorder. Are they the same in the bulk?

L. Balents Yes? ....

B. Halperin Well you have a gap, and this strong spin-orbit coupling, so I do not know what more, and if it is three-dimensional, there is presumably localization if it is really strong, but...

L. Balents Sure.

B. Halperin But I am not sure it is any different from from any other... I mean, once you get into the gap I would think it is just another state with spin-orbit coupling.

L. Balents Yes I mean this has been studied in two dimensions where you also have topological insulators and in that case we sort of know that the localization transitions themselves are not different between a two-dimensional topological insulator and a metal. In a sense the topological insulator is kind of only different in the way that it... they are only different if you measure one relative to an other, so the vacuum has a certain topology if you like and it is different from the topology of the solid then there are some phenomena at the interface but if our whole world is a topological insulator, from the inside we can not really tell.

B. Halperin I might add, in principle you can measure the band structure with a local measurement. It may be hard to do it, we do not really know how to do it, but in principle it could be done. But it is a high energy measurement so it is probably correct to say that no low energy measurement can distinguish the...

L. Balents Yes that is definitely the case because of the gap. The thought experiment that you can imagine is if our high energy colleagues would find us a magnetic monopole, we put that in our sample, then we can measure the charge accumulated around that monopole, so...

F. Wilczek I had a question also about topological insulators. I believe the effective theory of their low interactions is said to be $\theta = \pi$ electrodynamics. And that $\theta = \pi$ is singled out as a possibility because of time reversal symmetry. And of course, if time reversal symmetry is valid right out to the edge, then you get zero modes and a very characteristic phenomenology but this formulation in terms of θ electrodynamics also suggests an alternative which is that θ could continuously go from π to zero in one direction or another, breaking time reversal symmetry. Have people constructed microscopic models of that kind?

L. Balents To make sure I understand the question, so are you asking if one has models of bulk phases for which, by tuning some parameter, θ could be varied to take any value between zero and π. Is that ...

F. Wilczek No, π. But then you could have spontaneous breaking of time reversal in the boundary region, to allow it to go to zero.

L. Balents Oh at the boundary region. Yes, that is certainly something that has been considered. So we know that in the absence of interactions you would have gapless Dirac fermions at the surface. But if you now introduce interactions one possible thing that can happen is that those can magnetically order and then break time reversal just at the surface.

F. Wilczek So it is very analogous to the things we discussed in 1+1 dimensions a long time ago, yes.

B. Halperin As I recall there is a recent paper from Stanford, by Barkeshli and Qi, in which they actually consider what happens if you were to put a ferromagnetic moment on the surface of a topological insulator and it couples to the carriers at the surface and the material beneath. Some degree of broken time reversal should enter the material, and they consider a model which they parametrize as a θ that changes continuously and probably, it is a numerical calculation of what you are talking about.

E. Witten A variant of Frank's (Wilczek) question arises when you consider what is more standard which is explicit time reversal violation, for example due to a magnetic field. The massless gaps on the surface can disappear, and then what happens is that, from the point of view of low energy field theory, θ interpolates smoothly from π inside to zero outside, in a boundary layer.

B. Halperin Another question?

S. Kachru So my question is actually for Haldane. So you now have a formulation of the fractional quantum Hall effect that lets you keep track of more of the geometry of the flux attachment than for instance the Laughlin wave function, and my question is: do you foresee or have you already applied it to explain the sort of anistropic phases that show up in the 2D electron gas like the nematics or is there some obstruction to doing so?

D. Haldane We have not yet done any of that, but yes people have suggested that already and it is a fairly obvious thing. So in principle there are these two geometric features: the Landau orbit shape and the shape of the Coulomb interaction equipotentials. In fact, if you break rotational invariance by just tilting the magnetic field you have already gone into a kind of nematic phase. So in principle the Laughlin state will just adjust itself to have the right shape of its correlations to minimize the correlation energy. So whether that can happen spontaneously while you maintain a rotational symmetry is a good question and that could be a nature of the transition to the nematic states.

W. Phillips My question is for Sankar (Das Sarma) about graphene. You said that one of the limitations of graphene was the fact that there are these defects in there. Now people have suggested, even realized, graphene-like structures using optical lattices which will not have, at least, point-like defects like that. Will some of the things that would be inhibited by those defects in graphene, do you think it is possible, I mean I realize it is the early days for this sort of thing, but do you think it is possible that optical lattices will achieve some of those...

S. Das Sarma Yes, in fact that is an interesting direction. So the optical lattice people who are trying to make graphene, this is one of their motivations in my discussion with them. So what we have in graphene is not really defects in the sense of point defect. If you look at this TEM picture or STM pictures, graphene is very pure. All the atoms are there. But this is an exposed environment, so we have charged impurities... so these are worse than just point defects and an optical lattice of course would not have them. So in that sense your chance of accessing Dirac point physics is better in optical lattices.

B. Halperin But a lot of the Dirac point physics that you are most interested in come from the long range Coulomb interaction ...

S. Das Sarma Exactly, that is what I was going to say, that was my next comment. On the other hand, without Coulomb interaction, since it is not clear that there is anything interesting left, it is a double-edged sword.

W. Phillips Well, just to follow up on that, the question was raised earlier about whether you could have long-range interactions, and one of the answers was "well, if you have Rydberg atoms, or if you have dipolar molecules you could have dipolar interactions which are long-range", but they are not as long ranged as Coulomb. So the question is: will dipolar interactions allow you to see some of the interesting things that you would like, because the Coulomb interactions are always screened. So what is the deal there?

S. Das Sarma No, the whole point is that at a Dirac point there is no effective screening the way you are thinking about it. I have not looked at what would happen with dipolar interactions but I think some of these very interesting physics will not be there with dipolar interactions: we need true long-range interaction I believe.

B. Halperin Michael (Berry)?

M. Berry Yes, in keeping the spirit of these Solvay meetings and relevant to what has just been discussed about optical analogues of these Dirac points and Dirac cones as they call, it is worth pointing out that the very first physics associated with the conical intersection optics was Hamilton in 1830, and it is absolutely exactly a linear dispersion relation surrounding a conical intersection in momentum space, and it was the first prediction non-trivially using essentially the concept of phase space. Its experimental discovery, within a year of Hamilton predicting it, a discovery by Humphrey Lloyd, observation I should say, made Hamilton instantly famous. And it is worth pointing out that a lot of the physics (not the Coulomb stuff) was appreciated, and for example Humphrey Lloyd noticed what we now call the π-phase change of the eigenstates as you go around this conical intersection. And he said something which we might well emulate nowadays in our scientific writing. He said: "I have observed this sign change, it is what it is, in the polarization as you go around this conical intersection. It is not mentioned anywhere in Hamilton's paper, but I am sure he knew about it".

S. Sachdev Just a question to Sankar (Das Sarma). So you mentioned that Coulomb interactions are not that important in graphene, but there are lots of interesting experiments on bilayer graphene where it seems like a lot of interesting correlation physics is present. So if you want to summarize some of that?

S. Das Sarma Ok, Subir (Sachdev) gave me an opportunity, I had only one slide and five minutes. So, what Subir (Sachdev) is saying, I just talked about what is called monolayer graphene (just one layer of graphene). You can also have two layers of graphene and it turns out that a particular stacking of two-layer graphene, where you have these A-B atoms in a particular way which is the most standard kind of bilayer, is in some sense even more interesting than monolayer graphene. This is also a chiral material, but it is no longer linear, the low-energy dispersion is in fact quadratic (parabolic), but in high energies it becomes linear, so the actual band structure is hyperbolic. This system is very interesting because you can now open a gap at the Dirac point by just applying an electric field: basically, the two layers become asymmetric if you apply an electric field in this direction, by applying a gate. And there is a lot of interest in it, because if graphene is ever applied as a switch, this is the system that you are going to use, not monolayer graphene where you do not have a gap. And this system has very strong interactions: since this is parabolic, just by tuning density we can change what in QED would be the fine structure constant. You can make it (in principle) anything you want, you can make it a hundred if you want. So this is a very interesting system and I would say that much of the basic work in graphene has moved towards bilayer precisely for the reason that Subir (Sachdev) has just mentioned, that this is a strongly interacting system, various instabilities have been proposed. In fact almost all the instabilities that we discussed today have been proposed there, including mass generation, ferromagnetic instabilities, stripe instability, time-reversal invariance breaking, and it is a very beautiful playground. There are experimental indications, observations, of some of these instabilities. I know of four experimental papers. The only problem is that each of those four papers claim to see a different instability. And unfortunately, what is going on again, disorder is very important here, for the same reason because you still have charge impurities and the disorder energy scale with the current experiments is much larger than the interaction energy scale. So what you have to do is this very difficult problem in condensed matter, you have to somehow solve the problem of quantum phase transition with disorder and interaction. But I think that bilayer graphene is where interesting strong-correlation physics will be seen, because I think this material will become purer and purer.

B. Halperin Any questions or comments? Xiao-Gong (Wen).

X-G. Wen There is an issue that was raised about the possibility or impossibility of the emergent gravity from condensed matter systems. So I have a related question: Whether quantum gravity near flat space-time, at the low energy, can be viewed as a collection of harmonic oscillators or not?

B. Halperin Does someone want to answer that question?

I. Klebanov Does it not violate the Weinberg-Witten theorem, just emergent graviton...

X-G. Wen No, my question is that, we believe in gravitational waves, we believe the graviton, but whether that implies... you know, in condensed-matter when you have a wave (e.g. some magnon), we say that it is basically a collection of harmonic oscillators. Whether quantum gravity can be viewed in a same fashion? Whether gravitational wave can be viewed as a collection of oscillators?

I. Klebanov If any free field can be...

G. Gibbons I think that what makes gravity different from curved space is Einstein equivalence principle. If all the excitations in your system satisfy the same equations and follow the same metric, then you have an analogue of gravity. But it is easy to find analogue systems which you could describe in terms of curved space: for example, take a conductivity tensor, that defines a metric (it is the way the current flows), choose a different material or rather choose different carriers and then you will get a different metric. In order to really make the analogy with gravity you have got to show universality as far as the metric is concerned. And I think that this is probably going to miss in your model, because if you choose a different excitation it is not going to follow the same metric. But if it is, then you have a closer analogue.

X-G. Wen I am still asking my question: is that a collection of harmonic oscillators?

E. Witten I think we should just say *yes*. So approximately, gravitational waves are harmonic oscillator modes like other fields.

X-G. Wen Then there is an issue about this universality. The question is whether those modes can be gapless as a universal property? I don't know whether that is what you are talking about, that is the universality.

C. Bunster There is a book about Hector Rubinstein at the entrance, which is free, and I was reading one of the articles by Virasoro where he referred to the old string theory. Now I remember the excitement when the spin 2 was found in the spectrum and that it was realized that this theory was about gravity and it was not about what it was originally intended to be. And then, I was raised on the conviction that anything that looks like gravity *is* gravity. So I am confused by your discussion and I would like to have some clear statement as to whether this gravity-like thing that we are looking at is gravity or not.

D. Gross It was spin 2 and massless. Spin 2 does not have to be gravity. Massless, and finite charge, and Lorentz invariance.

J. Maldacena And Lorentz invariance. Wen's model was explicitly not Lorentz invariant.

C. Bunster So, is this gravity or not?

E. Witten Just to clarify, the finite charge I think means that there is a non-zero coupling to matter at low momentum.

D. Gross The only thing a massless spin 2 can couple to is energy-momentum. That is essentially the physics of it. And there is only one energy-momentum, and therefore only one gravity. And as far as I understand, Wen's stuff is not.

X-G. Wen Yes, it has no Lorentz invariance: the $\omega \propto k^3$ dispersion relation...

A. Polyakov There are rigorous definitions...

A. Polyakov But I am saying that gravity could be defined on a lattice, quite rigorously, at least in not too many dimensions, if the lattice is random. You have to assume that the lattice itself fluctuates.

G. Dvali I just missed the point. So is this excitation that you are finding, is it massless? Does it have long-range correlations, or...

X-G. Wen Yes, it is massless. The dispersion relation is $\omega \propto k^3$.

G. Dvali So it is long-range...

X-G. Wen It is gapless, it is not Lorentz invariant. So it is gapless.

G. Dvali No, this I understand, indeed. But it is sourced by energy, or you do not know?

X-G. Wen No it is $\omega \propto k^3$, so when k goes to zero, ω goes to zero.

G. Dvali No, but it is sourced by energy-momentum-tensor-type thing, or?

X-G. Wen It is a little bit like the emergent electromagnetism in fact. There is a sort of defect which is a source of this field. Yes, the source of this field is actually a defect.

G. Dvali A topological defect?

X-G. Wen Topological, yes.

B. Halperin I had a question for Duncan (Haldane), if I may. Maybe this should also go to Steven Girvin. So, this argument that S [structure factor] has to go with k^4 (momentum fourth power in the fractional quantized Hall regime, for the fractional quantum Hall effect), was at least originally derived for cases where you had Galilean invariant systems (momentum is conserved). You automatically have, for example, at $k = 0$, that all the weight is sucked up by the Kohn mode, which is inter-Landau level. I have not thought about it, but the question is: if I had a more general system, let's say on a lattice, where I did not have translational invariance, did not have Galilean invariance, and would not have Kohn's theorem, is it necessarily true that a quantized Hall state would have a k^4...

D. Haldane Yes, it has nothing in fact to do with Kohn's theorem. It could be derived completely after you have thrown away everything to do with the dynamical momentum of the system. And it is purely derivable within the quantum geometry...

B. Halperin So this has nothing to do with translational invariance, you could do this on a lattice, let us say, which is partially filled?

D. Haldane Yes, it arises from the quantum geometry. In fact it follows naturally from the metric formulation. But it does not require rotational invariance either.

B. Halperin Rotational invariance, I understand it would not require.

D. Haldane Yes, originally Girvin and MacsDonald, they put in rotational invariance, so it requires translational invariance in fact.

B. Halperin So let me thank all the speakers and all the contributors to this Session, and adjourn.

Session 5

Particles and Fields

Chair: *Howard Georgi*, Harvard University, USA
Rapporteur: *Frank Wilczek*, Massachusetts Institute of Technology, USA
Scientific secretaries: *Peter Tinyakov* and *Michel Tytgat*, Université Libre de Bruxelles, Belgium

Rapporteur talk by F. Wilczek: A Long View of Particle Physics

Abstract

2011 marked the hundredth anniversary both of the famous Solvay conferences, and of the Geiger-Marsden experiment that launched the modern understanding of subatomic structure. I was asked to survey the status and prospects of particle physics for the anniversary Solvay conference, with appropriate perspective. This is my attempt.

1. Origins: Understanding Matter

The intellectual origin of particle physics is quite straightforward: It arose out of the program of understanding the physical world through "Analysis and Synthesis" (Newton), or in modern terminology Reductionism. More specifically, it arose from the program of studying the smallest building blocks of matter and their properties, in the hope that those building blocks would obey simple laws, from which the nature and behavior of larger bodies could be inferred mathematically.

The success of that program was by no means logically guaranteed. Indeed, I think it is fair to say that only in the last hundred years has its slow-ripening fruit matured. But by now the strangeness, beauty, and richness of that fruit has far exceeded any reasonable, or even mystic, expectation. We out-Pythagoras Pythagoras.

Perhaps the most crucial development occurred almost precisely 100 years ago. I refer of course to the rapid sequence of events beginning with the Geiger-Marsden experiment (1911), advancing with Rutherford's interpretation of that result in terms of electromagnetic "Rutherford scattering" from heavy atomic nuclei in solar-system-like atoms (1912), and culminating in Bohr's inspired infusion of quantum ideas into subatomic dynamics (1913).

After the quantitative success of Bohr's model for hydrogen, and its many other semi-quantitative and qualitative successes, no doubt could remain about its central message: Atoms are held together by electromagnetic forces acting between small, massive nuclei and much lighter electrons, subject to rules of quantization. It was the work of a generation to create a physically satisfactory, mathematically coherent discipline of quantum dynamics. After more than a decade of indecisive skirmishing, breakthroughs by Heisenberg (1925), Schrödinger (1926), and others of their storied contemporaries rapidly established the basic outlines of quantum theory that we still recognize today.

The atomic nucleus was at first a tiny black box and a source of bizarre surprises, including notably the various emanations of radioactivity and the need for new binding forces. Chadwick's discovery of the neutron (1932), and the application of quantum principles, supported the rapid construction of a useful phenomenological description of nuclear phenomena, that we still use today.

With these achievements the "practical" goal of reductionism, for ordinary matter under ordinary conditions, was attained. An adequate foundation for condensed matter physics, materials science, chemistry, and (presumably) biology was in place. That foundation remains firm.

Though in some sense the original goal had been achieved, the intellectual situation was far from satisfactory. Theoretical nuclear physics, in particular, was a semi-empirical enterprise. It had not engendered governing equations or principles worthy to stand beside general relativity and covariance, or Maxwell's equations and gauge invariance. As we now know, in many ways the story was just beginning.

2. Phenomena: New Questions and Surprising Answers

The discovery of antimatter, and the successful measurement and calculation of effects of virtual particles, vindicated a radically conservative attitude toward the basic principles of quantum field theory. So did the experimental discovery of Yukawa's π mesons, posited to explain nuclear binding forces.

On the other hand many "gratuitous" new particles, starting with muons and K mesons, spawned a complex of flavor problems, that are still very much with us. And the complexity of nuclear forces, crowned by discovery of hordes of strongly interacting resonances, seemed to challenge the whole framework of quantum field theory.

Symmetry, harnessed to radically conservative quantum field theory, proved to be the most reliable and fruitful guide to decoding Nature's principles.

Universality and the $V - A$ structure of weak interactions led to the $SU(2) \times U(1)$ electroweak theory. Besides extending the gauge principle, this theory introduced two major dynamical innovations. One is the fundamental significance of fermion chirality. That concept erupted from the discovery of parity violation. The other is gauged spontaneously broken symmetry. That mechanism for generating gauge boson masses was implicit in the London-Landau-Ginzburg treatment of superconductivity, but the emphases of superconductivity theory were quite different. To recognize the massive photon interpretation of quanta *inside* the superconductor, and to give that interpretation a firm foundation in relativistic quantum field theory, were major achievements.

Patterns among resonances and Bjorken scaling led to the $SU(3)$ color theory of the strong interaction, quantum chromodynamics (QCD). Besides extending the gauge principle, this theory introduced two additional major dynamical innovations. One is the confinement phenomenon. It was shocking, in the early days of the quark model, to contemplate the existence of elementary quanta that do not appear in the physical spectrum. What could such "existence", which explicitly rejects physical existence, possibly mean? Yet we've come not just to live with confinement, but to understand that it is a natural consequence of gauge symmetry. (*Deconfinement* is the subtler case!) The other is asymptotic freedom, which both opens ultra-high-energy processes in cosmology and at accelerators to quantitative treatment, and enables a rigorous, completely nonperturbative approach to calculation of the spectrum, by means of a convergent discretization.

Both theories are, to an extraordinary degree, *embodiments* of ideal mathematical symmetry. For example, color gluons were introduced specifically to implement gauge covariant parallel transport, and their properties were derived without ambiguity from purely conceptual considerations, prior to any direct experimental evidence for them.

These two theories together constitute the Standard Model. The Standard Model overcomes the most unsatisfactory feature of the earlier, semi-phenomenological nuclear theory: It *is* based on equations worthy to stand beside Einstein's general relativity and Maxwell's electrodynamics.

With the mature result of reductionism before us, we can consider an important philosophical question: *Why* does reductionism work? Several specific aspects of physical theory underpin the success of that approach. Because of symmetry under translations of time and space, we can infer the laws through repeatable experiments, compiling work done over generations all over the world. Locality of the laws is also crucial, both to make repeatable experiments a practical possibility, and to guarantee that having completed the Analysis, we can perform the Synthesis. Relativistic quantum field theory explains the existence of many identical particles, as products of a common field, and thus supplies our elementary building-blocks. More subtle, but also crucial, is what I call *quantum censorship*: the feature of quantum physics, that complex systems can look like ideally simple ones, if we probe them only below their energy gap. Because the success of reductionism in fundamental physics depends on such specific, non-trivial features of physical law, we should not take the utility of that approach in other domains for granted.

3. Questions That the Standard Model Begs

3.1. *Questions from the Core*

The core of the Standard Model – i.e. the part embodying gauge symmetry – seems near to perfection. For just that reason, its remaining flaws stand in sharp relief. Most obviously, it has more moving parts than we'd like: 3 interactions (4, including gravity), each with its own coupling parameter; 6x3 fermion multiplets; and either one (or more) Higgs doublets, or some more complicated dynamics that produces equivalent effects at low energy.

In this accounting, only the enumeration of fermions requires further explanation. It is not appropriate, in a gauge theory, to regard particles related by gauge symmetry as truly independent entities. Thus the left-handed up and down quarks (related by electroweak $SU(2)$) of all three colors should be counted as one multiplet; the right-handed up quarks of all colors and the right-handed down quarks as two more; the left-handed electron and its neutrino as another; and the right-handed electron and neutrino as two others. The different families are not related by gauge symmetry, at least within the Standard Model, so their corresponding multiplets represent independent degrees of freedom.

In formulating the electroweak $U(1)$ couplings of fermions, phenomenology requires some peculiar-looking choices of fractions. If we require closure using the known degrees of freedom in the Standard Model, and demand absence of violations of gauge symmetry not only classically, but also quantum mechanically (anomaly cancellation), the choices are actually severely constrained; still, one might hope for more luminous, less arcane insight into the origin of those numbers.

3.2. *Loose Ends*

Once we look beyond its core, we find many loose-ended strands both within the Standard Model itself, and in its account of Nature:

The Higgs sector is poorly constrained theoretically and, at least for the moment, uncharted experimentally. (See the concluding Note.)

The astronomical dark matter is unaccounted for.

Flavor phenomena have been smoothly accommodated, but not comparably illuminated. There is no known theoretical principle that has, for the Yukawa couplings that encodes quark and lepton masses and mixings, anything approaching the power and coherence of the gauge principle for the core interactions. (It is amusing to note that this contrast, in different but recognizable forms, has been with us from the earliest days classical mechanics. Kinetic energies are simple and geometrical; potential energies are a never-ending work in progress.) The prediction of the third generation, and the brilliant success of Cabibbo-Kobayashi-Maskawa (CKM) phenomenology, are great achievements. But the *raison d'être* of family replication remains elusive, and the proliferation of theoretically unconstrained mass and mixing parameters is an embarrassment.

Within the complicated, opaque pattern of quark and lepton masses and mixings, a few features are so pronounced as to deserve *qualitative* explanation. These include:

Neutrino masses are very small, but (in at least two cases) not zero.

The θ parameter of QCD, which might introduce T violation into the strong interaction, is very small: $|\theta| \lesssim 10^{-10}$.

The top quark mass $M_t \approx 172$ GeV is much larger than the mass of any other quark or lepton. It is the only fermion for which the Yukawa coupling to the electroweak condensate, and presumably to the Higgs particle, is of order unity. In so far as the other couplings are small, their precise value might be complicated to calculate from fundamentals, but for the top quark there is less excuse.

The electron mass, and also the masses m_u, m_d of the light quarks, correspond to tiny Yukawa coupings, $y_{e,u,d} \sim 10^{-5} - 10^{-6}$.

The first two of these features have been connected to promising, though still speculative, fundamental ideas, as I'll elaborate below.

3.3. *Gravity*

Gravity, in the form of general relativity, can be brought into the standard model smoothly. Both quantitatively and qualitatively, however, the union is inharmonious.

Quantitatively: The observed strength of gravity – technically, the coefficient of the Einstein-Hilbert term – introduces a new large mass scale, the Planck mass $M_{\text{Planck}} \sim 10^{18}$ GeV. This is far larger than any other mass scale within the standard model. That disparity defines a *first hierarchy problem*. The observed cosmological term – technically, the coefficient of a pure volume term (and thus a universal pressure)

$$\Delta L = \lambda \int \sqrt{g} \, d^4 x$$

$$(\lambda)^{\frac{1}{4}} \approx 2 \times 10^{-3} \, \text{eV} \tag{1}$$

introduces a mass scale that is significantly smaller than most standard model mass scales, though it is comparable to probable neutrino masses. That disparity defines a *second hierarchy problem.*

QCD, through asymptotic freedom, advances the first hierarchy problem significantly. We can ask, specifically, why there is a big disparity between the QCD scale and the Planck mass. If we extrapolate logarithmic running of the coupling all the way to the Planck scale, we find that the QCD coupling is not terribly small there. (See immediately below, for a closely related discussion of coupling constant unification.) Reading it the other way: If we hypothesize a "generic, but not large" effective coupling at the Planck scale, then we will compute the proton mass to be many orders of magnitude smaller than the Planck mass, as is observed. Here of course I've assumed that the *quark* masses are negligibly small, as they are in Nature. We don't know why, so the first hierarchy problem is not completely solved.

The second hierarchy problem will be discussed extensively in the cosmology session of this meeting.

Qualitatively: The minimal implementation of general relativity within the standard model gives us an elegant, successful working theory of quantum gravity (modulo the hierarchy problems), but that minimal theory has not been formulated nonperturbatively, and at that level it almost certainly fails to exist. Moreover, there appear to be fundamental issues in black hole physics and in the treatment of cosmological singularities, that the minimal theory cannot address even qualitatively. Those problems will be discussed extensively in the string session of this meeting.

4. Approaches: "Modest" Improvements

4.1. *Unification and Supersymmetry*

At the level of quantum numbers, the interactions and multiplets of the Standard Model fit beautifully into a unified theory. The most attractive unification is based on the spinor representation of $SO(10)$, though variants are certainly possible, and have their advocates.

The renormalization group shows us how observed (low-energy) couplings might diverge from the (high-energy) equality that nonabelian symmetry requires.

Famously, if we construct our unified theory using only the degrees of freedom in the standard model, the resulting constraint among observed couplings doesn't quite work, while if we extend the theory to include the degrees of freedom required for approximate supersymmetry, at masses $\sim 10^2 - 10^4$ GeV, we get a much more successful relation. The observed low-energy couplings extrapolate to a common value g_U at a large mass scale M_U, with

$$g_U \approx .7 \quad (\alpha_U \approx \frac{1}{25})$$
$$M_U \approx 2 \times 10^{16} \, \text{GeV} \tag{2}$$

There is of course some play in these numbers, which arise in the minimal implementation of low-energy supersymmetry. Taking them at face value, however, we find many good features:

The unified coupling is not terribly strong. Thus the extrapolation of weak-coupling formulas for running of the couplings is internally consistent.

The unified coupling is not terribly weak. Thus M_U is plausibly associated with a scale for dynamical symmetry breaking. (Which after all is what it is!)

The unified scale is not too large. Thus grave uncertainties associated with quantum gravity at the Planck scale are sequestered.

The unified scale is not too small. Proton decay is sufficiently suppressed. More accurately, the calculable contribution to proton decay due to exchange of the new gauge bosons introduced by unification is sufficiently suppressed. Other less universal contributions, especially the contribution due to colored Higgsino exchange, are potentially dangerous, and constrain model-building.

The unified scale provides reasonable, though not perfect, input to a neutrino seesaw: $m_\nu \sim \frac{M_t^2}{M_U}$

One finds remarkably good, though not perfect, unification with gravity: $M_U \sim M_{\mathrm{Planck}}$. Since gravity is directly sensitive to energy it runs as a power when probed at high virtuality, even classically, rather than merely logarithmically due to quantum vacuum polarization, as for the gauge couplings. Remarkably: If we expand the running of couplings calculation to include gravity, we find approximate unification, even though gravity is *abysmally* weaker – by a factor $\sim 10^{-40}$! – than the other forces (measured, of course, at the level of elementary particles) at all practically accessible distances and energies.

To me the unforced fit of fermion families into spinor multiplets together with the accurate, multi-advantaged unification of couplings render unification, as quantitatively enabled by low energy supersymmetry, into a most compelling speculation.

Detailed implementations of low energy supersymmetry typically include a lightish mimic of the minimal standard model Higgs particle. A second doublet is mandatory, as well. Low energy supersymmetry also produces, in many but not all implementations, a dark matter candidate.

For all its attractiveness low energy supersymmetry both poses many theoretical challenges, and faces many experimental challenges. There is no consensus on the mechanism of supersymmetry breaking, and no existing mechanism seems entirely satisfactory. There is no evidence, so far, for any of the many additional contributions to flavor-changing processes, or T-violating processes, that low energy supersymmetry brings in. There is no reliable encouragement, so far, for supersymmetric dark matter candidates. And of course, most importantly, there is no evidence so far for any superpartners! In anticipation mixed with dread, we await the verdict of Nature.

Supersymmetry is, of course, a symmetry principle. It extends Poincare symmetry to include transformations into quantum dimensions, described by anticommuting coordinates. Two other popular symmetry-based suggestions for going beyond the Standard Model, that address some of the same questions as supersymmetry (but so far lack, as far as I know, comparable success stories), are technicolor and extra classical dimensions. They are based on additional gauge symmetry and on a more conventional extension of Poincare symmetry, respectively.

4.2. Θ *Problem and Axions*

The theory of the strong interaction (QCD) admits a parameter, θ, that is observed to be unnaturally small: $|\theta| < 10^{-9}$. That suspicious "coincidence" can be understood by promoting translation of θ to an asymptotic or classical quasi-symmetry, that is spontaneously broken.

The axion field a is established at the Peccei-Quinn transition, when a complex order-parameter field ϕ acquires an expectation value F:

$$\langle \phi \rangle = F e^{i\theta} = F e^{ia/F} \tag{3}$$

At the transition, which occurs (if at all) in the very early universe, the energy associated with varying θ is negligible, and differences from the minimum $\theta = 0$ can be imprinted. They store field energy that eventually materializes, with density roughly proportional to $F \sin^2 \theta_0$ today.

If no inflation occurs after the Peccei-Quinn transition then the correlation length, which by causality was no larger than the horizon when the transition occurred, corresponds to a very small length in the present universe. To describe the present universe on cosmological scales, therefore, we should average over $\sin^2 \theta_0$. One finds that $F \sim 10^{12}$ GeV corresponds to the observed dark matter density.

Since experimental constraints require $F \geq 10^{10}$ GeV, axions are almost forced to be an important component of the astronomical dark matter, if they exist at all. So it seems interesting to entertain the hypothesis that axions provide the bulk of the dark matter, and $F \cong 10^{12}$ GeV. That has traditionally been regarded as the default axion cosmology. A cosmic axion background with $F \cong 10^{12}$ GeV might be detectable, in difficult experiments. Searches are ongoing, based on the conversion $a \to \gamma\gamma(B)$ of axions into microwave photons in the presence of a magnetic field.

If inflation occurs after the Peccei-Quinn transition, things are very different. In that scenario a tiny volume, which was highly correlated at the transition, inflates to include the entire presently observed universe. So we shouldn't average. $F > 10^{12}$ GeV can be accommodated, by allowing "atypically" small $\sin^2 \theta_0$.

But now we must ask, by what measure should we judge what is "atypical"? In the large-F scenario, most of the multiverse is overwhelmingly axion-dominated, and inhospitable for the emergence of complex structure, let alone observers. Thus it is appropriate, and I would say necessary, to consider selection effects.

Fortunately, while θ_0 controls the dark matter density, it has little or no effect on anything else. It is hard to imagine a clearer, cleaner case for applying anthropic reasoning. The result of such an analysis is encouraging. Taken at face value, it suggests that in the large F axion cosmology the typical observer sees a ratio of dark to baryonic matter close to what we observe in our neighborhood (that is, in the universe visible to us!).

Very recently Arvanitaki and Dubovsky, elaborating earlier work by themselves and others, have argued that axions whose Compton wavelength is a small multiple of the horizon size of a spinning black hole will form an atmosphere around that hole, populated by super-radiance. That atmosphere can effect the gravitational wave and x-ray signals emitted from such holes, possibly in spectacular ways. Since

$$(m_a)^{-1} \approx 2\,\text{cm.}\ \frac{F}{10^{12}\,\text{GeV}}$$

$$R_{\text{Schwarzschild}} \approx 2\,\text{km.}\ \frac{M}{M_{\text{Sun}}} \tag{4}$$

this provides a promising window through which to view $F \geq 10^{15}$ GeV axions.

5. Experimental Frontiers

The central glory of particle physics has been its success in describing empirical facts using beautiful ideas. Can we keep it up, on the empirical side?

The LHC will be a beacon for investigating the mechanism of electroweak symmetry breaking, the possibility of low-energy supersymmetry, and a large class of dark matter candidates. Among more speculative prospects, I'd like to mention especially the possible existence of hidden sectors, i.e. standard model singlets. That possibility has, at least, the negative virtue of not endangering the observed quantitative accuracy of the standard model. Of course axions and the right-handed neutrinos of $SO(10)$ are standard model singlets, but the LHC is sensitive to other types, for example scalars that mix with the Higgs particle.

Barring discovery of some qualitatively new acceleration mechanism, accelerators at the high-energy frontier will continue to be engineering projects on an industrial scale. Whether there will be successors to the LHC is a question where politics and economics loom large. It's our proper duty, in any case, to integrate the fruit of our work into our surrounding culture; to keep our enterprise going strong, we'll need to do it well.

In the pioneering days of particle physics it was cosmic rays that gave access to the highest energies. Cosmic rays have, of course, many disadvantages and limitations as compared with accelerators, but they still do offer access to the highest energies, and perhaps satellite "whole Earth" monitoring can compensate for abysmal rates. In considering unconventional accelerator technologies, it may be worth keeping in mind the possibility of producing a terrestrial version of cosmic rays – that is, to sacrifice on intensity, focus, and energy resolution for the sake of cheapness and raw energy.

Other fundamental issues call for quite different kinds of experiments. The dramatic progress of atomic physics in recent years should empower new sensitive searches, through elementary electric dipole moments, for qualitatively new sources of T violation. Beautiful work has been done to look for feebly coupled light particles ("fifth forces"); this line should be pursued vigorously, and if possible extended to the monopole-dipole case, which remains nearly virgin territory. Finally, it is difficult to overstate the importance of the search for nucleon instability. Along with small neutrino masses, this is the second classic signature of unification. It is overdue. If found, it would open a unique window on the world at 10^{-29} cm.

6. Cosmic Questions: Way Beyond the Standard Model

(Since string theory has its own session, I'm avoiding that subject. Since cosmology has its own session, I'm mostly avoiding that, too.)

This final part will consist mostly of questions, though some of them are leading.

6.1. *Kinematics and Dynamics*

Quantum mechanics, like classical mechanics, is more accurately portrayed as a framework rather than as a world-model. Yet the kinematic structure of quantum theory is both considerably more elaborate and considerably tighter than that of classical mechanics. This suggests, perhaps, that the kinematic structure is not independent of of the dynamics. In any case, this is a sort of "unification" that would seem to me desirable, though it is rarely discussed. More concretely:

In our fundamental theories of physics we postulate commutation relations several times, in ways that seem only tenuously related, conceptually. We postulate the Heisenberg group for quantum kinematics, and separate symmetry groups for space-time transformations and for gauge structures. Are those groups destined to remain a direct product?

6.2. *Dynamics and Initial Values*

Can we transcend the distinction between laws of nature and initial values?

Let me remind you that many questions that were once existential have been subsumed into dynamics, and that this represents profound progress in our understanding of the world. Thus we now have *a priori* accounts (given the standard model) of what are the possible chemical elements, molecules, and nuclear isotopes, not to mention the hadron spectrum. Since the emerging standard model of cosmology needs only a few numerical inputs, it may not be silly to suspect that the "initial value" of the Universe might be deeply simple, and from there it's a small step to conjecture that it might not be independent of the laws of nature.

Is there a unique "wave function of the universe"? If so, what makes it unique? And if so, why does reality look so messy?

Hartle and Hawking have made a relevant proposal, the "no-boundary" prescription, and greatly advanced the dialogue on this issue, though their underlying microphysical framework, i.e. Euclidean quantum gravity, may be questionable. (Hartle reviewed that work nicely later in the conference.)

A plausible answer to that third question, at least, seems at hand. Reality will inevitably look messy *to us*, even if the total wave function of the Universe is completely symmetric and deeply principled, because we effectively sample only a small part of it, having decohered from the rest.

Once we begin to question the primacy of the initial value problem, it is hard to avoid asking:

Is it always appropriate to think from past, or present, to future? Might it be the future that's somehow profoundly simple instead, or as well? Or might time support interesting topology, with branchings and loops (enabled by singularities, or quantum fuzziness)?

Issues of a similar kind arise, in a special form, in the theory of eternal inflation.

6.3. *The Ubiquity of Spinors*

Spinors, like particle physics, are also almost exactly 100 years old (!), having been discovered by Élie Cartan in 1913. It is astonishing to realize that such a fundamental mathematical phenomenon, inherent in the nature of things and, according to modern understanding, at the very heart of Euclidean geometry, could have escaped human notice for so long. (One does find premonitions in Hamilton's quaternionic treatment of rotations, and possibly earlier.) Today spinors appear prominently at many frontiers of thought: in the description of space-time fermions, as the organization of standard model families and their unification, as supersymmetry generators, in the simplest nonabelian quantum statistics, and in the theory of quantum error correction.

Why are spinors so ubiquitous? Could the appearance of this common ingredient within such superficially diverse structures be hinting at new possibilities for unification?

6.4. *Information as Foundation?*

The primary ingredients of today's physical theory have been distilled and refined over a long process of coming to terms with the strange world that Nature presents to us. From real numbers to derivatives to operators in Hilbert space, they are often extremely sophisticated concepts, from an axiomatic perspective. One might hope to build the ultimate description of Nature on more logically primitive, less artificial foundations. To be specific:

Is information a foundational *physical* concept?

There are, I think, significant hints that it should be. *Action* is central to our formulation of fundamental physics, including both the standard model and most of its speculative extensions, using path integrals. With an assist from Planck's constant action becomes a pure number, as does entropy (with an assist, if you like, from Boltzmann's constant). Entropy, we understand, is deeply connected to (negative) information. Indeed Shannon's information theory bears, at several points, an uncanny resemblance to statistical mechanics and thermodynamics. In some contexts – most impressively, in black hole physics – the physical action can be interpreted as an entropy.

These analogies have been noted for decades. But they seemed to have limited promise, to me at least, because on the information side there did not appear to be richness of structure comparable to what we need on the physical side. Recent developments in quantum information theory have, however, unveiled a wealth of beautiful structure, and investigations of geometric or "entanglement entropy" in quantum field theory have established profound, natural connections to just that structure. So it may not be crazy to hope that we might go the other way, constraining or eventually deriving the physical action from hypotheses on entanglement entropy.

Note: My report also contained some brief remarks on issues related to the origin of mass within the standard model, and the implications of (tentative) Higgs particle phenomenology. Since that subject is especially topical now, and in flux, I decided to expand those brief remarks into a separate paper.[a]

Acknowledgments

This work is supported by the U.S. Department of Energy under cooperative research agreement Contract Number DE-FG02-05ER41360.

---

[a]F. Wilczek, Origins of mass, *Central Eur. J. Phys.* **10** (2012) 1021, arXiv:1206.7114 [hep-ph].

Discussion

H. Georgi Thank you Frank for a beautiful talk and for giving us so much time for discussions. The floor is now open for discussion. This can be questions for Frank, but this can also be ideas that were generally stimulated by the subject of the talk.

S. Dimopoulos You beautifully illustrated the success of the standard framework. I want to ask you about something that I know is there in your mind, which is the remaining 20 or so parameters of the Standard Model involving masses and mixing angles and neutrinos, which with the exception of the top quark Yukawa coupling seem to be all over the place. So, the Standard Model has three gauge parameters and the top Yukawa of order one, and everything else is all over the place: from 10^{-5} down to 10^{-22}. In particular, for the fermion masses, do you feel that there is room for anthropic reasoning, especially for the light quark and lepton masses of which we are made?

F. Wilczek Well, we should make a distinction (which is quite important) between anthropic reasoning and kind of giving up. If you believe in some kind of a multiverse picture where different regions of the Universe have effectively different fundamental constants, there could be some constants which are closely associated with the emergence of life to which you can apply anthropic reasoning. There could be other constants which are pretty irrelevant and are just completely random. But finally, if there are constants that are not random, that we can either determine uniquely from the dynamics somehow or at least relate to other things we know as opposed to giving up and saying: are they anthropic or are they random? And I don't think we should give up too easily. It's very tempting to give up or to say: Well, if I have not figured it out and I am so clever, it must be random... So I think the productive question to ask is can some of the things that appear to be qualitative, that deserve qualitative explanation get qualitative explanation, and also whether there might be some hidden regularity in this apparent chaos of masses and mixing angles. I've talked about one regularity that is striking, which is the smallness of theta-parameter which, in a precise sense, is related to complex part of quark masses. Another one you mentioned is the top quark. Although it was surprising historically because we had experience with lighter things, in retrospect it seems to be the most natural mass because it corresponds to Yukawa coupling that is of order unity. One modest improvement that could happen and is attractive to entertain is the idea that the bottom quark also has a natural order unity coupling constant. That could happen if you have – as in SUSY – two Higgs fields with the ratio of vacuum expectation values, the so-called $\tan(\beta)$, large, and that is something that has consequences and may very well be explored experimentally at the LHC. Another regularity that we see is that the different families seem to regularly go (at least, monotonically) up in mass. We might expect that the heaviest one is the easiest to analyze, because it is not subject to smaller effects, and just as there is the unification of couplings if you assume simple symmetry breaking patterns, you can relate masses of different particles. And as we know, there is a relationship that you can derive from unification between the tau mass and the bottom quark mass that works quite well, and if you extend it with this $\tan(\beta)$ prediction, you also get the relation to the top quark mass. So, there are bits and pieces that we can explain and chip away at, but at present it is nice to have a fall-back position that some of them are random and maybe even the light ones have an anthropic component.

H. Georgi Next I would like to recognize our chair who has some comments.

D. Gross This is an example of the discussion that is not a question, but resonates on Frank's (Wilczek) statement that particle physicists learned that symmetries are the secret of Nature and much of our success is exploring symmetries. But we have also learned that there are no global symmetries in Nature, and that in fact the symmetries we use are all expressions of redundancy. Introducing degrees of freedom to explain particles we observe and the forces, and then saying that we can make arbitrary rotations in the space of these degrees of freedom or changes of coordinates ... gauge symmetries are just expressions of redundant degrees of freedom, as for general coordinate invariance, which are actually dual to each other. So, why is that? Why do we need somehow, it seems, so much redundancy in our description of Nature? Can we eliminate it as in the S-matrix on-shell program. Is it necessary, useful? Or should we extend it, look for more redundancy? The symmetry ... I mean, may be that the principle you have learned in a hundred of years that the secret of Nature is not symmetry, but redundancy?

W. Zurek I was struck by your ambition to eliminate dichotomy between the laws and initial conditions. And it seems to me that one example where this happens is a view of the process of quantum measurement. Each time one measures, one ends up restarting the wave function with a new initial value, whatever came out of the measurement. And an attempt to understand measurements in terms of decoherence point to the ambition that you have expressed. But I am curious if you have other examples that you have had in mind where such a unification of laws and initial conditions could be.

F. Wilczek Well, the examples I gave are much more down to Earth. The things that used to be thought of as initial conditions like the abundance of the elements and the existence of different kinds of chemical materials were once initial conditions. Now we understand that they can be derived from fundamental principles. In a broad sense, also, modern cosmology derives from just a few hypotheses — it can be formulated that way. It seems – how should I say – that there are so few principles to unite that it does not seem out of a question that in a little more time we'll zip up the whole package.

J. Hartle You've mentioned the wave function of the Universe, and was it unique, and also the distinction between initial conditions and dynamics. Some current theories like the no-boundary proposal do provide some sort of unification between the dynamics and the state, because the same action that determines the dynamics also determines the state. But beyond that, to ask a question of whether there is a unique state presupposes the whole framework of quantum mechanics, which seems sort of redundant for the Universe. Why do we have a principle of superposition when the Universe only has one quantum state? So would you be willing to add to your questions the question of whether the present framework of quantum mechanics which we use in quantum cosmology even, will actually extend (or will need to be modified to extend) to the quantum mechanics of the entire Universe?

F. Wilczek Well, I certainly would not want to censor any questions. Maybe, you could question. I think this also relates to David's (Gross) comments. There is a tension always between the Occam's razor principle, between positivism of trying to get a description in terms of direct observations and not make hypotheses, which sounds like a great idea, except that history often suggests the opposite, that to understand the world we live in we have to imagine a much bigger world. I think that positivism has historically been fruitful and important when there is crisis, and something has to be given up. At the moment there are perhaps some conceptual crises in black hole physics, may be – I'd think that is not so clear – but really what I see is not a crisis but many loose ends, so at least for the moment, I think, the more fruitful strategy is like more symmetry, looking for bigger structures.

H. Georgi This is now the opposite of jeopardy: you have to state it in the form of something that is not a question... Misha (Shifman).

M. Shifman I have a very brief remark, rather technical: I want to do more justice to the Standard Model. Frank (Wilczek) mentioned that hypercharges are totally more or less random in the Standard Model and are determined from phenomenology only. That is not true because if you arbitrarily assign hypercharges you will get inconsistencies in the form of quantum anomalies in the Standard Model, so much of the assignment can be decided on the basis of the Standard Model alone.

F. Wilczek You are quite right.

H. Georgi Let us go on to other contributions.

S. Das Sarma I want to come down from philosophical questions to some factual issues. What will happen to the Standard Model if LHC sees no SUSY or even Higgs within this energy constraint? I mean, how much of what you discussed will change completely?

H. Georgi This is open to the floor but also we will be discussing this explicitly after the coffee break.

F. Wilczek I think it is premature to be so pessimistic. I do not even want to think about it.

G. 't Hooft I want to make a remark to Misha Shifman's observation. Today the Standard Model and also all its extensions to unified theories such as SO(10) have in common that they have a whole bunch of freely adjustable parameters, as Frank has said, and there is also the assumption that the gravitational force can be added to the theory as a sort of afterthought. But as I will explain more tomorrow, I think that there is a further constraint if you include gravity, and that is that gravity should describe consistently black holes without the black hole information paradox. And that would require the cancellation of more anomalies in the theory which in turn give you further constraints on the Standard Model interactions. I am not totally sure of this, it is something that has to be discussed further, but this would indicate that there are many more anomaly matching conditions, and that would then constrain interaction strengths, which would be very important if true. The problem is that the theory so far does not address at all most of the hierarchy problems, so we are not there yet, but the idea that all Standard Model interactions should be fixed from some mathematical principle is an intriguing prospect in future theories.

W. Phillips It was brought up, I think, on the first day that quantum opticians or atomic physicists have a very different view of symmetry breaking than do condensed matter physicists. Just to summarize what the difference is, a quantum optician will say that the thing that is called symmetry breaking in condensed matter physics is simply the result of quantum measurement, whereas from the point of view of a quantum optician it seems that condensed matter physicists invoke some sort of magic to produce spontaneously broken symmetry. But in fact once we talk to each other we find we don't really disagree with each other about any of these. But when I listen to a particle physicist talking about the spontaneously broken symmetry I get the feeling, perhaps just the feeling of the inadequacy of a poor quantum optician listening to all of this great stuff, that there is something more fundamental going on, than just a quantum measurement, because, after all, who performs the measurement if we are talking about the entire Universe? So what I am wondering is do in fact particle physicists and cosmologists see spontaneously broken symmetry as being something really fundamentally different from the way quantum opticians do as being just a result of a quantum measurement? Or can it all be worked out in the same framework so that we are all happy with the way we think about this? I think the answer is yes, and spontaneous symmetry breaking is interesting dynamics and that is one of the things that we really care about. But perhaps other people should answer as well.

F. Wilczek I am not sure I agree with Bill's (Phillips) characterization of – what was it? – the agreement between what he says quantum opticians think of it and condensed matter people and we think. When we talk about spontaneously broken symmetry, we are talking about something that happens when you have a very large number of particles. It is not just a question of measurement. We would talk about, say, a ferromagnetic transition being spontaneously broken symmetry, if you get very long-range correlations. Those long-range correlations would exist even if we did not measure it. And, on the other hand, if you go above the Neel temperature there is no such thing. I think that is much more analogous to the spontaneously broken symmetry that high-energy people are talking about.

H. Georgi I think that the way the people in particle physics view spontaneous symmetry breaking is quite similar to what is done in condensed matter physics. I think also in this case people in particle physics need an infinite volume universe in order to discuss the spontaneous symmetry breaking. The main difference, however, between particle physics and people doing condensed matter is that if you have a ferromagnet you may have many samples in which you can see the magnetization pointing in different directions, so you have many experimental validations of spontaneous symmetry breaking, while in the Universe we have only one experimental realization and we cannot change that for the moment.

G. Veneziano I wanted to come back to David's (Gross) observation about local symmetries and they being related to redundancy. I think the real physical gauge-invariant statement is that there are in Nature massless spinless particles and this is the way we can describe them through gauge theory. Of course, we can fix the gauge and then we talk about the gauge symmetry, it does not matter. The physics is determined by the fact that there are massless spin-one and spin-two particles in Nature. And here I would like to make an advertisement for string theory. String theory naturally gives these objects, automatically gives massless states which, by the way, are classically forbidden. So it is very important: quantum mechanics is essential in string theory to provide the *raison d'être* for the fundamental interactions. If I can add a question to Frank (Wilczek) I would like to know his opinion about quantum gravity. He seems to be happy enough that classical gravity unifies with quantum gauge theories near the grand unification scale, but is he happy to keep gravity classical? I am sure he is not. So, what does he plan to do about it?

H. Georgi Can we wait and discuss that in the session devoted to this subject? We have bunch of questions over there.

I. Antoniadis I would like first to make a comment on a Savas' (Dimopoulos) comment. I think that experimental indication that we have until now – not only previous experiments, but also from LHC – is that the Standard Model works very well not only qualitatively, but quantitatively, and up to now there is no hint at something beyond the Standard Model, because the Standard Model works very well. For me this is also an indirect hint that at least the Higgs sector should be discovered. I do not know about SUSY, but everything what is part of the Standard Model seems to describe the Nature very well. I would like to go back and play a bit of devil's advocate. Frank (Wilczek), you presented very nicely the idea of unification, and how it fits to the general picture, which I share, but sometimes I had the following feeling: I think unification is not so innocent as it seems, because it has a big price to pay. The price is a desert, a big constraint on what particle physics may be for 15 orders of magnitude above the energy that we explore. I think this is difficult to swallow. The other thing is that one can think what is the experimental indication for SUSY. As you presented, in the Standard Model there is no such indication, unification does not work. In order to make it work one has to postulate a new symmetry, but one has not observed it so far.

H. Georgi Let Frank (Wilczek) answer at the coffee break and just collect more comments.

F. Wilczek May I take 30 seconds? This calculation is very robust to certain kinds of perturbations. In particular you can add particles pretty freely as long as they form complete $SU(5)$ multiplets. So it does not require an absolute desert. It can be populated with singlets. I would call it quite a modest constraint.

F. Englert It's about the question of the relation of symmetry breaking in condensed matter and in elementary particle physics. I think the difference is essentially related for most of the case (superconductivity is an exception, if you wish) to the question raised by David (Gross), namely in the elementary particle physics really there is no symmetry breaking that we really know as a symmetry breaking, because it is local. In the ferromagnetism there is a global symmetry, that is a big difference. And in particular, for instance, just to point out the importance of the difference: if some theory of gravity can be quantized, just existence of Minkowski space is a symmetry breaking in the sense of elementary particle physics, because the fluctuation of is not zero. So I think there is a big difference actually, which is related to the redundancy which appears to be necessary in elementary particle physics. And I see very difficultly how this redundancy can be taken out, in particular, for instance, because of the topological effects like the Aharonov-Bohm effect. So I think this redundancy is fundamental, this is up to now the way we understand it. And that makes the symmetry breaking in elementary particle physics something completely different and actually not correctly named.

C. Bunster I would just like to point out that redundancy is intimately connected with locality, and that is why it appears to be unavoidable so far.

A. Polyakov I will resist the temptation to make philosophical statements. But I will just very briefly say a thing about this redundancy which David started to discuss. I think that the situation is dynamical, actually. We know in two dimensional models that when you have a gauge symmetry it can be realized in two ways: either integrating over the gauge fields reduces the number of degrees of freedom, or it adds the gauge bosons. In some cases, like, for example, $O(3)$ non-linear sigma model, you can introduce gauge fields, integrate it out and then you simply reduce the number of degrees of freedom. On the other hand, this gauge field can become dynamical, and then it's just a different phase. So, it could be both: it could be redundancy, or it could be ... I don't know what is the opposite of redundancy.

X-G. Wen It is hard to say where this gauge symmetry comes from in the Universe, but in condensed matter systems the emergence of gauge theory is really directly related to the long-range entanglement. So whenever you have a long-range entangled states and want to use local fields to describe long-range entanglement, that will be a very difficult. But somehow you manage to use the redundant local field to describe such entanglement. So once you have a long-range entanglement the collective fluctuations, or collective modes in the entangled state are naturally so-called gauge boson whose masslessness are really topologically protected, which means no local perturbation can give gauge boson a mass. So from this long-range entanglement point of view the masslessness of the vector boson seems natural just like masslessness of the Goldstone boson for symmetry breaking case. So long-range entanglement can be the reason for masslessness of the gauge bosons.

Prepared comments

G. Dvali: Classicalization as UV-Completion

One of the fundamental goals of high-energy physics is to understand the nature of UV-completion of the Standard Model. In the standard (Wilsonian) paradigm of UV-completion, the new high-energy physics comes in form of weakly-coupled quantum particles that become relevant degrees of freedom at scales shorter than the weak interaction length, approximately $10^{-16} - 10^{-17}$ cm. For example, a low scale supersymmetry is a typical representative of such a Wilsonian UV-completion. As it is well-known, the above energy frontier is currently being probed by the LHC experiments. Recently, a concept of non-Wilsonian self-UV-completion was introduced[a,b] in which no new weakly-coupled physics is required above certain cut-off energy scale M_* with the corresponding cutoff length $L_* \equiv \hbar/M_*$. In these theories low energy degrees of freedom (e.g. gravitons) naively become strongly interacting at distances shorter than L_* where perturbative expansion in $E L_*$ breaks down. However, in contrast to the Wilsonian picture, this breakdown does not imply the need for any new weakly-coupled physics that must be integrated-in at distances less than L_*. Instead, the theory cures itself in the following way. The would-be strongly-coupled particles get replaced by collective weakly-coupled degrees of freedom with an effective interaction strength suppressed by powers of $1/(L_*E)$. These collective degrees of freedom represent many-particle states of large wavelength. By large wave-length we mean the wavelengths that exceed L_*. This characteristic length is set by an energy-dependent scale which we denote by $r_*(E)$. The necessary property is the increase of $r_*(E)$ with E, so that $r_*(E) \gg L_*$ for $E \gg M_*$. As a result, the UV-theory when described in terms of collective degrees of freedom is weaker and weaker coupled and probes larger and larger distances with increasing E. This phenomenon describes the essence of what was termed as the non-Wilsonian self-completion.[a,b,c]

The intrinsic feature of the classicalization phenomenon, which makes this picture phenomenologically distinct from Wilsonian completion, is the efficient growth of the cross section at trans-cutoff energies as some positive power of EL_*,

$$\sigma \propto (EL_*)^\alpha, \quad \alpha > 0. \tag{1}$$

This growth can be understood as the result of creation of states composed of many soft quanta, which behave more and more classically at high-energies. These are so-called *classicalons*.

---

[a]G. Dvali and C. Gomez, arXiv:1005.3497 [hep-th].

[b]G. Dvali, G. F. Giudice, C. Gomez, A. Kehagias, *JHEP* **2011** (2011) 108; arXiv:1010.1415 [hep-ph].

[c]G. Dvali, C. Gomez, A. Kehagias, arXiv:1103.5963 [hep-th], *JHEP* **1111** (2011) 070.

Let us briefly describe their essence.

In the classical limit ($\hbar = 0$) classicalons represent static (usually singular) solutions of the classical equations of motion of characteristic radius $r_*(E)$ and energy (mass) E. These parameters appear as integration constants that can take arbitrary values. A well-known example of such solutions is the celebrated Schwarzschild black hole in classical general relativity. However, this geometric picture is only valid in an idealized classical limit. In reality nature is quantum and $\hbar$ is non-zero. A picture of quantum constituency of black holes and other classicalons has been suggested[c]. According to this picture, these objects represent self-sustained bound-states of many bosons of characteristic wave-length $\lambda = r_*(E)$ and occupation number,

$$N(E) = E\, r_*(E)/\hbar. \tag{2}$$

Due to their large wave-length and derivative coupling, these bosons interact extremely weakly, with the effective coupling constant

$$\alpha_{eff} = 1/N(E). \tag{3}$$

Thus, physics of classicalons in general, and black holes in particular, is a weakly-coupled large-N physics in 't Hooft's sense.[d] This property emerges as the result of maximal packing. The classicalons represent maximally packed states per given wave-length. The maximized occupation number density results into the oversimplification of the system and effectively converts it into a system with a single characteristics, N. In this way, classicalization replaces a would-be strongly coupled physics of few hard quanta at energy $E \gg M_*$ by an extremely weakly-coupled physics in which the same energy is distributed among many soft quanta of wavelength $r_* \gg L_*$. Thus, classicalization provides a quantum foundation for ideas about the existence of shortest length scale in quantum gravity.[e]

This picture defines the quantum N-portrait of black holes and other classicalons. The reason for the efficient production rate of these objects in high energy particle collisions is the exponential degeneracy of micro-states that over-compensates the usual exponential suppression of many-particle states. At high energies the cross-section grows as

$$\sigma \sim (\hbar\, N(E)/E)^2. \tag{4}$$

This cross section for large $N(E)$ can be interpreted as a geometric cross section

$$\sigma \sim r_*(E)^2. \tag{5}$$

[d]Gerard 't Hooft, *Nucl. Phys. B* **72** (1974) 461.

[e]D. Amati, M. Ciafaloni, G. Veneziano, *Phys. Lett. B* **197**, 81 (1987) and *Int. J. Mod. Phys. A* **3**, 1615-1661 (1988); D. J. Gross, P. F. Mende, *Nucl. Phys. B* **303** (1988) 407; G. 't Hooft, *Phys. Lett. B* **198**, 61-63 (1987).

The relations (2), (3), (4) and (5) describe the essence of classicalizing theories. Unlike in ordinary Wilsonian case, the high energy behavior of these theories , instead of probing short distances, in reality probes large distance physics, due to the fact that the high energy scattering is dominated by production of states with large occupation number of very soft quanta. Thus, deep-UV quantum behavior of classicalizing theories can be understood in terms of the classical IR dynamics of the same theory! For example, the behavior of deep-UV cross section can be derived by finding out the E-dependence of the $r_*(E)$ radius of a static source of mass E. The radius r_* can be defined as the shortest distance for which the linearized approximation is valid. This property simplifies the predictive power of the theory for high-energies, since the dependence of σ on center of mass energy E can be approximately read-off by solving the linearized classical equations of motion for a source of the same energy E.

However, we need to be extremely careful not to be mislead by this simplification. In order to understand properly the classicalon dynamics we must continuously monitor the information obtained in an idealized classical limit ($\hbar = 0$) by translating it into the language of the underlying quantum portrait. Without this guideline, the (semi) classical picture alone can lead us to wrong conclusions. This becomes obvious, ones we identify the correct classical and semi-classical limits. These limits correspond to taking

$$E \to \infty, \quad L_* \to 0 \quad r_* = \text{fixed} . \tag{6}$$

In addition, we may take $\hbar \neq 0$ or $\hbar = 0$ depending whether we want to be in semi-classical or classical treatment. For example, most of (if not all) the previous semi-classical analysis of the black hole physics is performed in this limit.

The quantum N portrait shows that none of these limits are correct approximations. This becomes very clear by realizing that in both limits, irrespective whether we keep $\hbar$ finite or zero, the occupation number of quanta becomes infinite, $N \to \infty$. This immediately tells us that all the subtleties of $1/N$ expansion become hidden. The typical example of the invalidity of this approximation is the application of the semi-classical limit for micro-black holes that can be produced at LHC. It is obvious that in reality these black holes correspond to the quantum states with $N \sim 1$. Thus, to apply to their properties the semi-classical limit ($N = \infty$) gives invalid predictions, such as thermality and democracy of their decay products. In reality, the micro-black hole if accessible at LHC will behave simply as unstable quantum particles.

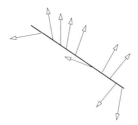

Fig. 1.   Non-Abelian strings have extra gapless modes (rotational). They also support confined monopoles (two-string junctions).

M. Shifman: SUSY-Based Methods at Strong Coupling

Among all of the theories discussed in high energy physics quantum chromodynamics has a special status: it is firmly established to be *the* theory of matter. Its Lagrangian is fully known. The complexity of QCD is due to the fact that the majority of physical phenomena that it describes occur at strong coupling.

The advent of QCD was a revolution in our understanding of field theory. We learned that what you see in the Lagrangian is not necessarily what can be detected; that the vacuum structure can be complex and the vacuum need not be unique; that small harmonic oscillations near vacuum are insufficient to explain strong dynamics; that there is a variety of diverse regimes (or phases) that can be implemented in YM field theory, such as Coulomb, Higgs, confinement, oblique confinement, conformal and more; that the dual Meissner effect presents a mechanism leading to confinement of quarks (color); that Wilson's operator product expansion (OPE) can be adjusted to perfectly fit QCD.

But let me first briefly outline the timeline spanning 40 years of continuous advances in QCD, putting emphasis on the milestone developments.

⋆ 1972-73: QCD/YM at strong coupling is established as *the* theory of matter (Gell-Mann; Gross, Wilczek, and Politzer);

⋆ 1974: The 't Hooft large-N limit revealed implying a 'stringy' picture behind QCD;

⋆ 1974-75: Dual Meissner effect conjecture suggested as a confinement mechanism (Nambu; Mandelstam; 't Hooft);

⋆ 1975: Instantons and monopoles discovered as important examples of nonperturbative effects (Polyakov et al.; Polyakov; 't Hooft);

⋆ 1980s: OPE-based (condensate, or SVZ) methods developed and applied to a large variety of problems of practical importance; advent of SUSY-based exact methods due to a generalized holomorphy of supersymmetric YM (e.g. the NSVZ β function);

⋆ 1994: Seiberg demonstrated "electromagnetic duality" in supersymmetric QCD (shortly after it was elevated to string theory); Seiberg and Witten found a breakthrough solution of $\mathcal{N} = 2$ SYM, analytically demonstrating, for the first time ever, the dual Meissner effect as the confinement mechanism;

⋆ 1998: The advent of string-gauge duality, or, holographic description (Maldacena; Klebanov and Polyakov; Witten);

⋆ 2000s: Planar equivalence between $\mathcal{N} = 1$ SYM and some nonsupersymmetric YM theories established (Armoni, Shifman, and Veneziano); non-Abelian strings (with orientational moduli on the world sheet constructed.

Comments before supersymmetry:

In the early days of (nonperturbative) QCD the advances were associated mainly with instanton studies and the proliferation of OPE-based methods in a large number of problems of practical interest, in particular, in heavy quark physics. Later progress was driven by a combination of large-N expansions and supersymmetry.

In gauge theories one of the most profound discoveries, which affected the way of thinking in the entire HEP community, was that of 't Hooft, who pointed out that $1/N$ is a (hidden) expansion parameter in QCD and Yang-Mills theories in general, corresponding to the expansion in topologies of the underlying Feynman graphs. Thus, there emerged a natural – albeit qualitative – correspondence between QCD and a string-like picture, with $g_s \sim 1/N$, where g_s is the string coupling constant. Moreover, the domain wall tension in super-Yang-Mills was shown to scale as $N \sim 1/g_s$, which served as a basis for identification of these domain walls with the string theory branes.

Instanton studies revealed a complex structure of the QCD vacuum and demonstrated the existence of a novel hidden parameter, the vacuum angle θ. This, in turn, paved the way to the emergence of various modifications of axions, which, in addition to the CP problem could comprise dark matter.

Comments on advent of supersymmetry:

The simplest super-YM theory is $\mathcal{N} = 1$ super-gluodynamics. Its fermion component is gluino, a Weyl field λ_j^i in the adjoint representation of $\mathrm{SU}(N)$. If we replace λ_j^i by a Dirac fermion $\psi^{[ij]}$ in the two-index (anti)symmetric representation we obtain a nonsupersymmetric ("orientifold") theory equivalent to its supersymmetric parent in the 't Hooft limit. At $N = 3$ the orientifold theory corresponding to $\mathcal{N} = 1$ super-Yang-Mills is just one-flavor QCD.

The pivotal (I could say, revolutionary) Seiberg-Witten solution paved the way to insightful analytic studies of the confinement mechanism. Confining strings following from the original solution are in fact Abelian: at distances of their formation only gluons from the Cartan subgroup play a role. In search of non-Abelian strings (i.e. such that all $N^2 - 1$ gluons play equal roles in string formation) people came across a few surprises. The search was successful (Auzzi et al.; Hanany and Tong; Shifman and Yung). Dynamics on the string world sheet in this case is nontrivial (e.g. $CP(N-1)$ models emerge, see Fig. 1). Being strongly coupled, the effective world-sheet models are solvable because they are two-dimensional. And – the most remarkable feature – the solution of these two-dimensional models provide us with unambiguous (exact) information on aspects of four-dimensional bulk theories. This remarkable phenomenon is now known as $2D-4D$ correspondence.

A few times in the last two decades many believed that the existing theory was at the verge of, if not the exact solution of QCD, at the very least, its solution in the planar limit (i.e. $N \to \infty$ with the fixed 't Hooft coupling). These high expectations never came true. The range of natural phenomena that are described by QCD is so diverse and complex, that such a universal solution seems unlikely (to me). And yet, our understanding of QCD and of more general non-Abelian gauge theories at strong coupling continues to grow.

N. Nekrasov: Quantization(s) and Gauge Theory

In recent years the new type of quantum symmetries have gradually emerged. Unlike the conventional realization of a symmetry algebra in the quantum model, where the symmetry is visible at the classical level, or, perhaps, becomes visible once some degrees of freedom are treated semi-classically, the generators of novel symmetries do not need to correspond to some infinitesimal transformations of the classical phase space.

One instance where such symmetries seem to be relevant is the landscape of supersymmetric vacua of theories with two dimensional super-Poincare invariance.[f] One finds that the vacua of a wide range of gauge theories with various matter content are in correspondence with the stationary states of quantum integrable systems such as spin chains. The latter possess a noncommutative algebra of symmetries which act irreducibly in the space of states and can be used to build the spectrum of the model, the so-called Bethe states. When translated to the gauge theory language this algebra acts not only on the vacua of a given theory, it relates the vacuum states of different gauge theories. The generators of the algebra are non-local from the quantum field theory point of view. In some cases they correspond to the domain walls, D-branes, Lagrangian correspondences,[g] or some other non-perturbative configurations, such as monopoles.

Sometimes there are dualities relating these non-local symmetries of one theory to more conventional local symmetries of another theory, its (Lang-

---

[f]N. Nekrasov, S. Shatashvili, arXiv:0908.4052, arXiv:0901.4748, arXiv:0901.4744.
[g]H. Nakajima, arXiv:math/9912158; M. Varagnolo, arXiv:math/0005277.

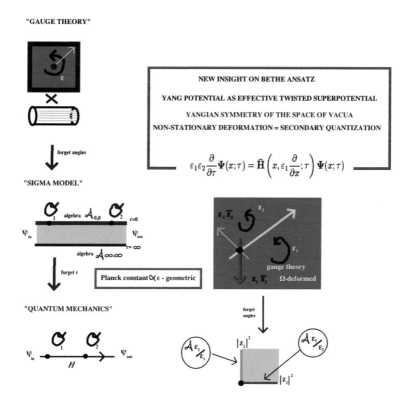

Fig. 1. Susy vacua, Bethe states, and $\varepsilon_1, \varepsilon_2$-quantizations.

lands) dual.[h]

Relation between the supersymmetric gauge theories and quantum integrable systems suggests an interesting extension of the usual paradigm of the quantum mechanics with the fixed Hilbert space of states, the algebra of Hermitian observables, and, sometimes, the quasi classical limit. The more general setup realizes separately the algebra of observables, and separately the representations of the algebra. This is done in the complexified approach to quantum mechanics,[i] where the path integral is replaced by the functional integral of a sigma model on the complexified phase space. A particular class of boundary conditions, the so-called canonical coisotropic branes[j] are used. Remarkably, these boundary conditions are the projections of the smooth boundary-less configuration in the four dimensional gauge theory.[k]

---

[h]E. Witten, arXiv:0905.4795.
[i]E. Witten, arXiv:1009.6032.
[j]A. Kapustin, D. Orlov, arXiv:hep-th/0109098.
[k]N. Nekrasov, E. Witten, arXiv:1002.0888; N. Nekrasov, A. Rosly, S. Shatashvili, arXiv:1103.3919.

The gauge theory has the two-parametric Ω-deformation[1] which leads to a two-parametric kind of deformation quantization. In the simplest cases the Ω-deformed gauge theory provides the realization of the so-called modular double of quantum group.[m]

The general picture is still missing. It is desirable to have a better understanding of these non-local symmetries, whether they act in any reasonable way on the full Hilbert space of the theory, whether they survive coupling to supergravity, the embedding to string theory. The two-parameter quantization leads to the refinement of the topological string theory[n] whose proper definition leads to the concept of the K-theoretic Donaldson-Thomas theory which seems to allow the computation of the Witten index of M-theory.[o]

I. Klebanov: Counting Degrees of Freedom in Conformal Field Theory

A deep problem in quantum field theory is how to define a measure of the number of degrees of freedom that decreases along any renormalization group (RG) trajectory and is stationary at fixed points. In two-dimensional QFT, an elegant solution to this problem was found by Zamolodchikov who used the two-point functions of the stress-energy tensor to define a "C-function" with the desired properties. At RG fixed points the C-function coincides with the Weyl anomaly coefficient c:

$$g^{ab}\langle T_{ab}\rangle = -\frac{c}{12}R .\tag{1}$$

In four-dimensional conformal field theory (CFT) there are two Weyl anomaly coefficients, a and c,

$$g^{ab}\langle T_{ab}\rangle = -\frac{a}{16\pi^2}\left(R_{abcd}^2 - 4R_{ab}^2 + R^2\right) + \frac{c}{16\pi^2}C_{abcd}^2 .\tag{2}$$

Cardy has conjectured that the a-coefficient decreases under RG flow and is stationary at fixed points. Since on the four-sphere the Weyl tensor C_{abcd} vanishes, the a coefficient can be calculated from the Euclidean path integral on S^4.

There have been many tests of the a-theorem for supersymmetric field theories where a is determined by the $U(1)_R$ charges. The Intriligator-Wecht principle of a-maximization, which states that at superconformal fixed points the R-symmetry locally maximizes a, has passed many consistency checks that rely both on field theoretic methods and on the AdS/CFT correspondence. For large N superconformal gauge theories dual to type

[1]N. Nekrasov, arXiv:hep-th/0206161.
[m]L. Faddeev, arXiv:math/9912078.
[n]A. Iqbal, C.Kozcaz, C.Vafa, arXiv:hep-th/0701156; M. Aganagic, M. Cheng, R. Dijkgraaf, D. Krefl, C. Vafa, arXiv:1105.0630.
[o]N. Nekrasov, A. Okounkov, talk by NN at the *Exact Methods in Gauge/String Theories*, PCTS, November 2011.

IIB string theory on $AdS_5 \times Y_5$, Y_5 being a Sasaki-Einstein space, a-maximization is equivalent to the statement that the Sasaki-Einstein metric on Y_5 is a volume minimizer within the set of all Sasakian metrics on this space. Recently, a proof of the a-theorem was constructed by Komargodski and Schwimmer.

Due to the abundance of fixed points in three-dimensional QFT and their relevance to observable phase transitions, it is of obvious interest to find a 3-d version of the 2-d c-theorem and of the 4-d a-theorem. Such a result would establish general restrictions on RG flows. However, because of the absence of anomalies in odd dimensions, the trace of the stress-energy tensor simply vanishes at the fixed points. A proposal for a good measure of the number of degrees of freedom in a 3-d Euclidean CFT is[p]

$$F = -\log Z_{S^3} \,, \tag{3}$$

where Z_{S^3} is the Euclidean path integral of the CFT conformally mapped to S^3. In a general QFT, F has divergent parts corresponding to renormalization of the terms $\int d^3 \sqrt{g}$ and $\int d^3 x \sqrt{g} R$ in the effective action. However, these counterterms are not Weyl invariant; therefore, in any CFT the finite part of F is well-defined. If a theory flows from a UV CFT to an IR CFT, then the conjectured F-theorem states that $F_{IR} < F_{UV}$. The conjecture also states that F is stationary at RG fixed points and is positive for unitary theories. The stationarity of F at fixed points follows from the vanishing of all one-point functions in a CFT on a three-sphere. This argument also implies that F is constant along lines of fixed points.

In recent literature the F-theorem has been subjected to a variety of tests. Some of them are possible because, as established by Kapustin, Willett and Yaakov, and by Jafferis, the path integral for $\mathcal{N} \geq 2$ supersymmetric CFT on a three-sphere localizes to a finite dimensional integral. For a small number of colors such integrals may be calculated analytically or numerically, while for large N using expansion around saddle points. For $\mathcal{N} = 2$ supersymmetric CFT, where one generally finds anomalous dimensions of operators, they are fixed by the Jafferis F-maximization principle which states that the superconformal R-symmetry locally maximizes F. This is analogous to the Intriligator-Wecht a-maximization in four dimensions. For the large N CFT's that have dual AdS_4 descriptions, these results have been compared successfully with the predictions of the AdS/CFT correspondence. Both the $N^{3/2}$ scaling of F and its normalization have been shown to agree. This matching of F seems analogous to the well-established AdS/CFT Weyl anomaly matching for even-dimensional CFT.

[p]D. Jafferis, The exact superconformal R-symmetry extremizes Z, arxiv.org/1012.3210; D. Jafferis, I.R. Klebanov, S. Pufu and B. Safdi, Towards the F-theorem: $\mathcal{N} = 2$ theories on the three-sphere, *JHEP* **1106** (2011) 102, arxiv.org/1103.1181.

Some tests of the F-theorem have also been carried out for non-supersymmetric theories.[q] They include RG flows produced by slightly relevant operators. It is also possible to show that F decreases for flows produced by double-trace operators in large N theories. Finally, is also not hard to calculate F for free CFT's. For example for a free conformally coupled scalar, $F = \frac{\log 2}{8} - \frac{3\zeta(3)}{16\pi^2} \approx 0.0638$.

N. Seiberg: Some Thoughts About Quantum Field Theory

We have heard here beautiful talks, which demonstrated the effectiveness of quantum field theory both in condensed matter physics and in particle physics.

So, instead of using my five minutes to talk about additional recent developments in the field, I thought I should take a broader perspective and ask the following question. Given the spectacular progress during the past several decades, where is the field heading?

In particular, should we look for a new formulation of quantum field theory? I would like to present a number of arguments suggesting that this is indeed the case:

- Physicists love Lagrangians. A Lagrangian is the standard starting point of a classical field theory, which we later quantize. (More technically, it is a UV free field theory deformed by a relevant operator.) But several arguments suggest that this might not be the only good starting point.

  - First, some interesting quantum field theories are intrinsically quantum mechanical. They have no semi-classical limit and hence they have no Lorentz invariant Lagrangian. This happens even in free field theories in $2k + 2$ dimensions when self-dual fields are present. More interesting examples involve isolated fixed points of the renormalization group, which do not arise as the IR limit of a free theory. The most prominent example of this is the celebrated six-dimensional $(2, 0)$ theory, but there are also many other examples. Without a natural Lagrangian, we do not have a clear formulation of these theories. So how should we define them?

  - A lot of the developments during the past decades circled around dualities. These are situations in which several different Lagrangians describe the same physics. I will not review the subject here, but will simply mention the two classes of examples. In gauge/gauge duality different Lagrangians with different gauge

---

[q] I.R. Klebanov, S. Pufu and B. Safdi, *F*-theorem without supersymmetry, *JHEP* **1110** (2011) 038, arxiv.org/1105.4598.

symmetries lead to the same physics. In gauge/gravity duality a local quantum field theory is equivalent to a gravitational theory in another spacetime. If a given theory does not have a unique Lagrangian, what is its fundamental description?

- Next, many exact solutions of quantum field theory rely on algebraic methods – they do not use the Lagrangian formulation (even when it is known). Examples are integrable field theories and two-dimensional conformal field theories. In these cases a Lagrangian exists, but it is not being used to solve the theory.

- Finally, we have learned that scattering amplitudes of quantum field theory exhibit magical properties. These properties allow an effective computation of the scattering matrix, which is much more powerful than the more traditional Feynman diagrams techniques. This magic clearly points to another viewpoint of the scattering matrix and perhaps even of the whole theory.

In summary, some theories do not have a Lagrangian, some theories have more than one Lagrangian, and in some situations a Lagrangian exists but it is not useful in solving the theory.

- Despite a lot of effort, a complete, satisfactory and rigorous formulation of quantum field theory is still lacking. This is one of the reasons that mathematicians have such hard time understanding quantum field theory. I believe that a new formulation, based on a new perspective, would be more palatable to mathematicians and more physically insightful.

- Finally, I would like to point out that string theory has stimulated us to explore certain non-gravitational theories, which are clearly not local quantum field theories. Examples of such theories, which go beyond our standard framework, are field theories on a non-commutative spacetime and little string theory. In these situations the notion of local operators and in particular the energy momentum tensor is confusing. Furthermore, these theories exhibit nontrivial mixing between physics in the UV and physics in the IR. This makes the standard Wilsonian renormalization group setup puzzling.

These points do not prove that we need to reformulate the theory. But I feel that a better perspective of the theory would be helpful.

I should emphasize that I have no doubt that our existing formulation of the theory and our current understanding are correct. They simply do not do justice to some of the marvelous properties of this beautiful subject.

Hopefully, by the next Solvay meeting, a new formulation of the theory will be known and it will uncover deep new insights.

Discussion

H.Georgi Thank you. The floor is now open for discussion.

L. Randall I have a question for Nati Seiberg. Could it be that Lagrangians are some sort of effective theory and, do we actually know that the underlying theory will be unique in anyway? It seems that the examples you gave are very different and it could be that Lagrangians are good effective theories for many different underlying formulations.

N. Seiberg I tried to emphasize that Lagrangians are useful.

L. Randall The question is more about the uniqueness of The Underlying Formulation.

N. Seiberg If you say "The", then it should be unique.

L. Randall One could imagine that there are many.

N. Seiberg For a given theory, there might be many. There could be several formulations, but one may be more fundamental than the others.

H. Georgi Does somebody want to comment on that?

H. Ooguri Actually I have a question for Gia Dvali. I want a clarification of the notion of UV completeness. It appears to be different from what I am used to. I have in mind the example of Einstein gravity, that he mentioned. It is true that for scattering in Einstein gravity at large energies compared to the Planck scale, it is possible to describe the process semi-classically. However, if the energy is near the threshold, namely at the Planck scale, then Einstein gravity alone cannot describe what will happen, as higher loop effects will be highly relevant. So the outcome of the scattering near the threshold (near the Planck scale) depends critically on what the underlying theory is. So what does he mean by UV completion?

G. Dvali Thank you. This is an excellent comment. You are correct. Einstein's gravity is unique in that sense. Let's compare with another effective theory, like Fermi's theory. Fermi theory does not prove that there are new degrees of freedom around the cut-off. In Einstein gravity you know for sure that the only propagating degrees of freedom cannot be Einstein's graviton. Why? Because we also control the large distance sector of the theory. We know that there are big black holes. We can make a thought experiment with a big black hole, wait that it evaporates down to the Planck size at which point it becomes quantum. At that point inevitably, if you have a theory that "classicalizes" in the UV, the same theory tells you that there must be new propagating degrees of freedom that emerge at the quantum scale, in this case propagating micro-black holes. Now, of course, what we can conclude is that if we do deep high-energy scattering in gravity, since we are dealing with long wavelengths, the configurations we are producing are classical, up to corrections from these contributions around the Planck scale which are presumably exponentially small. These corrections may be very important, for instance for unitarity.

H. Georgi Please finish this up so that we can go on with other things...

E. Witten I have a question for Seiberg. A new formulation of QFT will be general, or will it just apply to a good class of theories?

N. Seiberg I don't know. It might be valid everywhere, but perhaps more effective in some situations than others. I wish I knew what it is...

E. Verlinde I usually think of QFT in the way Wilson told us, with short distance degrees of freedom inducing effective operators, like order or disorder operators, and making them into fields following the RG. Then you conclude that every QFT is an effective QFT. Then there simply has to be a UV cut-off. Every QFT is like that. Do we need to talk about really fully-defined QFT, while it is against the philosophy of QFT? QFT is totally logical when you think the Wilsonian way, it is just an effective description of Nature.

D. Gross About effective theories. The problem is that nobody has a starting point for some of these more fundamental theories. Now there is a problem with thinking of QFT as purely effective QFT. It is great when discussing large scale phenomena. But it you want a picture to arbitrarily short distances, then you to worry about irrelevant operators, and there is an infinite number of effective operators, and that is Terra Incognita, totally undefined. Nathan (Seiberg) gave a good example of that which is non-commutative gauge theories, which in a sense is a controllable irrelevant deformation of this theory in the UV. I think he was thinking of discussing QFT without limits, and then we have now an effective QFT approach.

M. Shifman I have a comment regarding Seiberg's talk and David Gross's remark. If your task is to build a theory of everything, you are in a much harder position than if you try to describe the surrounding world at our scale. The comment to Nathan Seiberg is the following. In the 60's physicists made an attempt of abandoning the Lagrangian approach and developing the S-matrix approach. There were some successes, and some important spin-offs, like string theory. But the theory of the surrounding world did not go along this way. Moreover the useful Lagrangians went far beyond this matrix theory approach. For instance it would be very difficult to guess from the S-matrix elements the existence of the magnetic monopole. You need some redundancy in the description of the degrees of freedom in order to say this.

A. Polyakov This discussion reminds me of a famous remark by Landau: "Lagrange should be buried with all honors that he deserves!" Anyway, I think that field theories should be formulated as an operator product expansion. As a matter of fact, the equations of motion that we use should be read from the right to the left. You take for instance $\Box \phi = \phi^3$, you should understand it as of OPE of ϕ^3 which contains $\Box \phi$. It's like Hebrew. That's for me a good way of formulating field theory. Another comment is more concrete and new. In conformal field theories, you have a series of non-unitary CFT. There are very useful in condensed matter, to describe polymers for instance. The question is whether we can interpret the negative probabilities that arise in such conformal field theories? My statement is that yes, you can. There should be interpreted as theories on a metastable vacuum.

N. Seiberg I want to address Erik Verlinde's comment. As David Gross said, theories with a cut-off have an infinite number of parameters. So they may be approximately useful, but cannot be a complete description. Furthermore, in some of the examples I presented, they are not even useful. One example is the $(2, 0)$ theory for which we do not have a useful cut-off description. More interesting is the example of non-commutative field theories. These theories do not work in the standard Wilsonian approach. It is correct to say that they are controllable irrelevant deformations of the theory, but it is also misleading as there is no separation of UV and IR. This tells us that there is something we are not understanding.

E. Silverstein This is a comment/question for Nathan Seiberg. Something I would like to understand better is the role of magnetic matter. Some very interesting fixed points have both electric and magnetic matter. It would be nice to have an effective theory which would make clear which of magnetic or electric flavour make sense in the UV and how they contribute to renormalization. Do you have ideas about that?

N. Seiberg I am not sure which example you have in mind, but whenever we try to describe simultaneously electric and magnetic matters, we are led to some confusion. As for theories at fixed points, these are just CFTs, they should not be thought of theories with interacting electrons and monopoles.

L. Randall If I may, I think that the question of Eva Silverstein is whether, in the new formulation you refer to, if it exists, should you suspect that it would include both electric and magnetic degrees of freedom?

N. Arkani-Hamed An obvious comment. The usual structure of QFT and Lagrangians, etc, is designed to make locality and unitarity manifest, and this introduces all the gauge freedom we have been talking about. But we have learned that there are many other properties of theories that we would like to make manifest but which are make invisible in the usual formulations, like dualities, and hidden symmetries. These are reasons to suspect that other formulations should exist. A good example is for instance the 6D (2,0) for which we do not have a Lagrangian formulation, but it explains many dualities. What is the moral of this theory for string theory is the question I would like to throw out.

H. Georgi We should stop at this point. These are great questions and we might want to come back to them on Saturday. Now let's go on to one of the most important scientific projects of all times, the LHC.

Prepared comments

G. Giudice: What have we Learned from Higgs Searches at the LHC?

Next December the LHC experimental collaborations will release their results on the Higgs searches based on about 5 fb^{-1} of data. This should be enough for exclusion at 95% CL in the mass range between about 130 and 600 GeV or for discovery in a more limited range. The LHC integrated luminosity is expected to at least double next year. This means that by the end of 2012 discovery will be possible for any Higgs mass. So the Higgs saga, which started 47 years ago, will possibly be over less than two months from now and will almost certainly be over within a year. But the question is: Are we going to learn anything fundamental from these searches?

I am raising this question because some physicists consider the Higgs boson as the *chronicle of a discovery foretold*. Certainly it is the most "expected" discovery that the LHC can make, but nevertheless the discovery (or exclusion) of the Higgs boson will tell us something fundamental about nature. One of the key questions addressed by the LHC is: What is the force responsible for electroweak symmetry breaking, the fifth force of nature?

As beautifully reviewed by Frank Wilczek, one of the greatest successes of particle physics was to reduce all forces to a gauge principle. This gauge principle guarantees elegance, robustness against deformations of the theory, and predictability. If the Higgs boson is discovered, there will be some evidence for a non-gauge fundamental force. Indeed, gauge interactions are the only renormalizable interactions until you introduce scalars. In the Standard Model a self-scalar interaction is used to trigger electroweak breaking and Yukawa interactions are used to communicate the information to fermions. It is exactly this departure from the pure gauge paradigm, this non-gauge nature of the fifth force, that is causing all the troubles that we have with the Higgs mechanism, from a theoretical point of view:

(1) Lack of uniqueness of the Higgs sector (many deformations of the minimal Higgs sector are equally possible; we can easily change the number of fields and the allowed or forbidden interactions).

(2) Lack of predictivity for the particle masses and mixings (each particle mass or mixing corresponds to a free coupling and there is no symmetry relating them).

(3) Naturalness problem (again, lack of symmetry forbidding certain terms in the scalar potential).

The question of whether the Higgs boson is a truly fundamental particle, or a composite, or something hiding some notion of a symmetry principle is a key issue. This question can be answered experimentally by precise measurements of the Higgs couplings and by investigation of the WW scattering channel, even in the presence of the Higgs boson. So the first point I want to make is that precision measurements in Higgs physics are not a moot issue, but an interesting research program that addresses a fundamental question regarding the nature of the fifth force.

Point number two. The measurement of the Higgs mass, which will be carried out at the LHC, can give us crucial information about underlying theories. Let me illustrate this point with only two examples, but I could make others.

The measurement of the Higgs mass is a unique probe of the scale of super-symmetry breaking, even if this scale lies well above what is predicted by naturalness or what can be directly tested in collider experiments. Indeed, supersymmetry determines the matching condition for the Higgs quartic coupling at the supersymmetry-breaking scale, thus allowing us to calculate the Higgs mass. It is then possible to predict the Higgs mass as a function of the supersymmetry breaking scale, which can vary from the weak scale all the way up to the Planck scale[a]. This has been done in two cases: 1) when the Standard Model degrees of freedom survive below the cutoff scale, thus retaining the motivation of supersymmetry as an ingredient of the funda-mental theory, but disregarding naturalness; 2) when a particular subset of supersymmetric particles are kept light (the so-called Split Supersymme-try), thus potentially explaining dark matter, gauge-coupling unification, but not naturalness. Once the Higgs mass is experimentally determined, we will be able to tell if the measured value is consistent with the idea that supersymmetry is broken at some super-heavy scale. Or we could experi-mentally exclude such possibilities, under some reasonable assumptions. It is remarkable how the Higgs mass measurement can teach us something even about energy scales well beyond what is probed directly in collider searches.

The second example concerns the stability of the electroweak vacuum. If the Higgs mass is found to be in the range between about 115 and 130 GeV, than the Standard Model vacuum is potentially unstable. Either new physics intervenes at high scales modifying the Higgs potential, or thermal and quantum fluctuations in the early universe did not destabilize the vacuum. In either case, we will learn something fundamental about physics well beyond the reach of present colliders.

Moral of my presentation: The Higgs story is an old one, but it can still offer many interesting lessons for fundamental physics. The discovery or exclusion of the Higgs boson is not likely to be the last word on our under-standing of electroweak symmetry breaking, but rather just the beginning.

S. Dimopoulos: Naturalness and the Higgs

The Higgs boson is the last missing ingredient of the Standard Model (SM)

---

[a]G.F. Giudice and A. Strumia, *Nucl. Phys. B* **858** (2012) 63 [arXiv:1108.6077].

- Where is the Higgs?

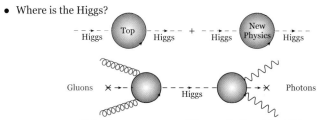

Naturalness may suppress Higgs production at the LHC

- Where is Supersymmetry?

	1980s	1990s	2000s	2010s
Squark mass limit	50 GeV	100 GeV	300 GeV	1000 GeV

If all colored sparticles degenerate and lightest sparticle stable
Missing Energy at the LHC

Lighter Sparticles if:

Lightest Sparticle Unstable: No Missing Energy at LHC
Only 3rd sparticle family light: Stop Mass ~300 GeV

Fig. 1.   Top: The Higgs and Naturalness. Bottom: The Missing Superpartner Problem.

and it has been the focus of collider searches in the past 30 years. It explains electroweak (EW) symmetry breaking and the particle masses but at the price of a huge fine-tuning needed to keep the scale of weak interactions far below the Planck scale. Any solution to this hierarchy problem requires new degrees of freedom at the EW scale to cancel the quadratic divergences of the Higgs as shown in the top of Fig. 1. These particles also affect the main Higgs production mechanism at the Large Hadron Collider (LHC) via two-gluon fusion, as well as the Higgs decay to two photons (see Fig. 1) which is one of the channels that are easily accessible at the LHC for a light Higgs. So, if there are new degrees of freedom at the weak scale that tame the SM quadratic divergences, they will alter the Higgs properties measured at the LHC.[a] Therefore a natural Higgs at the LHC behaves differently from the SM Higgs. Conversely, the more the Higgs is SM-like the more tuned or unnatural the theory will be. This fact highlights the importance of accurately measuring the Higgs couplings to SM particles.

S. Dimopoulos: The Missing Superpartner Problem

The confirmation of the supersymmetric prediction of gauge coupling unification in the early '90s by LEP and SLC gave a tremendous boost to

---

[a] A. Arvanitaki and G. Villadoro, A non standard model Higgs at the LHC as a sign of naturalness, *JHEP* **1202** (2012) 144 [arXiv:1112.4835 [hep-ph]].

naturalness and weak-scale SUSY, and led to the expectation of a major discovery by LEP2. Unfortunately this was not to be. By the late nineties the absence of the Higgs showed that simple versions of supersymmetric theories were tuned at the $\leq 10\%$ level, as shown in the bottom portion of Fig. 1. This "little hierarchy problem" left open the possibility that naturalness was realized in more involved theories. The LHC has significantly changed the prospects for naturalness. The direct superpartner limits, which are now beginning to exceed ~ 1.5 TeV for all colored sparticles except the stop and sbottom, pose a serious problem for naturalness. Nevertheless, there are still ways out for naturalness. One involves models of SUSY where the amount of missing energy in LHC events of sparticle production is reduced. This automatically relaxes the bounds on SUSY as missing energy is one the main SUSY search strategies at the LHC. One of the most appealing class of theories where naturalness constraints from the LHC are relaxed are the theories of "Natural Supersymmetry".[b] In these models only the third family particles – which contribute the most to the Higgs mass – are light, whereas the superpartners of the first two generations can be several TeV without spoiling naturalness. The stop quark in such theories can be as light as the top mass and can be accessible in the near future.

N. Arkani-Hamed: Naturalness And Its Discontents

The central drama of the electroweak scale has to do with questions of naturalness: does the hierarchy problem have a natural solution, or is there fine-tuning for electroweak symmetry breaking, as we perhaps also see for the cosmological constant? Supersymmetry has long been the most attractive candidate for natural physics beyond the standard model, given the spectacular successes of gauge-coupling unification and dark matter. And yet there has been a growing sense of unease, since fully natural supersymmetric theories could have already been discovered at LEP. A Higgs heavier than 115 GeV, together with the absence of signals for stop squarks so far at the LHC, are starting to significantly squeeze the idea of naturalness for low-energy supersymmetry. But if this idea is off the mark, why did it seem we were on the right track, with gauge-coupling and unification and dark matter?

Back in 2004, together with Dimopoulos I proposed the idea of "split supersymmetry", which resolves this tension. In this model, the scalars of supersymmetry are much heavier than the fermions, which can be lighter since they carry an additional R symmetry, and are motivated to be close to the TeV scale by dark matter considerations. The simplest version of this

---

[b]S. Dimopoulos and G. F. Giudice, *Naturalness constraints in supersymmetric theories with nonuniversal soft terms*, *Phys. Lett. B* **357** (1995) 573 [hep-ph/9507282].

idea has parametrically a one-loop splitting between scalars and gauginos, with scalars between 100's to 1000's of TeV. This predicts a higgs mass between $\sim 120 - 140$ GeV. If the gauginos are discovered, a smoking gun for this idea would be moderate (millimeter range) displacement for gluino decays, which would give sharp evidence for heavy scalars and a part in 10^6 tuning for electroweak symmetry breaking.

In this decade, the LHC should experimentally settle the question of whether the weak scale is fine-tuned or not. After three decades of speculation, it is time for the truth.

Discussion

L. Randall I would like to make a few general points. One is that I think it is true that a lot of people are upset at not finding SUSY this year. But as Nima (Arkani Hamed) has correctly pointed out, I believe, the possibility of a strongly interacting sector was at the heavy end already. If there was anything that was lighter it would be the weakly interacting particle that has not been searched for yet. There is another general lesson which is that, given the constraint of naturalness, it does seems to point to a spectrum that is split. I think that this is the most optimistic scenario. In fact the spectrum seems to be heavy just given the experimental constraints. The optimistic scenario is that there will be some additional light particles in the theory as well. One last thing is the possibility of strong interactions. Of course we tend to keep under the lamp of theories we can solve. But I think we have learned a bit about strongly interacting theories recently, both from warped extra dimensions and from dualities. In both cases, it might be these are right by themselves, but also they point to something general. I think that they do point to one thing, which is that if a composite model works, it is because of some mixing between composite and fundamental particles, and both strong and weak interactions would play a rôle. This would be compatible with having both a light and heavy spectrum. It does seem that if we do find things at the LHC, it would be the tail-end of a more larger theory, that will not be completely exposed there. I think it is a very nice scenario.

I. Antoniadis There is a part of the Higgs doublet that has already been discovered, which is the longitudinal component of the gauge fields. If the Higgs were composite, we should have already some indication of this and this is not seen. I think that the experimental situation points rather to a linear realization of the symmetry, and what is missing is a physical particle of spin zero, which is the Higgs scalar.

H. Georgi OK, I don't agree with that but we go on with Frank Wilczek.

F. Wilczek This is a comment for the previous session. To me the greatest triumph of QFT of recent years, and which has not yet been mentioned, is the calculation of the hadron spectrum based on lattice gauge theory, which really uses the full, non-perturbative content of the theory and gets the right answer. So in the spirit of "shut up and calculate" we should also mention that.

I. Klebanov I have a question for Savas Dimopoulos. The cancellation you talk about, has it been checked in simple models? It really looks like different diagrams. When you talk of cancellation, it means that the coefficients are the same in the diagrams. Does it happen for the stop?

S. Dimopoulos Yes, the cancellation is being checked. If you ignore A-terms, then this is not a cancellation. The two diagrams both contribute to the beta-function and both fermions and scalars give you the same sign for the beta-function. Now for moderate to large A-terms, there is a sign reversion and they cancel. So what we are doing is plotting the parameter space of A *vs* stop mass for which this cancellation would happen and reduce the Higgs production cross section by a factor of two.

H. Georgi I'd like to follow a bit about what Lisa Randal said about composite and fundamental particles. I would not say that the evidence is overwhelming from electroweak precision tests that we cannot have strong EW symmetry breaking. The problem is really flavor and what one does with the global symmetries that are present in the effective theory of flavor, including all the gauge interactions. Gauge symmetries are not very efficient in breaking global symmetries, so you really have to have a theory that does everything at once. Start with a theory with no global symmetries, and hope that somehow the SM will emerge. That's way beyond our abilities at the moment. So I don't think that it means that it is impossible, but it is certainly something we cannot quite imagine and this is perhaps the reason why so many people are thinking of fundamental particles.

N. Arkani-Hamed If we do not find the Higgs with the expected cross-sections, there are two obvious things that could be going on. It may be that the Higgs mixes with some gauge singlet particle. That would change many things. One could easily reducing the production cross-sections by a factor of a few. One could also imagine exotic decay modes. I personally think that we will see a "vanilla" Higgs.

G. Guidice If I may reply to Antoniadis, I don't agree with that point of view. I think that, even if we discover the Higgs, it does not mean that it is necessarily a fundamental particle. For instance, it could be some remnant of some other sector. The nice thing is that we can address this question experimentally, through precise measurements of Higgs couplings.

L. Randall I would like to follow on the problem of flavor because it is an underestimated problem of getting at the same time electroweak symmetry breaking and the right mixing in the quark and lepton sectors. One thing that happens in warped geometries is interesting. It is an example of a composite theory where flavor works. The composite degrees of freedom are in the IR, and they mix with fundamental degrees of freedom. It is because of this mixing that it works, so perhaps this is pointing to something more general.

H. Georgi Alright, let us thank all the speakers.

Session 6

Quantum Gravity and String Theory

Chair: *Joseph Polchinski*, University of California at Santa Barbara, USA
Rapporteurs: *Juan Maldacena*, Institute for Advanced Study, Princeton, USA and
Alan Guth, Massachusetts Institute of Technology, USA
Scientific secretaries: *Riccardo Argurio*, Université Libre de Bruxelles, Belgium
and *Ben Craps*, Vrije Universiteit Brussel, Belgium

Rapporteur talk by J. Maldacena: The Quantum Spacetime

Abstract

This is a very general overview of the quantum mechanics of spacetime, starting from some generalities and then moving on to more current developments in string theory.

1. Classical Spacetime Dynamics

Our current view of spacetime is based on the theory of general relativity, which states that spacetime is a dynamical object. It can support propagating "vibrations", or gravity waves.

General relativity had two surprising predictions: Black holes and the expanding universe. These predictions were so surprising that even Einstein had trouble with them. In fact, Einstein said to Lemaître (maybe here in Brussels): "Your math is correct, but your physics is abominable". I like this phrase because it is similar to what string theorists are sometimes told.

It is interesting that the evidence for these more surprising aspects of general relativity came earlier, and is stronger, than that for the more straightforward gravity waves.

2. Quantum Spacetime

Since nature is quantum mechanical, and spacetime is part of the dynamics, we should quantize spacetime. For example, you can think about the gravitational field of quantum superposition of a particle at two different locations.

Quantizing linearized gravity waves is as easy as quantizing the free electromagnetic field. It is just a collection of harmonic oscillators. But the theory is not free. Including interactions perturbatively, one finds the unique structure of General Relativity by postulating that gravitons interact with the energy of gravity waves in a self consistent fashion. In other words, one can derive general relativity by postulating relativistic massless spin two particles interacting in a self consistent way.

Also if we have a curved background we can quantize fields moving in it, including the quantization of the gravitational field.

3. Two Surprising Predictions

This approximate approach gives two surprising predictions:

1) Black holes emit Hawking radiation. They have a temperature and an entropy.
2) Inflation produces the primordial fluctuations.

Both change the classical behavior of the system in surprising and physically important ways.

Experiments have essentially confirmed the inflationary predictions. This can be viewed as a confirmation of quantum gravity. Though, only of the non-dynamical modes of the geometry which are modified by the scalar field (or inflationary clock). Seeing primordial, or inflation generated, gravity waves would be a more direct test that spacetime geometry should be quantized.

Again, quantum effects lead to surprises. We have quantum mechanics at the longest observable distances!. And they are crucial for understanding the universe. Without them we would have a uniform universe. With them we have an essentially unique quantum initial state, and the complexity of the world arises through the measurement process or decoherence. At least for the fluctuations.[a]

4. Quantization at Low Energies

Let us go back to quantizing spacetime. The effective coupling among is proportional to the square of the typical energy of the interacting gravitons. $g_{eff}^2 = E^2 G_N = E^2 M_{pl}^2$.[b] And, this is the size of the quantum gravity effects.

---

[a]Namely, we cannot currently predict the values of the constants of nature: number of gauge groups, ranks, gauge couplings, other couplings, etc. However, given these, the rest of the properties of the universe can be computed from the quantum state produced by inflation. Primordial fluctuations produce structure, which produces stars, etc...

[b]In units with $\hbar = c = 1$.

Gravity can be quantized as a low energy effective theory to any order in perturbation theory. We should introduce new parameters at each order in perturbation theory. This is similar to the Fermi theory of weak interactions. This works fine for low energies. It is OK during inflation, for example.

However, it fails completely when the energy is comparable to the Planck mass. Or when we require non perturbative precision. This is not just a problem of resuming the perturbation theory, but the fact that we have an infinite number of undetermined constants, which makes the theory ill defined.

5. UV Completion in Field Theory

In quantum field theory, such effective field theories have a UV completion. In the case of the low energy field theory describing a condensed matter problem, this completion can be a lattice model, or the Schroedinger equation. For the fermi theory of weak interactions, it is the electroweak theory.

Going to short distances we find the local degrees of freedom which are the "fundamental" description of the theory.

6. UV Completion in Gravity?

Could gravity be UV completed in a similar way?. We expect that the answer is no.

We start with a local classical lagrangian, so that one might expect a picture similar to the field theory one. But a big difficulty arises because we cannot devise a thought experiment that would allow us to explore short distances. If we collide high energy particles, we form black holes which get bigger as we increase the energy.

But the problem does not just appear when we go to the Planck scale, it also shows up at long distances in the form of information bounds, that are believed to hold for any quantum gravity theory. These bounds say that the total quantum information, or number of q-bits, we can store in a region of space is given by the area in Planck units. $S \leq \frac{\text{Area}}{l_p^2}$.[1] You might be familiar with the fact that the entanglement entropy in quantum field theory has a similar expression, with l_p replaced by the UV cutoff. In QFT we can have q-bits in the interior which are not entangled with the exterior, so that total entropy in a region can be bigger than the area. This is not so in gravity. For example, if we have a dilute gas of particles, the entropy is naively expected to grow like the volume, thus, if have a sufficiently big gas of particles we would naively violate the entropy bound. So, what is wrong with such a large sphere of gas? Well, when it is about to violate the bound, it collapses into a black hole!

7. Perturbative String Theory

String theory is sometimes presented as UV completion of gravity, but it is not a completion in the same sense. It is a theory that perturbatively constructs the S-matrix. For an introduction see refs. 2, 3.

We introduce a new length scale l_s, where new massive particles appear, and a dimensionless interaction constant g_s governing the quantum corrections. $G_N \sim g^2 l_s^2$.

These massive particles can be viewed as the oscillation modes of a string. There is a massless spin two particle, so we recover gravity at low energies. The amplitudes do not increase with energy and the quantum corrections are finite and calculable.

The simplest examples are ten dimensional and supersymmetric. It has no parameters. The coupling is the vacuum expectation value of some field.

8. Unification

In string theory we replace the classical notion of geometry by the new notion of stringy geometry, which is different at short distances. This stringy geometry has some very surprising features. For example, if we turn one of the dimensions into a circle of radius R, the physics is equivalent to that of a circle of radius $R' = l_s^2/R$. So, as we try to shrink a dimension, a new dimension grows to large size.

String theory also provides a unified description of matter and spacetime. Gravitons, gauge bosons, Higgs bosons, fermions, all come from the same string.

By going from ten to four dimensions on a compact six dimensional manifold we get gauge fields, chiral matter, and presumably the particle physics we see in nature.

9. Beyond Perturbation Theory?

There is a large amount of evidence that there is an exact theory whose approximation is perturbative string theory. This exact theory is usually also called "string theory", though it might not contain discernible strings, if we are at strong coupling.

At weak coupling, one can compute non-perturbative effects by considering D-branes. For example, the leading low energy correction to the scattering of gravitons in ten dimensions can be computed exactly.[4]

The very strong coupling behavior of one theory is believed to be dual to other string theories, or to an eleven dimensional theory. And all string theories are connected by such dualities. These strong/weak coupling dualities give rise to mathematical identities which are very non-trivial. These can be checked to be true, giving evidence for the dualities.[3]

10. Beyond Perturbation Theory

Many conceptually important problems in gravity seem to lie beyond perturbation theory:

1) Initial cosmological singularity, origin of big bang.
2) Graviton scattering at Planckian energies.
3) Describing black holes in a unitary fashion.

In fact, one gets confused at the very start. It is hard to define precise observables. In ordinary quantum mechanics, the position of a particle, or a spin projection, are well defined observables whose expectation values we can compute with arbitrary precision.

In gravity, nothing that can be measured by an observer living and measuring in a finite region seems to have this quality, since the observer has a finite number of degrees of freedom to store this information. Related to this, we can always have a quantum fluctuation of the metric where this region is completely absent.

In spacetimes with a simple asymptotic shape we can define precise observables. These include asymptotically flat space and asymptotically AdS space. Quantum fluctuations are suppressed at long distances and we can make precise measurements. Such as the measurement of the S-matrix in the flat space case. Knowing that they are well defined is nice. But, can we calculate them?

11. Non Perturbative Quantum Spacetimes

The existence of D-branes in string theory made it possible to discover some non-perturbative descriptions of some spacetimes. The examples include:

1) Matrix theory, which describes the S-matrix of some flat spacetimes.[5]

2) Gauge gravity duality, describing AdS spaces (and other spacetimes with a timelike boundary).[6]

In both cases we extract the spacetime physics by doing a computation in a well defined quantum mechanical system with no gravity. We will discuss this in more detail in the case of the gauge/gravity (or gauge/string) duality.

12. Hyperbolic Space

A few words about a central player in this story. This is hyperbolic space the simplest and first example of a negatively curved space. The two dimensional version has the metric $ds^2 = \frac{dr^2 + r^2 d\varphi^2}{(1-r^2)^2}$. It has a boundary at $r = 1$, and this boundary is at an infinite proper distance. It also has higher dimensional generalizations. Including time we get Anti-de-Sitter space. Thinking of the radial direction as time, we get de-Sitter (or ordinary inflation). We will discuss the Anti-de-Sitter case.

13. Quantum Hyperbolic Space

We can get Anti-de-Sitter from string theory via a compactification on a suitable internal space. We have $AdS_d \times M^{10-d}$, with M^{10-d} a compact manifold. The string theory defines its quantum geometry perturbatively.

The gauge/gravity duality says that the physics of this quantum space is the same as the physics of an ordinary quantum field theory on the boundary. There are various examples. The simplest one involves a gauge theory similar to quantum chromodynamics, but with more supersymmetries.

The spacetime becomes classical and described by Einstein gravity when the gauge theory has a large number of colors and it is strongly coupled.

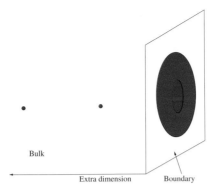

Fig. 1. In the boundary we have two excitations of different sizes. In the bulk they both correspond to particles of the same size but at different distances from the boundary.

14. Emergent Space

It is very easy to understand how the extra dimension emerges. Normally, in order to specify the state of an excitation we need to give its position. In a scale invariant theory, we also have to give its size. Thus, an excitation is a certain blob of some size. Blobs with different sizes are related by a scale transformation. Thus, we have an extra coordinate: the size of the blob.

In the interior, these blobs correspond to a particle of a fixed size that is at various positions along the extra direction. Changing the size of the boundary blob corresponds to changing the radial position of a particle in AdS. In Fig. 1 the red blob is smaller. It is described by a particle closer to the boundary. This picture is backed up by the conformal group representation theory.

15. Black Holes in AdS

Any state in the interior has a corresponding state on the boundary. What about a black hole in the interior. It corresponds to a thermal system on the boundary. Its entropy, which is equal to the area of the horizon in gravity, corresponds to the ordinary statistical entropy in the boundary theory. The long distance dynamics of the black hole is related to hydrodynamics on the boundary theory. In fact, one can get the ordinary Navier Stokes equation from Einstein's equation by looking at the long wavelength excitations of such black holes. This picture makes it possible to compute transport properties of strongly coupled theories (see talks by Horowitz and Sachdev[7,8]). A wave falling into the black hole corresponds to dissipation and thermalization on the boundary theory.

16. The Information Problem

Since the information problem was the main historical reason for thinking about the gauge/gravity duality, let me describe it in some more detail, since thinking about it will probably lead to important new insights.

In a nutshell, the problem is the following. You form a black hole from a pure state. The black hole evaporates in a thermal fashion. In the end you got thermal radiation and the information about the pure state was lost.[9]

This is incompatible with quantum mechanical evolution from the point of view of the outside observer.

17. The Information Problem

The perturbative gravity description seems to lose track of information. It seems lost to all orders in perturbation theory. This was most clearly seen in two dimensional models in the early 90's.[10]

My opinion, which is not shared by everyone, is that this is not a problem by itself, since one needs non-perturbative accuracy in order to really test whether information is lost or not. Thus, to check whether information is lost one needs a non-perturbative method for computing the evolution of the state.[11]

In conclusion, the Hawking argument does not show that information is lost, since it is not accurate enough. This can be understood without resorting to any duality. However, it does raise the important question of whether it is lost or not. The gauge gravity duality shows that information is not lost, since the boundary theory is unitary.

However, it does not give a clear bulk description for how information is recovered.

Also, one would like to get a description of the black hole interior from the point of view of the gauge theory.

18. Lessons

Let us mention some other lessons of this description of quantum spacetime.

- Spacetime is emergent. This means that spacetime is not fundamental, there isn't an operator in the theory which is the "space shape" operator, as there is a position operator in for a quantum mechanical particle. Note also that in string theory even the dimension of spacetime is an approximate concept. As we vary the parameters we can go from ten dimensional to eleven dimensions.

- Holographic bounds are obeyed, and are essential for making sense of the relationship.

- The boundary conditions at the AdS boundary specify the system, they specify the Lagrangian of the dual theory. Thus, in a sense, this is a realization of Mach's principle. The shape of space far away determines they physics of the system in the interior.

One might wonder whether any theory has a gravity dual, or which are the theories that have gravity duals. As we reduce the number of colors we might get a more and more strongly coupled theory in the bulk, one that deviates in a strong way from Einstein'st theory. If one is willing to accept such theories as possible states in quantum gravity, then one concludes that quantum gravity includes all quantum systems.

But, what are the particular systems that have a weakly coupled and Einstein gravity dual?. We have two necessary conditions, we need a large number of degrees of freedom, and strong interactions. How generic such theories are, is not understood. We just know a variety of examples. These include an example involving simple matrix harmonic oscillators with an anharmonic coupling term. This is an ordinary non-relativistic quantum mechanical system that has a gravity dual.

As in condensed matter theory one would like to classify all conformal field theories. This goal might be too ambitious, due to the large expected number of theories. Maybe the particular class with gravity duals with a macroscopic spacetime, could be classified. The greeks classified regular polyhedra. Now, we can certainly classify highly supersymmetric field theories. How should we think about the less supersymmetric ones...?

We saw that approaching the AdS boundary meant going to short distances in the boundary theory. But the AdS boundary is infinitely far away in the bulk. This implies that there is a UV/IR connection, which relates long distances in the gravity theory to short distances on the boundary theory.

There is close connection between the de-Sitter, inflationary computations, and the present anti de Sitter discussion. In fact, there is a simple analytic continuation that allows one to compute the inflationary perturbations from the anti-de-Sitter ones.

One can say that the scale invariance seen in cosmic fluctuations is related to the scale invariance of critical phenomena, that we encounter in conformal field theories. For inflating universes, at the moment this is just a hint, and not a complete and proper connection. But one would hope to be able to apply some of these ideas to cosmology.

In fact, the Euclidean field theory partition function is equal to the Hyperbolic space analog of the Hartle Hawking, or no boundary, wavefunction, originally proposed for the de Sitter case.

19. String Theory and the Real World

As we mentioned, string theory has four dimensional vacua that have features similar to those of nature: gauge fields, chiral matter, inflation, etc.[12] This I would call top down unification. In fact, there seem to be so many vacua.[c][13,14] that one with the right cosmological constant is very likely to exist.

In addition, we have seen a kind of different kind of unification. In fact, the physics that governs QCD is the same as the one governing spacetime. The precise QCD string is not known, but there is a lot of evidence that it should be part of what we now call "string theory". These strings have certainly been experimentally observed. Let me make a historical analogy. Newton observed that the force that

---

[c]More properly, we should call them long lived metastable states.

makes an apple fall is the same as the one moving the heavens. We can now say that the quantum spacetime, is the same "stuff" as the one that gives the apple its mass: the strong interactions.[d]

20. Some Unsolved Problems

I would now like to close with some of the big challenges that we are facing in our field.

One important problem is to describe spacetimes with cosmological, or big bang, singularities. To describe, means to be able to give probabilities for what comes out of the singularity. Related to this, is the description of the interior of black holes. Also vaguely connected to this is the measure problem in eternal inflation (see Guth's talk[15]).

It is also important to make some concrete prediction, that we could experimentally check for the spectrum of particles/inflation/susy/dark energy from string theory. The simple predictions of string theory involve high energies and are hard to check experimentally. Current experiments involve low energy physics and are hard to predict from the theory.

Hopefully we will have more surprises and unexpected predictions that will be simpler to check. Maybe predictions about questions that we now think are unrelated to quantum spacetime!.

Acknowledgments

This work was supported in part by U.S. Department of Energy grant #DE-FG02-90ER40542.

References

1. J. D. Bekenstein, Generalized second law of thermodynamics in black hole physics, *Phys. Rev. D* **9**, 3292 (1974). G. 't Hooft, Dimensional reduction in quantum gravity, gr-qc/9310026. L. Susskind, The World as a hologram, *J. Math. Phys.* **36**, 6377 (1995) [hep-th/9409089].
2. M. B. Green, J. H. Schwarz and E. Witten, *Superstring Theory. Vol. 1, Vol. 2*, Cambridge Monographs On Mathematical Physics (Cambridge, Uk: Univ. Pr.) (1987) 469 P.

---

[d]Let us make a comment of how to think about the application of the known examples of the gauge gravity duality to the study of quantum gravity, strong interactions and condensed matter. Let us suppose that, instead of physicists trying to understand nature, we were doctors trying to cure cancer. Now in order to learn about cancer, one might want to study fruit flies. Obviously, fruit flies and humans are different. In the same way the field theories that have gravity duals are not the same as the ones in nature. Nor the gravity theories that have known field theory duals are the ones in nature. (At least the currently understood ones.) They are analogous to the fruit fly, they have components or aspects which we believe are the same as those in natural systems, but are easier to understand, and to "take apart". They are "model organisms" for the study of quantum gravity or strongly interacting quantum systems.

3. J. Polchinski, *String Theory. Vol. 1, Vol. 2*, (Cambridge, UK: Univ. Pr.) (1998) 531 p.

4. M. B. Green, S. D. Miller, J. G. Russo and P. Vanhove, Eisenstein series for higher-rank groups and string theory amplitudes, arXiv:1004.0163 [hep-th].

5. T. Banks, W. Fischler, S. H. Shenker and L. Susskind, M theory as a matrix model: A conjecture, *Phys. Rev. D* **55**, 5112 (1997) [hep-th/9610043].

6. J. M. Maldacena, The large N limit of superconformal field theories and supergravity, *Adv. Theor. Math. Phys.* **2**, 231 (1998) [*Int. J. Theor. Phys.* **38**, 1113 (1999)] [hep-th/9711200]. E. Witten, Anti-de Sitter space and holography, *Adv. Theor. Math. Phys.* **2**, 253 (1998) [hep-th/9802150]. S. S. Gubser, I. R. Klebanov and A. M. Polyakov, Gauge theory correlators from noncritical string theory, *Phys. Lett. B* **428**, 105 (1998) [hep-th/9802109].

7. G. Horowitz, Contribution to these proceedings.

8. S. Sachdev, Contributions to these proceedings.

9. S. W. Hawking, Breakdown of predictability in gravitational collapse, *Phys. Rev. D* **14**, 2460 (1976).

10. A. Strominger, *Les Houches Lectures on Black Holes*, hep-th/9501071.

11. J. M. Maldacena, Eternal black holes in anti-de Sitter, *JHEP* **0304**, 021 (2003) [hep-th/0106112].

12. L. Ibañez and A. Uranga, *String Theory and Particle Physics: An Introduction to String Phenomenology*, (Cambrigde U. Press).

13. R. Bousso and J. Polchinski, Quantization of four form fluxes and dynamical neutralization of the cosmological constant, *JHEP* **0006**, 006 (2000) [hep-th/0004134].

14. S. Kachru, R. Kallosh, A. D. Linde and S. P. Trivedi, De Sitter vacua in string theory, *Phys. Rev. D* **68**, 046005 (2003) [hep-th/0301240].

15. A. Guth, Contribution to these proceedings.

Discussion

J. Polchinski Thank you Juan for finishing ahead of time. We will now have a discussion and I remind you that this is not a question and answer period, but like the famous Solvay meetings from the past, this is a chance for everybody in the room to discuss these issues.

W. Phillips I want to ask about the thing you called the information problem, the idea that you start with a pure state, that collapses into a black hole, then you let it evaporate, and it seems like you have lost something. A quantum optician would say, big deal! I start with an atom in a pure state, I let it radiate into the vacuum, I trace over the radiation and I no longer have a pure state. Is there something more than that going on? It is like the difference between an open system and a closed system. Of course if I have an open system I do not expect it to have a unitary evolution.

J. Maldacena The big difference is that in the case of quantum optics we know what the unitary description is, we know that there is a Schrödinger equation which is unitary. In the case of black holes, the bulk evolution would suggest that we do not have a unitary evolution, and raises the question of whether there is or there is not a unitary evolution. People suggested seriously that quantum gravity as viewed by an outside observer, would not be unitary. This is the information problem.

W. Phillips What I am wondering is that if you assume it is unitary, then is the problem simply solved by saying this is not a closed quantum system?

J. Maldacena The point is that you can assume it is unitary, and it might satisfy you if you just assume it, but we are in the business of describing quantum space-time, so we would like to show it from the theory. We have string theory, it is a quantum description, we would like to show it is unitary.

J. Polchinski Let me also just give an answer and then move on. The other distinctive difference between a black hole and a lump of coal, say, or a collection of atoms, is that there is a horizon. Part of the quantum state has fallen behind the horizon and, unlike a lump of coal where it can diffuse out later, it can never get out. It is the presence of the horizon that really makes this a problem, a tension between locality, causality and quantum mechanics.

A. Sen We have seen some impressive tests for AdS/CFT at the planar level. My question is, to what extent has it been tested beyond the planar level, to see some genuine quantum gravity effects like, say, virtual gravitons propagating in the loop.

J. Polchinski Anyone can answer!

J. Maldacena There are certainly finite N effects coming from D-branes and so on.

A. Sen Can we really see a graviton loop from the gauge theory analysis?

J. Maldacena You can certainly calculate such things from the gauge theory, but not directly compare them to gravity. I should mention that in some examples of AdS_3, you can calculate the one-loop partition function in Anti-de Sitter space and it agrees with the expected properties in the gauge theory.

S. Sachdev You mention that any quantum field theory has at least a strongly coupled gravity dual, but then subsequently you said we only classify CFTs that have gravity duals. So, is there a tension here, and is there any example of a CFT that does not have a gravity dual?

J. Maldacena If you define the gravity duals sufficiently generally, I would say any theory has one, because it is defined by the CFT. An example of field theories that do not have a weakly coupled gravity dual could be the Ising model. That will not have a weakly coupled gravity or even string dual. If you take any weakly coupled theory, if it has a large number of colors, it might have a string dual, but not one that would be described by Einstein's theory. In order to have a description by Einstein's theory, you need strong coupling for example.

V. Mukhanov I want to go back to this information paradox, because people discussed it for already ten years, and I think nothing changed since then. Some people think that there is a paradox, some people think that there is no paradox. Now, on your transparencies, I have seen that you wrote without question mark, that these things can be checked by making perturbations in field theory. Is there lost information or not? The question is, why it was not done until now, what is the obstacle and when can it be done?

J. Maldacena As I mentioned, these perturbative checks were most clear in two dimensions. Taking into account backreaction, you can do an expansion in the coupling, and check that information is lost, in that approximation. I think that answers part of your question. How to describe how information is preserved from the bulk point of view is not understood.

V. Mukhanov Before you will resolve the problem of the singularity within the black hole, you will not be able to do anything, non-perturbatively even.

S. Das Sarma My comment is on the same point. I of course know very little about the subject, but it seems to me that to conclude that there is information loss based on perturbative theory, even to all orders, does not make much sense. There are many examples, also trivial examples, where perturbation theory at every order gives you something, but the non-perturbative result is completely different. For example, you may have poles that get converted to branch cuts, you may have no poles become poles. I can give many examples like that in low-energy field theory that I work with. Why is this such a problem, why are people worried, if this is a perturbative result only? We just have to discover the right theory.

J. Maldacena The point is that this is a laboratory for understanding how quantum gravity works, because this perturbative feature is, as Polchinski just said, due to the presence of the horizon. Somehow the non-perturbative description, if you use the duality, manages to give you the unitary description. Understanding in more detail how this works from the bulk point of view would be helpful, as it is helpful in other systems to understand this transition between one limit and another limit. This is an example where the long time limit and the classical limit are not commuting with each other.

D. Gross The reason it is a problem is that if information were lost in any corner, one consequence would be the modification of quantum mechanics, which as we have discussed is awfully hard to modify. A second point, as Hawking suggested, is that if real black holes were to eat up information, then virtual black holes might, and quantum mechanics might be fundamentally wrong once gravity is brought into account. So this is, though most people think it is no longer, a fundamental challenge to quantum mechanics, in fact the only interesting one in a hundred years.

S. Wadia There is a simple way of understanding information loss in *AdS*/CFT correspondence, because the long wavelength description of ripples on the horizon can be described by fluid dynamics with viscosity. If you have viscosity, we know in fluid dynamics there is information loss, but nobody bothers about it because you understand why there is information loss, namely you coarse grain the microscopic degrees of freedom. I do not think there is a big mystery here, except that recovery of information, even in fluid dynamics, is not a trivial question. Having said that, I would like to ask one of these questions that you mentioned, that you need a non-perturbative calculation actually to understand the recovery of information. Can you comment on that?

J. Maldacena You can set up some problem where you do a calculation, and you try to find the difference between a theory that loses information and a theory where you preserve information. A simple example is when you send a particle into a black hole, and then you see whether the state, after a long time, has changed or not, from the thermal state. You can see that it is perfectly consistent with unitary evolution, that this two-point function decreases and decreases until it gets very tiny, of a size which is non-perturbative in the effective coupling, which in this context is a size of order e^{-S} where S is the entropy of the black hole. So you need that kind of precision in order to be able to tell whether information is lost or not.

I. Antoniadis I would like to ask a different question. Maldacena discussed two aspects of string theory, one mostly as a quantum theory of gravity, where using *AdS*/CFT one could have a way to describe non-perturbative aspects of the theory, and the other is a possible manifestation of string theory to describe the real world. Does it make sense to ask the question of which is more fundamental than the other? In other words, suppose that string theory is realized in Nature, and the LHC or the next accelerator finds strings. Take the simplified case in which the string scale is different from the Planck scale, so that we do not need to go to the Planck scale.

J. Maldacena I think string theory has already been discovered. We already have strings in Nature, they have been seen experimentally, they have been used in describing the results of the LHC. In that sense they have already been seen. And if we see them again at the LHC, then we will see them again.

I. Antoniadis I think this is similar to the question of how you view a theory. We had a discussion on the second day, how for instance gauge theories are seen by condensed matter physicists, and how they are seen by particle physicists. So I think there is a difference if the theory is fundamental or not.

J. Maldacena There is the question of whether string theory describes quantum gravity or not, and I think the only way to answer this question is to get to the Planck scale, at least this is the only way I can see. If we see strings at the LHC, we would not be testing quantum gravity, they could be strings similar to the strings of QCD. Even if you form black holes, and we could say we have already formed some kind of black holes at RHIC, and maybe we could form something that is better approximated by a black hole at the LHC for example, but that would not necessarily test string theory as a theory of 4-dimensional quantum gravity, it might be that the graviton is still not described by a string. It is logically possible, though I am not sure.

G. 't Hooft As Maldacena clearly described, string theory is very good at generating scattering matrix amplitudes, but it does not work so well off mass shell. This to me is very reminiscent of a situation we had in the 60s, when there were big attempts trying to generate all interactions among the elementary particles, that we now know as the Standard Model, by starting from the scattering matrix and then trying to figure out conditions that it has to obey. The program never materialized because people did not understand the local structure of the theory, which now in the case of the Standard Model we understand very well. I think there is some similarity here, that string theory is somehow missing a local description of what is going on. (I see Gross shaking his head.)

J. Maldacena I tried to emphasize that we will not get this, and that this is a virtue rather than a defect.

M. Douglas Coming back to AdS/CFT, what are the prospects for a microscopic explanation and derivation of the duality? For example, since the early days, it has been thought that AdS/CFT has a lot in common with the renormalization group, and that the extra radial direction is the renormalization group scale. Can we hope to really make that precise? Of course there have been numerous difficulties in really making it precise, but then, if we could, would not that answer any of these other questions, such as the origin of the space-time, the question of which field theories have gravity duals, and the like?

E. Silverstein I will try to make this as a statement rather than a question. Maldacena briefly discussed the possibility of analytically continuing the AdS/CFT dictionary to de Sitter space, but it is worth emphasizing the vast difference that has to do with the causal structure of de Sitter space, and the horizons that exist in the cosmological version. Also the fact that de Sitter space, from various points of view, seems likely to decay ultimately into a decelerating FRW phase. I think that an equally valid starting point for holography in de Sitter is to work within a causal patch of the space-time, discussing what a given observer can actually see operationally, and to also take into account the fact that de Sitter seems to decay into a late time decelerating phase. That is part of the system that we are really trying to formulate in quantum gravity.

Prepared comments

J. Hartle: Local Observation in Eternal Inflation

In the outline for his talk Prof. Maldacena raised the issue of whether the no-boundary theory of the universe's quantum state predicts enough inflation and whether it predicts eternal inflation. As he put it —"great for small deformations, completely wrong for the constant modes". It does predict enough inflation and I will devote a few minutes to explaining how.

The no-boundary wave function is a theory of the universe's quantum state. The wave function is not a state on a spacelike surface *in* some background spacetime. It is not an initial condition. Rather it predicts probabilities *for* spacetimes. In particular it predicts probabilities for the possible four-dimensional classical histories of geometry and matter fields that the universe may exhibit. It predicts probabilities both for backgrounds and for the fluctuations on backgrounds.

We do not observe four-dimensional histories of spacetime. Our observations are limited to a small part of our past light cone located somewhere on a surface of the observed density and Hubble constant in a vast universe. To get probabilities for observation we have to sum probabilities for histories over what we *don't* observe. It's like summing over unobserved spins in a scattering experiment. In particular, in any one spacetime we have to sum over the unknown location of our light cone on the constant density surface. Assuming that we are rare on this surface, that sum weights the probabilities for a space time by the volume of that surface. It is by the volume weighted probabilities for our observations that the theory is tested.

$$p(\mathcal{O}) \propto \int_{\mathcal{O}} \delta h_{ij} \delta \phi |\Psi[h_{ij}, \phi]|^2 V(h_{ij})$$

Volume weighting is not a choice, but an inevitable consequence of calculating probabilities for observation. It's just quantum mechanics.

The NBWF does and excellent job of predicting small sub-horizon fluctuations. That is because it is the cosmological analog of the ground state constructed from a Euclidean functional integral. Fluctuations begin in their ground state in a given background, and that, as is well known, is consistent with observations. The question is rather about the superhorizon structure.

By themselves the NBWF probabilities favor histories with a small number of efolds and little reheated matter. That is the puzzle. But we are interested in the probability for the number of efolds *in our particular spacetime* — *the one we are in!* These are the volume weighted probabilities and they favor a large number of efolds. In a larger universe there are more places for us to be. That is how the NBWF successfully makes predictions for both large and small wavelengths.

Volume weighting drives us the the largest amount of inflation. It favors eternal inflation. Perturbation theory suggests that eternally inflating spacetimes are highly inhomogeneous. That makes the calculation of predictions by traditional methods difficult. This afternoon Stephen Hawking will explain how to make more secure predictions for our observations in the eternally inflating universes favored by the NBWF.

G. Horowitz: Gravity and Condensed Matter Physics

We may be witnessing the birth of a new field: one that was unimaginable just five years ago. This field is based on surprising connections between gravity and condensed matter physics. It is an outgrowth of the gauge/gravity duality that emerged from string theory and was discussed by Maldacena. It says that a theory of gravity in a $d+1$ dimensional space with certain boundary conditions is equivalent to a theory without gravity living on the d dimensional boundary of that space.

This duality was originally applied to particle physics, but recently, it has also been applied to condensed matter. Among other things, this provides a new tool for calculating transport properties of strongly coupled systems at finite temperature. This is because a state of thermal equilibrium at temperature T is dual to a black hole with Hawking temperature T. Thermal and electrical conductivity can be obtained by studying perturbations of the black hole. In particular, dissipation is just the result of waves falling into the horizon.

In the few years since people have started to explore this gravity/condensed matter duality, it has been shown that gravity can indeed reproduce various aspects of condensed matter systems including superconductivity and superfluidity, Fermi surfaces, and Fermi and non-Fermi liquids. These arise due to new properties of black holes that could have been discovered 30 years ago if people asked the right question. I will briefly describe these properties.

Consider a charged black hole coupled to a charged scalar field with boundary conditions required by this duality. Then one finds that at high temperature the black hole is stable with no scalar field outside the horizon, while at low temperature it becomes unstable toward forming a static nonzero scalar field. This is surprising. Ever since the 1970's there has been the idea that black holes have no "hair". This means that matter fields outside the horizon usually radiate out to infinity or fall into the black hole. Nevertheless, with the boundary conditions required by this duality, scalar hair indeed forms at low temperature. This is a second order phase transition and is the gravity dual of forming a charged condensate in a superconductor. Furthermore, by perturbing the black hole and translating the results to the dual field theory, one finds that the DC conductivity diverges and the optical conductivity is suppressed at low frequency and low temperature, all standard properties of superconductors. One can add magnetic fields and show that these "holographic superconductors" are type II. One can even construct gravitational duals of Josephson junctions and reproduce the standard relation between the current across the junction and the phase difference in the two superconductors on either side of the gap.

Fermi surfaces arise from another surprising property of black holes. If one studies a Dirac fermion propagating in the background of a charged black hole, then one finds that there is a static, normalizable mode only for one particular momentum, which is the gravitational dual of the Fermi momentum. This is because this mode produces a singularity in the retarded Green's function of a fermionic operator in the dual theory. Studying the dispersion relation near this momentum one finds either Fermi or non-Fermi liquids, depending on the charge and mass of the Dirac fermion. One can even get a marginal Fermi liquid which was proposed earlier to obtain a resistance proportional to temperature which is found in strange metals.

I realize that most condensed matter physicists want to see a definite prediction that they can test in the lab, and we don't have that yet. But I am still amazed that we can take general relativity, do calculations with black holes and get anything like what is seen in condensed matter systems.

In the early stages of a new field, it is easy to be optimistic about the future. So looking ahead (this list is ordered by increasing optimism): (1) Perhaps one can use this technique to classify states of matter at zero temperature (like the compressible states discussed by Sachdev). (2) One may be able to gain insight into high temperature superconductors. (3) Even more ambitiously, one can hope to someday apply the duality in reverse and use condensed matter physics to help answer fundamental questions in quantum gravity.

G. 't Hooft Spontaneously Broken Local Conformal Symmetry and the Black Hole Information Paradox

Local conformal symmetry exists in a somewhat hidden way already in classical and quantized Einstein-Hilbert gravity. Let us write the metric tensor $g_{\mu\nu}$ as

$$g_{\mu\nu} = \kappa^2\omega^2\hat{g}_{\mu\nu} ; \qquad \kappa^2 = \frac{4\pi G}{3}, \tag{1}$$

where $\omega(\vec{x}, t)$ is treated as a dynamical field describing the overall conformal factor, while $\hat{g}_{\mu\nu}$ is demanded to obey some gauge constraint (besides the four gauge constraints that fix the freedom of the space-time coordinates). One could for instance choose $\det(\hat{g}_{\mu\nu}) = 1$, but there are more interesting gauge choices. The normalization of the parameter κ differs by a factor 6 from the usual one; the present normalization gives the following expression for the Einstein Hilbert lagrangian:

$$\mathcal{L} = \sqrt{-\hat{g}}\left(\tfrac{1}{2}\hat{g}^{\mu\nu}\partial_\mu\omega\partial_\nu\omega + \tfrac{1}{12}\hat{R}\omega^2 + \mathcal{L}^{\text{matter}}(\hat{g}_{\mu\nu}, \omega)\right) . \tag{2}$$

We now propose that not $g_{\mu\nu}(\vec{x}, t)$ but $\hat{g}_{\mu\nu}(\vec{x}, t)$ be used to describe the metric of space-time, while $\omega(\vec{x}, t)$ can be fixed in different ways. The *vacuum state* however, requires ω to be constant: $\omega = 1/\kappa$, and we interpret this as saying that the vacuum spontaneously breaks local conformal invariance.

A local conformal transformation may be described as

$$\hat{g}_{\mu\nu} \to \alpha^2(\vec{x}, t)\, \hat{g}_{\mu\nu} \,, \quad \omega(\vec{x}, t) \to \alpha^{-1}\omega \,, \quad \phi(\vec{x}, t) \to \alpha^{-1}\phi \,, \quad \text{etc.,} \qquad (3)$$

where $\phi(\vec{x}, t)$ is any other type of scalar fields, and of course we can continue including fermions and gauge fields. Note that, even if the matter lagrangian in Eq. (2) contains mass terms and dimensionful couplings, the explicit occurrence of the ω field in there makes everything conformally invariant.

If this conformal symmetry would be an exact one, it could be used to shine new light[a] on the black hole information difficulty, which is the problem to identify its micro states as features of the horizon, in spite of the fact that he horizon is just a coordinate artifact. Without such an identification, quantum gravity seems to exhibit explicit nonlocal features, and this causes considerable complications in the interpretation of the dynamical laws. To see what conformal invariance can do, just consider the mass M of a black hole; it is not conformally invariant and therefore, different observers may attach different mass values to a black hole. Observers going in can reach the horizon, and they will have to conclude that, according to their metric, the black hole cannot lose its mass when the Schwarzschild time parameter tends to infinity. Outside observers however, may observe that a black hole shrinks due to Hawking radiation, so they would conclude that, at time infinity, the mass actually disappears altogether. It may be that these two observers simply use different gauges to fix $\hat{g}_{\mu\nu}$.

We observe that the metric tensor $T_{\mu\nu}$ may be used to fix the gauge. Under a conformal transformation, the Ricci curvature changes, and therefore, the metric tensor transforms non-trivially:

$$\hat{T}_{\mu\nu} \to \hat{T}_{\mu\nu} + C(D_\mu \partial_\nu \alpha - \hat{g}_{\mu\nu} D^2 \alpha) + \mathcal{O}(\partial \alpha)^2 \,. \qquad (4)$$

It has ten components. One of these can be used to fix the gauge.

Consider radial light cone coordinates for a black hole:

$x^\pm = r + 2M \log(r - 2M) \pm t$.

Instead of the "standard gauge", $\omega = 1/\kappa$, one could impose:

$$\hat{T}_{--} = 0 \,, \quad \textit{black hole} \text{ gauge} \qquad (5)$$

$$\text{or} \qquad \hat{T}_{++} = 0 \,, \quad \textit{white hole} \text{ gauge} \,. \qquad (6)$$

In the "black hole gauge", no matter particles are seen to emerge alongside the future event horizon. This is a gauge in which no Hawking particles are observed, so it is used by ingoing observers. Matter falling in the black hole of course then is observed. If, however, the white hole gauge is chosen, no ingoing matter can be discerned, while the Hawking particles are seen as real matter.

---

[a]G. 't Hooft, arXiv:1009.0669[gr-qc]; 1011.0061[gr-qc]; 1104.4543.

The transition from the black hole gauge to the white hole gauge is then seen as a gauge transformation. It is this gauge transformation that identifies outgoing matter in terms of matter falling in. Thus, it is this transformation that generates the microstates.

The procedure does not come without a price. We observe that, when a quantum field theory is renormalized, more often than not conformal symmetries are broken *explicitly*. There are two kinds of anomalies. One is the anomaly that occurs when the background metric $\hat{g}_{\mu\nu}$ itself is curved. It is an anomaly that itself is conformally invariant, so that it is proportional to the Weyl tensor squared. The coefficient in front of it depends on the number of scalar fields (including the ω field), spinor fields and vector fields present. All these fields contribute with the same sign, and this means that we have to find new mechanisms to remove this anomaly, possibly through gravitino fields, or some even more drastic modification of the theory.

The second anomaly may be more interesting: it is the anomaly associated to ordinary scale transformations, indeed, it is the same anomaly that produces non-trivial β functions. All dimensionless couplings of the theory run logarithmically as a function of the scale. Since now these anomalies are not allowed, we must demand that all β functions vanish:

$$\beta_j(\lambda_i,\ \kappa m_i,\ \kappa^2 \Lambda,\ \cdots) = 0 \ . \tag{7}$$

These are not the ordinary β functions, since the ω field is involved. Indeed, the masses m_i are there as well, and also one finds the cosmological constant Λ in there, all made dimensionless by the appropriate factors of κ.

Solving this equation turns out to be cumbersome. The solutions depend on the details of the field theoretical algebra of the matter model. One can choose this algebra in such a way that the suspected solution gives couplings very close to zero, but even then, the equations are intertwined quartic polynomials, and the question is whether there are any solutions at all with physically allowed values of the parameters. None has been found yet. An *existence proof* may be constructed starting from $\mathcal{N} = 4$ super Yang-Mills theory, but this theory might be too featureless to be itself of interest physically. Perhaps interesting non supersymmetric modifications of the theory can be considered.

N. Arkani-Hamed: Locality and Unitarity

The usual formulation of quantum field theory is built on the two pillars of Locality and Unitarity. The standard apparatus of Hamiltonians, Lagrangians and path integrals are designed to make these two fundamental principles manifest. This is however associated with the introduction of a large amount of unphysical redundancy in our description of physics. Even for the simplest scalar field theories, there is the freedom to perform field-redefinitions. Starting with massless particles of spin one and higher, we are forced to use larger gauge redundancies.

Over the past few decades, there has been a growing realization that these redundancies hide amazing physical and mathematical structures lurking within the heart of quantum field theory. This has been seen dramatically at strong coupling, in gauge/gauge and gauge/gravity dualities. The past decade has uncovered remarkable new structures in quantum field theory even at weak coupling, in properties of scattering amplitudes in gauge theories and gravity. The computation of simple tree amplitudes, of relevance for calculating Standard Model backgrounds at Hadron colliders, can run to hundreds of pages of algebra, and yet the final answers can be expressed in a few terms. It is startling to see that even in this familiar territory, where we have in principle "understood" the physics for sixty years, the commitment to a particular, gauge-redundant description of the physics blinds us to astonishingly simple and beautiful properties seen in physical observables of the theory.

All of this strongly suggests a new formulation of quantum field theory where the principles of locality and unitarity are not the stars of the show, but instead emerge as derived concepts from more primitive principles, likely involving new mathematical structures.

Along these lines, recently a new mathematical structure in algebraic geometry has been found to lie at the heart of scattering amplitudes in maximally supersymmetric gauge theories in the planar limit. The amplitude to all loop orders is associated with a contour integral over the space of k planes in n dimensions–the Grassmannian $G(k, n)$–with a very simple and special measure. This connection has allowed an explicit recursive determination of the integrand of the scattering amplitudes to all loop-orders, giving a formulation of the physics where the words "space-time", "Lagrangian", "path Integral" and "gauge redundancy" make no appearance. While locality and unitarity are not primary, the Grassmannian does make a hidden, infinite-dimensional "Yangian" symmetry of the scattering amplitudes manifest. Remarkably, this structure in the Grassmannian turns out to have been studied by mathematicians in the past five years, where it is known as the "positive Grassmannian", a special example of a "cluster variety". The backbone of these ideas turns out purely combinatorial, based on a new way of thinking about permutations. Furthermore, at least in simple examples, the full amplitude has a striking geometric interpretation as the volume of a certain polytope. Also, carrying out the loop integrals to arrive at the final results gives rise to results far simpler than what would be expected from usual methods; the simplicity has been made manifest using deep ideas ultimately related to number theory and Grothendik's theory of "motives".

It is remarkable and surprising to see ideas from algebraic geometry, combinatorics and number theory playing a central role in the very basic physics of particle scattering. The hope is that pursuing these ideas will lead to a deeper understanding of the origin and meaning of space-time and quantum mechanics.

Discussion

I. Klebanov I would like to comment on *AdS* duals of condensed matter physics. A set of fixed points that are ubiquitous in condensed matter physics are Wilson-Fisher fixed points for $O(N)$ symmetric field theories. For example, for $N = 1$ this is the famous 3-dimensional Ising model, and then there are various higher N generalizations. There is building evidence that at least for very large N the dual theory in AdS_4 is something known as a higher spin gauge theory. It is a very complicated theory involving interactions of the infinite number of massless higher spins in 4-dimensional anti-de Sitter space. This is a kind of opposite limit to the Einstein gravity approximation. If I recall Maldacena's nice description of how Einstein's gravity arises, it basically comes from sending the gauge theory to very strong coupling. Now one could ask the question of what happens if you instead start reducing the coupling, then both gravity and gauge theory sides tend to get complicated. But then what if you go all the way to the free theory, after all the simplest conformal field theories are free. Then there is growing evidence that there is suddenly a different type of simplification which involves this higher spin gauge theory. It is very hard to study on the bulk side of the duality, but it seems people are getting better and better at it, so this could be another controlled corner of AdS/CFT.

M. Fisher This is a question for Horowitz. My understanding in the gravity/conformal field theory duality is that some of the gravity theories have a healthy ultra-violet completion, ultimately in string theory I guess. I have two questions. Does one have a case of a dual superconductor on the gravity side which has an ultra-violet completion, and more generally should one worry in those cases in the duality when one does not have an ultra-violet completion on the gravity side?

G. Horowitz The original work in constructing these holographic superconductors was done in a sort of bottom-up approach, where we just basically put in the minimal ingredients that we thought were needed from a gravity description to get a superconductor, and worked with that. But it was shown, within a year, that one could get these minimal ingredients from a consistent embedding in string theory. So the answer is yes, you can do this in a way which we believe has a well-defined UV completion.

M. Fisher Should it be important to try to do that in all the cases where one has a dual description, for example in the Fermi surface situation?

G. Horowitz I guess there are arguments both ways. I am not sure that is a requirement. One could be looking at some low-energy effective description. Maybe other people have different views on this.

N. Seiberg This is a question that I have asked some of the experts in the audience, but since I have got different answers I thought I would bring up the question again. We know of all these beautiful backgrounds for string theory, say AdS labeled by the parameter N, or we know the $c = 1$ matrix model which describes two-dimensional strings. Now, how should we think of all these backgrounds, are these different ground states of the same theory, or are these distinct theories? In other words, if we have the asymptotic behavior in one of them, do we or do we not see in the bulk remnants of the other theories? Usually in field theory, if we go to high energies, we do not care which ground state we picked. Maldacena made the point that almost anything is a string theory, every quantum mechanical system is a string theory, so let us take the harmonic oscillator. This is a nice quantum mechanical system, does it include in it all of the richness of *AdS*, or maybe flat space?

J. Maldacena I do not know, probably not. You could take a spin system, which has a finite number of states, so that is even more dramatic. My first inclination is to say probably not. Maybe quantum gravity has some kind of wave function which, when you evaluate inner products, you are getting the partition functions of these various theories, spin 1/2 and so on. It is not clearly understood I think.

E. Rabinovici 't Hooft mentioned the importance of conformal or scale invariance as a guiding principle, so I would like, like Cato the elder, to mention that conformal invariance has a unique property, that in theories where the symmetry is global, the vacuum energy does not depend on the value of any expectation values of the fields involved, when they are generated by spontaneous symmetry breaking, and I think that is a very important property. Now one signature of such a spontaneous breaking in a global context would be the presence of the Goldstone boson, the dilaton. I would like to hear from Dimopoulos about the new tests of the equivalence principle, which you mentioned, what new light they could shed on the properties of the dilaton. The question to 't Hooft, if he would be willing to be more explicit on how the dilaton is swallowed in his way of having a breaking of local conformal invariance.

G. 't Hooft There certainly is not a massless particle for the same reason the Higgs theory has no massless Goldstone particle, because it is a local symmetry which is spontaneously broken, and not a global one. I am hardly changing anything in the ordinary Einstein-Hilbert theory, I just declare that the conformal part of the metric tensor is also a dynamical field, which we usually simply gauge to one. But you can let it flop freely, then you have to make another gauge constraint on the theory. That is the added value of the system, provided there are no anomalies. The only new thing is the cancellation of the anomalies, which is now a new requirement, which was not there previously, but the theory is not in any other sense different from ordinary canonical gravity, which has no massless dilaton running around. The dilaton is swallowed just like in the Higgs mechanism.

B. Altshuler I have a question for Horowitz. Condensed matter systems are not always in thermal equilibrium, there is an *H* theorem that tells that entropy is increasing, and it is not always trivial. For instance, consider the Gross-Pitaevsky equation for a Bose condensate, and the Ginzburg-Landau equation for a superconductor. Stationary, they are the same, but dynamics in Gross-

Pitaevsky is non-dissipative, and dynamics in Ginzburg-Landau is dissipative, and at the same time there is a cross-over between BCS and BEC descriptions. My question is, if you have a non-equilibrium system in condensed matter which increases entropy, how would you discuss it in terms of this correspondence with gravity and black holes?

G. Horowitz That is one of the great things about this correspondence, that there is a very simple gravity description of the increase in entropy that we see in non-gravitational systems. That just is Hawking's area theorem for black holes, which says that the area of a black hole always increases. It has been known since the early 70s that for a black hole with matter around it, when the latter falls in, the area will increase. With the Hawking-Bekenstein entropy formula, which makes the entropy proportional to the area, you automatically get entropy increase from a gravitational perspective.

B. Altshuler How would you do it universally? Relaxation times are not universal properties of equilibrium systems.

G. Horowitz You can calculate relaxation times. You can perturb a black hole, and watch how the waves decay away, and the black hole settles down to another stationary state. That is given by the quasi-normal mode frequencies. You can also go far from equilibrium, and actually form a black hole from collapse of anything. That is a very non-equilibrium evolution towards a thermal state. So all of these things can be studied.

E. Witten Just to make a quick clarification. I think that the non-universality of the relaxation rates, probably corresponds in the black hole to the fact that you have not just gravity, but gravity together with other fields. The details of how a black hole grows depend slightly on the other fields that are present.

N. Nekrasov I have a comment on the discussion between Seiberg and Maldacena, about spin systems representing a space-time geometry. If you consider a spin chain even with a finite number of degrees of freedom, but sufficiently large, that could be the space of vacua of a supersymmetric gauge theory in two dimensions, which is actually a deformation of a superconformal theory. If you send the number of spins to infinity, that would have an AdS_3 dual.

G. Dvali I have a question to Horowitz. You mentioned these theorems from the 70s, but those were proven for primary black hole hair. Obviously we can just introduce a scalar and couple it to the square of the Riemann tensor, and that will develop a hair. That is secondary hair, and it is perfectly fine with those theorems. So, in this particular situation, you are talking about this type of secondary hair for the charged scalar, or is it something different?

G. Horowitz No, it is important that this be what you call primary hair. As you say, it has been known for a long time, that if you couple a scalar to something, it has to be there because it is sourced. Here you definitely want this to be primary hair, because you want there to be a phase at high temperature, where the scalar field is simply absent. It cannot be there because of coupling to the Maxwell field or curvature. There are consistent solutions in which it is zero.

Even at low temperature, there is a solution where the scalar field remains zero, it just turns out that that solution is unstable, and the black hole wants to generate the scalar hair. The stable solution, the one with lowest free energy, has non zero scalar profile outside the horizon.

G. Dvali I thought that if you couple ϕ^2 to Riemann squared, it had similar properties.

G. Horowitz If it is an even power, that would be another way of doing this. But the simplest models just have minimal coupling and it works fine.

E. Verlinde I have a statement. The picture in AdS/CFT seems to be always that there is one emergent dimension and this to me seems very special. Because, if you think from a D-brane perspective, originally, you have to construct many dimensions. The reason why this happens in AdS/CFT is because of the way it is constructed. Namely, you go to very low energies, and you take the near horizon limit, in such a way that you stack all the D-branes together, and this is why there is one special direction which is emergent. But clearly in our space, it is not like this one dimension that is emerging. There is a much more general story about how space-time should emerge, and AdS/CFT is just a special case, a very special case and not the prototype example of what emergent space-time should be like.

E. Silverstein I had a brief comment on Fisher's question about the UV completion. In my view it will be important, and it already has been seen to be important. If you ask these questions about entropy, at low energies, the kind of instabilities that we have seen that lead to zero entropy at zero temperature, have to do with stringy effects, things like D-brane condensation in the bulk of the space or Kondo lattice models of Kachru and others, that involve the UV completion of the theory. The kind of 2KF singularities that you are looking for, that you described in your talk, probably come from stringy effects as opposed to GR effects. And finally the phenomenology of non-Fermi liquids depends on dimensions of operators, the dynamical critical exponents, and so on, and these things are just taken as free parameters in the low-energy theory, but are discrete and determined in any UV complete example. So in my view it is going to become increasingly important to control the stringy effects.

H. Ooguri I would like to respond to Verlinde's comment. Perhaps what I am going to say may be tangential to that. I just want to point out two examples where more than one dimension can be generated. One is a matrix model, where you start with one-dimensional quantum mechanics and generate the eleven dimensions. The other example is Dijkgraaf-Vafa type matrix integrals, where an eigenvalue distribution generates the entire 6-dimensional Calabi-Yau manifold from zero dimensions. So, maybe Verlinde has in mind something else, but there are certainly examples where many dimensions are generated from lower dimensional models.

E. Verlinde I meant those examples. It is not just AdS/CFT, that is what I am saying.

G. Parisi I have just a simple question. When you [Horowitz] say that in the superconductor transition, this can be related to the transition of the black hole hair, what happens to the critical exponent? Because the critical exponents are not trivial for the superconductor transition. Can one hope to have any way to grasp the values of the critical exponents doing this type of correspondence?

G. Horowitz The exponents you get in terms of how the condensate turns on when you lower the temperature a little bit below T_c, in these models are the simple mean field exponents. But there are interesting modifications of this, for example you can change the theory in a way that drives the critical temperature to zero, to actually model a quantum critical point, and you can get non trivial exponents in those examples.

S. Kachru I have a comment on Verlinde's and Ooguri's comments. One of the things that still bothers me most about talking about emerging space-time is that in examples we have in string theory, and the matrix theory example is a great one where from dynamics of 0-brane you get 11-dimensional space, the space-time that emerges is a moduli space of supersymmetric quantum mechanics. And as soon as you get rid of the supersymmetry in this matrix theory example it looks like space-time itself disappears. Now, obviously a crucial property of our own emergent space is that there is no supersymmetry but we have macroscopic space-time. So it seems to me we are still missing some very fundamental way of understanding how space-times emerge without all the extra bells and whistles of supersymmetry.

Rapporteur talk by A. Guth: Quantum Fluctuations in Cosmology and How they Lead to a Multiverse

Abstract

This article discusses density perturbations in inflationary models, offering a pedagogical description of how these perturbations are generated by quantum fluctuations in the early universe. A key feature of inflation is that rapid expansion can stretch microscopic fluctuations to cosmological proportions. I discuss also another important conseqence of quantum fluctuations: the fact that almost all inflationary models become eternal, so that once inflation starts, it never stops.

1. Introduction

I have been asked to describe quantum fluctuations in cosmology, which I find a fascinating topic. It is a dramatic demonstration that the quantum theory that was developed by studying the hydrogen atom can be applied on larger and larger scales. Here we are applying quantum theory to the universe in its entirety, at time scales of order 10^{-36} second, and it all sounds incredibly fantastic. But the shocking thing is that it works, at least in the sense that it gives answers for important questions that agree to very good precision with what is actually measured. In addition to discussing the density perturbations that we can detect, however, I want to also discuss another important aspect of quantum fluctuations: specifically, quantum fluctuations in cosmology appear, in almost all our models, to lead to eternal inflation and an infinite multiverse. This is a rather mind-boggling concept, but given our success in calculating the fluctuations observed in the cosmic microwave background (CMB), it should make good sense to consider the other consequences of quantum fluctuations in the early universe. Thus, I think it is time to take the multiverse idea seriously, as a real possibility. The inhomogeneities that lead to eternal inflation are nothing more than the long-wavelength tail of the density perturbations that we see directly in the CMB.

2. Origin of Density Perturbations During the Inflationary Era

The idea that quantum fluctuations might be the origin of structure in the universe goes back at least as far as a 1965 paper by Sakharov.[1] In the context of inflationary models, the detailed predictions are model-dependent, but a wide range of simple models give generic predictions which are in excellent agreement with observations. In this section I will give a pedagogical explanation of how these predictions arise, based on the time-delay formalism that was used in the paper I wrote with S.-Y. Pi.[2] This formalism, which we learned from Stephen Hawking, is the simplest to understand, and it is completely adequate for the dominant perturbations in

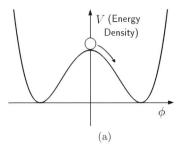 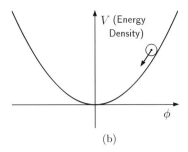

Fig. 1.   (a) Potential energy function for new inflation. (b) Potential energy function for chaotic inflation.

single-field, slow-roll inflation.[a] More sophisticated approaches are needed, however, to study multifield models or models that violate the slow-roll approximation, or to study extremely subdominant effects in single-field, slow-roll models. Even for multifield inflation, however, some of the simplicity of the time-delay formalism can be maintained by the use of the so-called δN formalism.[11,12] There are a number of reviews[12–14] and textbooks[15–18] that give a much more thorough discussion of density perturbations in inflationary models than is appropriate here.

Inflation[20–22] takes place when a scalar field has a large potential energy density. A straightforward application of Noether's theorem[19] gives the energy-momentum tensor of a canonically normalized scalar field as

$$T^{\mu\nu} = pg^{\mu\nu} + (p+\rho)u^\mu u^\nu \;, \tag{1}$$

where

$$\rho = -\frac{1}{2}g^{\mu\nu}\partial_\mu\varphi\partial_\nu\varphi + V(\varphi) \;, \tag{2}$$

$$p = -\frac{1}{2}g^{\mu\nu}\partial_\mu\varphi\partial_\nu\varphi - V(\varphi) \;, \tag{3}$$

$$u^\mu = \left(-g^{\lambda\sigma}\partial_\lambda\varphi\,\partial_\sigma\varphi\right)^{-1/2} g^{\mu\rho}\partial_\rho\varphi \;, \tag{4}$$

---

[a]The original work on density perturbations arising from scalar-field-driven inflation centered around the Nuffield Workshop on the Very Early Universe, Cambridge, U.K., June-July 1982. Four papers came out of that workshop: refs. 3, 2, 4, and 5. Ref. 5 introduced a formalism significantly more general than the previous papers. These papers tracked the perturbations from their quantum origin through Hubble exit, reheating, and Hubble reentry. Earlier Mukhanov and Chibisov[6] had revived Sakharov's idea in a modern context, studying the conformally flat perturbations generated during the inflationary phase of the Starobinsky model.[8] They developed a method of quantizing the metric fluctuations, a method more sophisticated than is needed for the simpler models of refs. 2–5, and gave a formula (without derivation) for the final spectrum. For various reasons the calculations showing how the conformally flat fluctuations during inflation evolve to the conformally Newtonian fluctuations after inflation were never published, until the problem was reconsidered later in refs. 9 and 10. The precise answer obtained in ref. 6, $Q(k) = \sqrt{24\pi G}M\left(1 + \frac{1}{2}\ln(H/k)\right)$, has not (to my knowledge) been confirmed in any modern paper. However, the fact that $Q(k)$ is proportional to $\ln(\text{const}/k)$ has been confirmed, showing that the 1981 paper by Mukhanov and Chibisov did correctly calculate what we now call n_s (as was pointed out in ref. 7).

where $\partial_\mu \equiv \partial/\partial x^\mu$. So, as long as the energy of the state is dominated by $V(\varphi)$, Eq. (3) guarantees that the pressure is large and negative. Einstein's equations imply that negative pressure creates repulsive gravity, so any state whose energy is dominated by the potential energy of a scalar field will drive inflation. There are two basic scenarios — one where φ starts at the top of a hill (new inflation[21,22]), and one where it starts high on a hill and rolls down (chaotic inflation[23]); see Fig. 1. Either scenario is successful, and for the density perturbation calculation we can treat them at the same time. We use comoving coordinates, with a background metric describing a flat Friedmann-Robertson-Walker (FRW) universe:

$$ds^2 = -dt^2 + a^2(t)\,\delta_{ij}dx^i\,dx^j \ , \tag{5}$$

where $a(t)$ is the scale factor. Objects moving with the expansion of the universe are at rest in this coordinate system, with the expansion described solely by $a(t)$: when $a(t)$ doubles, all the distances in the universe double. In this metric the Klein-Gordon equation for a scalar field is given by

$$\ddot{\varphi} + 3H(\varphi,\dot{\varphi})\dot{\varphi} - \frac{1}{a^2(t)}\nabla^2\varphi = -\frac{\partial V(\varphi)}{\partial\varphi} \ , \tag{6}$$

where an overdot indicates differentiation with respect to time t, and ∇^2 is the Laplacian, $\sum_i \partial^2/\partial(x^i)^2$, with respect to the coordinates x^i. The equation is identical to the Klein-Gordon equation in Minkowski space, except that there is a drag term, $3H\dot{\varphi}$, which can be expected, since the energy density must fall if the universe is expanding. In addition, each spatial gradient is modified by $1/a(t)$, which converts the derivative to the current scale of spatial distance. The Hubble expansion rate $H \equiv \dot{a}/a$ is given by the Friedmann equation for a flat universe,

$$H^2 = \frac{8\pi}{3}G\left(\frac{1}{2}\dot{\varphi}^2 + V(\varphi)\right) \ . \tag{7}$$

In this language, the repulsive effect of the negative pressure that was mentioned above can be seen in the equation for the acceleration of the expansion,

$$\ddot{a} = -\frac{4\pi}{3}G(\rho + 3p)a \ . \tag{8}$$

If $p = -\rho$, as one finds when $V(\varphi)$ dominates, this equation gives $\ddot{a} = (8\pi/3)G\rho a$.

Using an assumption called the slow-roll approximation, which is valid for a large range of inflationary models, we can ignore the $\ddot{\varphi}$ term of Eq. (6) and the $\dot{\varphi}^2$ term in Eq. (7). In addition, at sufficiently late times the Laplacian term can be neglected, since it is suppressed by $1/a^2(t)$. We are then left with a very simple differential equation,

$$3H(\varphi)\dot{\varphi} = -\frac{\partial V}{\partial\varphi} \ , \tag{9}$$

which has a one-parameter class of solutions. That one parameter is itself trivial — it is a time offset. Given one solution $\varphi_0(t)$, the general solution can be written as $\varphi_0(t - \delta t)$, where δt is independent of t. Since the differential equation (9) has no spatial derivatives, δt can depend on position, so the most general solution can be written as

$$\varphi(\vec{x}, t) = \varphi_0\big(t - \delta t(\vec{x})\big) . \tag{10}$$

Since we are interested in developing a first order perturbation theory, we can expand about $\varphi_0(t)$,

$$\varphi(\vec{x}, t) \equiv \varphi_0(t) + \delta\varphi(\vec{x}, t) = \varphi_0(t) - \dot{\varphi}_0(t)\, \delta t(\vec{x}) , \tag{11}$$

so

$$\delta t(\vec{x}) = -\frac{\delta\varphi(\vec{x}, t)}{\dot{\varphi}_0(t)} . \tag{12}$$

Even though the numerator and denominator of the above expression both depend on time, the quotient does not. Thus at late times (within the inflationary era) — times late enough for Eq. (9) to be accurate — the nonuniformities of the rolling scalar field are completely characterized by a time-independent time delay.[b] It is

---

[b]The description of the perturbations at late times by a time-independent time delay $\delta t(\vec{x})$ is in fact much more robust than the approximation that $\ddot{\varphi}$ can be neglected. It is a consequence of the Hubble drag term, and will hold at sufficiently late times in any single-field model for which the slow-roll approximation is valid for more than a few e-folds. To see this, consider Eq. (6), with H taken to be an arbitrary function of φ and $\dot{\varphi}$. We will neglect the Laplacian term, since it is suppressed by $1/a^2(t)$, and we are interested in late times. Then, for each value of $\vec{x}$ there is a two-parameter class of solutions to this second order ordinary differential equation. To see the effect of the damping, suppose that we know the unperturbed solution, $\varphi_0(t)$, and a nearby solution, $\varphi_0(t) + \delta\varphi(t)$, where $\delta\varphi(t)$ is to be treated to first order. $\delta\varphi$ can depend on $\vec{x}$, but we suppress the argument because we consider one value of $\vec{x}$ at a time. We then find that $\delta\varphi(t)$ and $\dot{\varphi}_0(t)$ obey the same differential equation. If we construct the Wronskian $W(t) \equiv \dot{\varphi}_0\, \delta\dot{\varphi} - \ddot{\varphi}_0\, \delta\varphi$, we find that

$$\dot{W} = -3\left(H + \frac{\partial H}{\partial\dot{\varphi}}\dot{\varphi}_0\right)W ,$$

the solution to which is

$$W(t) = W_0 \exp\left\{-3\int_{t_0}^{t} dt\left(H + \frac{\partial H}{\partial\dot{\varphi}}\dot{\varphi}_0\right)\right\} .$$

Thus $W(t)$ falls off roughly as e^{-3Ht} or faster ($\dot{\varphi}_0\partial H/\partial\dot{\varphi} > 0$), and so can be neglected after just a few e-folds of expansion. Then note that

$$\frac{d}{dt}\left(\frac{\delta\varphi}{\dot{\varphi}_0}\right) = \frac{W(t)}{\dot{\varphi}_0^2} ,$$

while in the slow-roll regime $\dot{\varphi}_0^2$ is approximately constant — from Eq. (9) one can show that the fractional change in $\dot{\varphi}_0^2$ during one Hubble time (H^{-1}) is approximately $2(\epsilon - \eta)$, as defined in Eqs. (14) and (15). Thus the time derivative of the ratio $\delta\varphi/\dot{\varphi}_0$ falls off as e^{-3Ht} or faster, implying that the time delay rapidly approaches a fixed value. The time-delay description remains accurate throughout the reheating process, even though the slow-roll conditions will generally fail badly at the end of inflation, when the scalar field starts to oscillate about the bottom of the potential well. The time delay is maintained because $\delta\varphi$ and $\dot{\varphi}_0$ continue to obey the same linear differential equation. Thus if $\delta\dot{\varphi}/\delta\varphi = \ddot{\varphi}_0/\dot{\varphi}_0$ at the end of the slow roll period, then $\delta\varphi$ will remain proportional to $\dot{\varphi}_0$ for all later times. This argument, which generalizes an argument in ref. 2, is in contradiction with ref. 24, where it is argued that the time delay persists to the end of inflation only under very stringent assumptions about the potential. The argument of ref. 24, however, is useful to define a dimensionless measure of the time delay,

$$\delta N = H \delta t , \tag{13}$$

which can be interpreted as the number of e-folds of inflation by which the field is advanced or retarded.

To justify the slow-roll approximation, we must adopt restrictions on the form of the potential energy function $V(\varphi)$. The slow-roll approximation is equivalent to saying that the field φ evolves approximately at the drag-force limited velocity, where the drag force equals the applied force, with inertia playing only a negligible role. (This would be called the terminal velocity, except that it can change slowly with time.) From the first two terms of Eq. (6) one can see that the velocity approaches the drag-limited value with a time constant of order H^{-1}. Thus, for the field to evolve at the drag-limited velocity, it is essential that neither the drag coefficient nor the applied force changes significantly during a time of order H^{-1}. Thus we want to insist that $H^{-1}|\dot{H}| \ll H$, and that $H^{-1}|(\partial^2 V/\partial\varphi^2)\dot{\varphi}| \ll |\partial V/\partial\varphi|$. Using Eq. (9) to approximate $\dot{\varphi}$, these two conditions can be expressed in terms of the two slow-roll parameters[15]

$$\epsilon \equiv \frac{1}{16\pi G} \left(\frac{V'}{V} \right)^2 \approx -\frac{\dot{H}}{H^2} , \quad 0 < \epsilon \ll 1 , \tag{14}$$

$$\eta \equiv \frac{1}{8\pi G} \frac{V''}{V} \approx -\frac{V'' \dot{\varphi}}{HV'} \approx \epsilon - \frac{\ddot{H}}{2H\dot{H}} , \quad |\eta| \ll 1 , \tag{15}$$

where a prime denotes a derivative with respect to φ. Note that these slow-roll conditions do not by themselves guarantee that φ will evolve at drag-limited velocity, because a large initial velocity will take time before it approaches the drag-limited value. But the slow-roll conditions do guarantee that for times long compared to H^{-1}, φ will evolve at very nearly the drag-limited velocity.

To proceed, I will make two approximations that will simplify the problem enormously, but which are nonetheless extremely accurate for single-field slow-roll inflation. First, we will neglect all perturbations of the metric until the time when inflation ends. That is, until inflation ends we treat the scalar field as a quantum field in a fixed de Sitter space background. Thus, we will be ignoring the fluctuations in the energy-momentum tensor of the scalar field, since we are not allowing them to perturb the metric. However, we will calculate the fluctuations in the scalar field itself, as described by the time delay $\delta t(\vec{x})$. Since the scalar field is driving the inflation, the time delay $\delta t(\vec{x})$ measures the variation in the time at which inflation ends at different places in space. The amount of energy that is released at the end of inflation is much larger than the energy-momentum tensor fluctuations during inflation, so the spatial variation of the timing of this energy release becomes the dominant source of the density perturbations that persist at later times. To describe

---

really a discussion of the validity of Eq. (9), but we have seen that the time delay is preserved even when Eq. (9) fails.

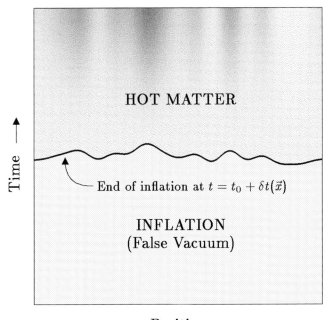

Position

Fig. 2.   Schematic illustration of our approximations: (1) before the end of inflation, shown as a wiggly line, the spacetime is the unperturbed, exponentially expanding flat (de Sitter) space corresponding to the "false vacuum" state of the scalar field; (2) inflation ends on a sharp line, at which the matter is immediately transformed into thermal radiation.

this release of energy, we make our second approximation. We will treat the ending of inflation as instantaneous. We will assume that the potential energy of the inflaton field is converted instantaneously into the thermal radiation of effectively massless particles, beginning the radiation-dominated era of cosmological history.

I will give heuristic justifications for these approximations, but I am not aware of a more rigorous justification that can be explained without developing an understanding of what happens when these approximations are avoided, and then the problem requires a much more detailed analysis. Such analyses have of course been done and are even described in textbooks. The answer that we will obtain agrees with the textbooks,[15,16,18] for the single-field slow-roll case, down to the last factor of $\sqrt{\pi}$. While the textbooks corroborate the answer that we will obtain, the methods are sufficiently different so that very little light is shed on the approximations described here. In a future publication,[25] I will attempt a more detailed justification.

For a given theory we can calculate $\varphi_0(t)$ by solving an ordinary differential equation, so Eq. (12) reduces the problem of calculating $\delta t(\vec{x})$ to that of calculating the fluctuations of the scalar field, $\delta\varphi(\vec{x}, t)$. This is a problem in quantum field theory, albeit quantum field theory in curved spacetime. The calculations closely resemble the familiar quantum field theory calculations for Minkowski space, but there is one point in the calculation where cosmology rears its head. While Minkowski space has a well-defined vacuum state, which is the starting point for most calculations, it is less clear what quantum state should be used to describe the fields evolving in de Sitter space, where the time dependence prevents the existence of a conserved total energy. In principle the quantum state is determined by the initial conditions for the universe, about which we know very little. However, while we do not know the quantum state of the early universe, there is a very natural choice, corresponding at least locally to the concept of a vacuum state. To understand this choice, recall that we are interested in an exponentially expanding space, the de Sitter spacetime of inflation, so to a good approximation $a(t) \propto e^{Ht}$, where H is constant. If we now treat the inflaton field $\varphi(\vec{x}, t)$ as a quantum operator, we can as usual consider its Fourier transform:

$$\varphi(\vec{x}, t) = \frac{1}{(2\pi)^3} \int d^3k \, e^{i\vec{k}\cdot\vec{x}} \tilde{\varphi}(\vec{k}, t) \,. \tag{16}$$

For a free quantum field theory in Minkowski spacetime, $\tilde{\varphi}(\vec{k}, t)$ would be the sum of an annihilation operator term for particles of momentum $\vec{k}$ and a creation operator term for particles of momentum $-\vec{k}$, each corresponding to a de Broglie wavelength $\lambda = 2\pi/|\vec{k}|$. For an FRW spacetime, since the Fourier transform is defined in terms of the *comoving* coordinates $\vec{x}$, the physical wavelength for a mode $\vec{k}$ is not constant, but is given by

$$\lambda_{\text{phys}}(t) = a(t)\frac{2\pi}{|\vec{k}|} \,. \tag{17}$$

In other words, each mode is stretched as the universe expands. Thus, if we follow any mode backwards in time, it will have a shorter and shorter wavelength and a higher and higher frequency. The Hubble expansion rate H is approximately constant during inflation, so at very early times H is very small compared to the frequency, and hence is negligible. Thus, any given mode behaves at asymptotically early times exactly like a mode in Minkowski space, so the "natural" initial state is to simply start each mode in its Minkowski vacuum state in the asymptotic past. This is called the Bunch–Davies vacuum,[26] and it is identical to what is also called the Gibbons–Hawking vacuum.[27] Gibbons and Hawking developed their description of the vacuum in a completely different formalism, based on the symmetries of the de Sitter spacetime (Eq. (5) with $a(t) \propto e^{Ht}$), but the two vacuum states are identical, as one would hope. The Bunch–Davies / Gibbons–Hawking vacuum is taken as the starting point for all standard calculations of density perturbations.

To discuss the spectrum of fluctuations of a spatially varying quantity such as $\delta\varphi(\vec{x}, t)$, which is assumed to be statistically homogeneous and isotropic, cosmologists define a power spectrum $P_{\delta\varphi}(k,t)$ by[c]

$$\langle \delta\tilde{\varphi}(\vec{k},t)\,\delta\tilde{\varphi}(\vec{k}',t)\rangle \equiv (2\pi)^3 P_{\delta\varphi}(k,t)\,\delta^{(3)}(\vec{k}+\vec{k}')\,, \tag{18}$$

or equivalently

$$\langle \delta\varphi(\vec{x},t)\,\delta\varphi(\vec{y},t)\rangle = \frac{1}{(2\pi)^3}\int d^3k\, e^{i\vec{k}\cdot(\vec{x}-\vec{y})} P_{\delta\varphi}(k,t)\,. \tag{19}$$

When $\delta\varphi(\vec{x},t)$ is a quantum field operator, the power spectrum is nothing more than the equal-time propagator, which can be calculated straightforwardly once the vacuum is specified, as described in the previous paragraph. From Eq. (6), $\delta\tilde{\varphi}(\vec{k},t)$ can be seen to obey the equation

$$\delta\ddot{\tilde{\varphi}} + 3H\delta\dot{\tilde{\varphi}} + \frac{k^2}{a^2}\delta\tilde{\varphi} = -\frac{\partial^2 V}{\partial\varphi^2}\delta\tilde{\varphi}\,. \tag{20}$$

To make use of this equation, we consider its behavior around the time of Hubble exit, $t_{\mathrm{ex}}(k)$, when the wavelength is approximately equal to the Hubble length, defined more precisely by

$$\frac{k^2}{a^2(t_{\mathrm{ex}})} = H^2\,. \tag{21}$$

We assume that the slow-roll conditions of Eqs. (14) and (15) are valid within several Hubble times (H^{-1}) of t_{ex} (but it is okay if they are violated later during the period of inflation, as described in footnote b). For $t \gtrsim t_{\mathrm{ex}}$, the k^2/a^2 term of Eq. (20) becomes insignificant. From Eq. (15) we see that the right-hand-side of Eq. (20) has magnitude $3\eta H^2\,\delta\tilde{\varphi}$, where $\eta \ll 1$. Thus for times up to and including $t_{\mathrm{ex}}(k)$ and a little beyond, we can neglect the right-hand-side and treat $\delta\varphi(\vec{x},t)$ as a free, massless, minimally coupled field in de Sitter space. It is then straightforward to show that[d]

$$P_{\delta\varphi}(k,t) = \frac{H^2}{2k^3}\left[1 + \left(\frac{k}{a(t)H}\right)^2\right]\,. \tag{22}$$

The time delay should be calculated at a time slightly beyond t_{ex}, say by a few Hubble times, when $k^2/a^2 \ll H^2$, when we can neglect the k^2/a^2 terms in both Eqs. (20) and (22). Using Eqs. (9), (12), and (14), one finds several useful expressions for $P_{\delta N}(k)$:

$$P_{\delta N}(k) = \frac{H^2}{\dot{\varphi}_0^2}P_{\delta\varphi}(k) = \frac{H^4}{2k^3\dot{\varphi}_0^2} = \frac{2\pi G H^2}{k^3\epsilon} = \frac{9}{2}\left(\frac{8\pi G}{3}\right)^3\frac{V^3}{k^3 V'^2}\,. \tag{23}$$

---

[c]Conventions vary, but here we follow the conventions of refs. 15 and 18. The quantity $\Delta f(\vec{k})$ defined in ref. 2 is related by $\Delta f(\vec{k})^2 = k^3 P_f(k)/(2\pi)^3$. Another common normalization, called $\mathcal{P}(k)$ in ref. 15 and $\Delta^2(k)$ in ref. 18, is given by $\mathcal{P}(k) = \Delta^2(k) = k^3 P(k)/2\pi^2$, so $\Delta f(\vec{k})^2 = \mathcal{P}(k)/4\pi$. According to ref. 15, $P(k)$ and $\mathcal{P}(k)$ are both called *the spectrum*. In the context of the curvature perturbation $\mathcal{R}$, to be defined below, ref. 28 defines yet another normalization that remains in common use, $\delta_H \equiv \frac{2}{5}\mathcal{P}_{\mathcal{R}}$.

[d]One source from which this equation can be deduced is ref. 26, but note that Eq. (3.6) is misprinted, and should read $\psi_k(\eta) = \alpha^{-1}(\pi/4)^{1/2}\eta^{3/2}H_\nu^{(2)}(k\eta)$.

Since the time delay is evaluated a few Hubble times beyond $t_{ex}(k)$, the quantities V, V', and H appearing in the above expressions should all be evaluated at $\varphi_0(t)$ for $t \approx t_{ex}(k)$. The distinction between $t_{ex}(k)$ and a few Hubble times later is important only at higher order in the slow-roll approximation, since the change in φ_0 over a Hubble time is of order ϵ.

Note that the fluctuations in δt are inversely proportional to ϵ, which is a consequence of the presence of $\dot{\varphi}_0(t)$ in the denominator of Eq. (12). We are now in a position to test the consistency of the first of our approximations (see Fig. 2), the approximation of neglecting the fluctuations in the scalar field energy-momentum tensor. The major source of these fluctuations is the fluctuation in potential energy caused by the fluctuations in φ. Thus $\delta\rho/\rho \approx V'\delta\varphi/V$, so

$$P_{\delta\rho/\rho}(k) \approx \left(\frac{V'}{V}\right)^2 P_{\delta\varphi}(k) = \frac{8\pi G H^2 \epsilon}{k^3}\left[1 + \left(\frac{k}{a(t)H}\right)^2\right].\tag{24}$$

Thus, the fractional fluctuations in the mass density are of order ϵ^2 times the dimensionless fluctuations $P_{\delta N}(k)$ of the time delay, so we expect that we can neglect the mass density fluctuations if we are interested in calculating the dominant term in the small ϵ limit.[e,f]

3. Evolution of the Density Perturbations Through the End of Inflation

Thus, we have reduced the calculation of the density fluctuations to the schematic description of Fig. 2, with the power spectrum of the time delay given by Eq. (23).

---

[e] One might worry that the second term in square brackets in Eq. (24) becomes large at early times, since $a(t) \propto e^{Ht}$. But note that the factors of H cancel, so the term is independent of H, and that $k/a(t) = k_{phys}$; this term is just the short distance divergence that would also be present in a Minkowski space background. It leads to divergences, but those divergences must be canceled by the prescription for regularizing $T_{\mu\nu}$. The density fluctuations of cosmological relevance, which are finite and can be treated classically at late times, arise from the first term in square brackets.
[f] The fact that $P_{\delta\rho/\rho}(k) \propto \epsilon$ is a strong argument to justify the neglect of metric fluctuations, but some experts may not be convinced without seeing a more complete formulation in which metric fluctuations can actually be calculated. In ref. 25 I will show how the time-delay formalism can be embedded in a complete first-order calculation in synchronous gauge. (Synchronous gauge is most useful here, since the time delay is time-independent when measured in proper time. Other time coordinates will obscure the underlying simplicity.) Following the notation of ref. 16, the metric is written as

$$g_{00} = -1, \quad g_{0i} = 0, \quad g_{ij} = a^2(t)\left[(1+A)\delta_{ij} + \frac{\partial^2 B}{\partial x^i \partial x^j}\right].$$

Defining a new auxiliary field χ (not used in Ref. 16) by

$$\chi \equiv \frac{1}{2}\left(3\dot{A} + \nabla^2\dot{B}\right) - 3\left[\frac{\partial H}{\partial\varphi_0}\delta\varphi + \frac{\partial H}{\partial\dot{\varphi}_0}\delta\dot{\varphi}\right],$$

the scalar field equation of motion in this gauge can be written as

$$\ddot{\varphi} + 3H(\varphi_0, \dot{\varphi}_0)\delta\dot{\varphi} + 3\dot{\varphi}_0\left[\frac{\partial H}{\partial\varphi_0}\delta\varphi + \frac{\partial H}{\partial\dot{\varphi}_0}\delta\dot{\varphi}\right] - \frac{1}{a^2}\nabla^2\delta\varphi + V''(\varphi_0)\delta\varphi + \dot{\varphi}_0\chi = 0.$$

χ can be found from the source equation

The role of quantum theory in this calculation is finished — it determined the power spectrum of the time delay. The rest of the calculation is general relativity and astrophysics. Here I will continue the derivation through the end of inflation, which carries it far enough to compare with the literature and to at least qualitatively understand the observational consequences.

To understand the implications of $\delta t(\vec{x})$ in the cosmic evolution as described by Fig. 2, it is convenient to switch to a new coordinate system in which the end of inflation happens at t_0 everywhere. Just define $\vec{x}' = \vec{x}$ and $t' = t - \delta t(\vec{x}')$. Clearly $dt = dt' + \partial_i \delta t \, dx'^i$, so the transformation of the metric of Eq. (5) becomes $ds^2 = g'_{\mu\nu} dx'^\mu dx'^\nu$, where

$$g'_{00} = -1 \,, \quad g'_{0i} = g'_{i0} = -\partial_i \delta t \,, \quad g'_{ij} = a^2 \big(t' + \delta t(\vec{x}') \big) \delta_{ij} \,, \tag{25}$$

with an inverse metric, to first order in δt, given by

$$g'^{00} = -1 \,, \quad g'^{0i} = g'^{i0} = -\frac{1}{a^2} \partial_i \delta t \,, \quad g'^{ij} = \frac{1}{a^2 \big(t' + \delta t(\vec{x}') \big)} \delta_{ij} \,. \tag{26}$$

In the primed coordinates the phase transition happens sharply at $t' = t_0$, with a sudden change from $p = -\rho \equiv -\rho_{\rm inf}$ in the inflationary phase to $p_{\rm rad} = \frac{1}{3}\rho_{\rm rad}$ in the radiation phase. During the inflationary phase the energy-momentum tensor is simply $T_\mu{}^\nu = -\rho_{\rm inf}\delta_\mu^\nu$, while after the transition it is given by Eq. (1). The energy-momentum tensor must be covariantly conserved, which means that

$$D_\nu T_\mu{}^\nu = \partial_\nu T_\mu{}^\nu + \Gamma^\nu_{\nu\lambda} T_\mu{}^\lambda - \Gamma^\lambda_{\nu\mu} T_\lambda{}^\nu = 0 \,, \tag{27}$$

where D_ν denotes a covariant derivative. The affine connection coefficients Γ will not contain any δ-functions, so $\partial_\nu T_\mu{}^\nu$ cannot contain any δ-functions either; thus

$$\frac{\partial}{\partial t}(a^2 H \chi) = \dot{H}\nabla^2 \left(\frac{\delta\varphi}{\dot{\varphi}_0} \right) \,.$$

The full metric can be recovered by using

$$\nabla^2 A = 2a^2 H \chi \,,$$

and then using the definition of χ to find $\dot{B}$. Note that the terms involving partial derivatives of H reproduce to first order the dependence of H on φ and $\dot{\varphi}$ in Eq. (6), so these "metric perturbations" are taken into account by the time-delay calculation. The metric perturbations that are ignored are those proportional to χ, and they can be seen to be small in slow-roll inflation. The source term on the right is proportional to $\dot{H}$, which is of order ϵ. The term enters the scalar field equation of motion with a prefactor of $\dot{\varphi}_0$, which contributes another factor of $\sqrt{\epsilon}$ to the suppression. At later times near the end of inflation, when the slow-roll condition might fail badly, χ is strongly suppressed by the factor $1/a^2$. For the slow-roll solution with $\delta\varphi/\dot{\varphi}_0 \approx -\delta t(\vec{x})$, the source equation can be solved to give

$$\chi(t) \approx \frac{H(t_{\rm ex}) - H(t)}{a^2(t)H(t)} \nabla^2 \delta t(\vec{x}) \,,$$

where a constant of integration was chosen so that $\chi(t) \approx 0$ at Hubble exit. Thus, this formulation gives a solid underpinning to the intuitive idea that, at late times, each region can be treated as an independent Robertson-Walker universe. Each independent universe follows essentially the same history, differing from the other universes by only a time offset.

$T_\mu{}^0$ must be continuous at $t' = t_0$. This implies that

$$T_0{}^0{}_{\mathrm{rad}} = p_{\mathrm{rad}} + u^0{}_{\mathrm{rad}}\, u_{0,\mathrm{rad}}\, (\rho_{\mathrm{rad}} + p_{\mathrm{rad}}) = T_0{}^0{}_{\mathrm{inf}} = -\rho_{\mathrm{inf}} \ , \tag{28}$$

$$T_i{}^0{}_{\mathrm{rad}} = (\rho_{\mathrm{rad}} + p_{\mathrm{rad}})\, u_{i,\mathrm{rad}}\, u^0{}_{\mathrm{rad}} = T_i{}^0{}_{\mathrm{inf}} = 0 \ . \tag{29}$$

The second of these equations can only be satisfied if $u_{i,\mathrm{rad}} = 0$, because $u^0{}_{\mathrm{rad}}$ cannot vanish, as u^μ is timelike, and $(\rho_{\mathrm{rad}} + p_{\mathrm{rad}}) = \frac{4}{3}\rho_{\mathrm{rad}}$ cannot vanish without violating the first equation. Thus, the radiation fluid is necessarily at rest in the frame of reference in which the phase transition occurs simultaneously. Requiring $u^2 = -1$, one finds that

$$u_{0,\mathrm{rad}} = -1 \ , \quad u_{i,\mathrm{rad}} = 0 \ ; \quad u^0{}_{\mathrm{rad}} = 1 \ , \quad u^i{}_{\mathrm{rad}} = \frac{1}{a^2}\partial_i \delta t \ . \tag{30}$$

Given the above equation, Eq. (28) leads immediately to $\rho_{\mathrm{rad}} = \rho_{\mathrm{inf}}$; the energy density is conserved across the transition. Since the Einstein equations are partial differential equations that are second order in time derivatives, the metric and its first time derivative will be continuous across $t' = t_0$. So we have now found all the information needed to give a well-defined Cauchy problem for the evolution of the model universe, starting at the beginning of the radiation-dominated era. At this point the perturbations of interest have wavelengths vastly larger than the Hubble length, but during the subsequent evolution the Hubble length will grow faster than the perturbation wavelength, so later the perturbations will come back inside the Hubble length. The description of the perturbations through the time of Hubble reentry was given in refs. 2–5, and in many later sources, but for present purposes we will stop here.

At this point we can discuss the validity of the second of our key approximations, the approximation of an instantaneous phase transition. The actual transition, during which the scalar field rolls down the hill in the potential energy diagram and then oscillates about the minimum and reheats, very likely takes many Hubble times to complete. However, we need to keep in mind that the modes of interest exited the Hubble horizon some 50 or 60 Hubble times before the end of inflation, which means that at the end of inflation their physical wavelength is of order e^{50} to e^{60}, or 10^{21} to 10^{26}, times the Hubble length. Thus, even if the phase transition takes 10^{10} Hubble times, during this time light would be able to travel less than 10^{-10} wavelengths. Thus the phase transition is effectively instantaneous, on the time scale that is relevant for influencing a wave with the wavelengths under consideration. (Of course the reheat energy density that we calculated, $\rho_{\mathrm{rad}} = \rho_{\mathrm{inf}}$, was an artificiality of the instantaneous approximation. But the calculation can easily be adjusted to account for a lower reheat energy density, which can be found by doing a more accurate, homogeneous calculation. At the end of inflation we would still obtain, in the primed coordinate system, a radiation fluid that is at rest, with a uniform energy density.)

Note that the method used here depended crucially on the assumption that all parts of the universe would undergo the same sequence of events, so that the only difference from one place to another is an overall time offset. If there were more than one field for which the quantum fluctuations were relevant, then this would not be true, since a fluctuation of one field relative to the other could not be described as an overall time offset. Thus, multifield inflation requires a more sophisticated formalism.

4. Simplifying the Description

The description we have given so far, specifying the metric and the matter content, is sufficient to calculate the rest of the history, but it is rather complicated. Furthermore, it is equivalent to many other complicated descriptions, related by coordinate transformations. It is therefore very useful to find a coordinate-invariant way of quantifying the density perturbations. One convenient approach is motivated by considering the Friedmann equation for a universe with spatial curvature, a universe which might be closed, open, or in the borderline case, flat:

$$H^2 = \frac{8\pi}{3} G\rho - \frac{k}{a^2} . \tag{31}$$

Here k is a constant, where positive values describe a closed universe, and negative values describe an open one. We are interested in describing a perturbation of a homogeneous background universe that is flat, $k = 0$. Thus $\rho(\vec{x}, t)$ will on average equal $\rho_0(t)$, the value for the background universe, but it will fluctuate about this average. H is normally thought of as part of the global description of the universe, but it has a locally defined analog given by

$$H_{\mathrm{loc}} \equiv \frac{1}{3} D_\mu u^\mu , \tag{32}$$

where D_μ is the covariant derivative and u^μ is the fluid velocity. (Here we will deal only with a radiation fluid, but in a multicomponent fluid u^μ can be defined in terms of the total energy-momentum tensor, as described on p. 225 of ref. 16.) We can then define

$$K(\vec{x}, t) \equiv a^2(t) \left[\frac{8\pi}{3} G\rho(\vec{x}, t) - H_{\mathrm{loc}}^2(\vec{x}, t) \right] . \tag{33}$$

This quantity has a property called gauge invariance, which means that its value for any coordinate point $(\vec{x}, t)$ is not changed, to first order in the size of the perturbations, by any coordinate transformation that is itself of order of the size of the perturbations. In this case, the gauge invariance follows from the fact that, apart from the factor $a^2(t)$ which is irrelevant for this issue, $K(\vec{x}, t)$ a quantity that is coordinate-invariant, and which vanishes for the background universe. (Note that coordinate invariance by itself is not enough; if the quantity varied with time in the background solution, then its value at $(\vec{x}, t)$ would change if t were redefined by a small amount.) K also has the convenient property that it remains constant as long as spatial derivatives can be neglected, because it is exactly conserved for the case of a homogeneous universe.

We can calculate $K(\vec{x}, t)$ just after the end of inflation, using the primed coordinate system but dropping the primes. We have $H_{\text{loc}} = \frac{1}{3}D_\mu u^\mu = \frac{1}{3}g^{-1/2}\partial_\mu(g^{1/2}u^\mu)$, where $g \equiv -\det(g_{\mu\nu}) = a^6(t + \delta t(\vec{x}))$ and u^μ is given by Eq. (30). This gives $H_{\text{loc}} = H_{\text{inf}} + \nabla^2\delta t/3a^2$, which gives

$$K = -\frac{2}{3}H_{\text{inf}}\nabla^2\delta t\,, \tag{34}$$

where H_{inf} is the Hubble expansion rate during inflation, which we have treated as a constant.[g] As in Eq. (6), ∇^2 denotes the Laplacian operator with respect to the coordinates x^i. Then, given the power spectrum of Eq. (23) for $\delta N = H\delta t$, we find a power spectrum for K given by

$$P_K(k) = \frac{4k^4}{9}P_{\delta N}(k) = \frac{2kH^4}{9\dot{\varphi}_0^2} = \frac{8\pi GH^2 k}{9\epsilon} = 2\left(\frac{8\pi G}{3}\right)^3\frac{V^3 k}{V'^2}\,. \tag{35}$$

In the literature K is seldom used, but instead it is much more common to use a variable called the curvature perturbation $\mathcal{R}$, for which the usual definition is somewhat complicated.[h] However, it is shown in ref. 28 that

$$K = -\frac{2}{3}\nabla^2\mathcal{R}\,, \tag{36}$$

so

$$P_{\mathcal{R}}(k) = \frac{9}{4k^4}P_K(k) = P_{\delta N}(k)\,, \tag{37}$$

where $P_{\delta N}(k)$ is given by Eq. (23). This answer agrees precisely with the answers obtained in refs. 15, 16, 18, and 2.[i]

The WMAP seven-year paper[29] quotes $\mathcal{P}_{\mathcal{R}}(k_0) = (2.43 \pm 0.09) \times 10^{-9}$, where $k_0 = 0.002$ Mpc^{-1} and $\mathcal{P}_{\mathcal{R}}(k) = k^3 P_{\mathcal{R}}(k)/(2\pi^2)$. (The WMAP papers use $\Delta_{\mathcal{R}}^2(k)$

---

[g]At this point one might worry that the approximation of instantaneous reheating might be crucial to the answer we obtained. If we had used a more realistic picture of slow reheating which leads to a lower reheat energy density, we might expect that our method would give $K = -\frac{2}{3}H_{\text{reheat}}\nabla^2\delta t$, which could be much smaller. Although it is not obvious, however, the reheat energy density does not affect K, which is conserved for wavelengths large compared to the Hubble length. Thus, the value we obtained here could have been calculated before reheating began, and is equal to the value that holds long after reheating, whether reheating is fast or slow. To understand the evolution of K when H changes, however, requires a more detailed calculation. Because of the factor of $a^2(t)$ in Eq. (33), the value of K is in fact sensitive to quantities that are suppressed by factors of $1/a^2$. To see the conservation of K for long wavelengths, one needs to include the contribution of the auxiliary field χ defined in footnote f. This issue will be discussed in more detail in ref. 25.

[h]See, for example, p. 246 of ref. 16.

[i]To compare with ref. 16, note that $(2\pi)^3|\mathcal{R}_q^0|^2$ in this reference corresponds to $P_{\mathcal{R}}(q)$, and is given on pp. 482 and 491. To compare with ref. 15, note that $\mathcal{P}_\zeta(k)$, given on p. 406, corresponds to $k^3 P_{\mathcal{R}}(k)/(2\pi^2)$, as described on p. 89, and that M_{Pl} is the reduced Planck mass, $1/\sqrt{8\pi G}$. To compare with ref. 18, use $P_\zeta(k) = P_{\mathcal{R}}(k)$, where $P_\zeta(k)$ is given on p. 170. As explained in ref. 16, ζ and $\mathcal{R}$ refer to slightly different quantities, but they agree for wavelengths long compared to the Hubble length. To compare with ref. 2, note that $S = K/(a^2 H^2)$, and that the equation for $S(t' = 0)$ describes the conditions just after the end of inflation, with the scale factor $R(t) = e^{\chi t}$, $\chi = H_{\text{inf}}$. The quantum fluctuations are quantified in this paper by $\delta t = \delta\varphi/\dot{\varphi}_0$, with $\Delta\varphi^2 = k^3 P_{\delta\varphi}(k)/(2\pi)^3 = H^2/(16\pi^3)$, in agreement with Eq. (22) of the current paper.

for $\mathcal{P}_\mathcal{R}(k)$.) If these fluctuations come from single-field slow-roll inflation, we can conclude from Eqs. (37) and (23) that at the time of Hubble exit,

$$\frac{V^{3/2}}{M_{\mathrm{Pl}}^3 V'} = 5.36 \times 10^{-4} \; , \tag{38}$$

where $M_{\mathrm{Pl}} = 1/\sqrt{8\pi G} \approx 2.44 \times 10^{18}$ GeV is the reduced Planck mass.

5. Deducing the Consequences, Comparing with Observation

From Eqs. (37) and (23) we can also deduce how the intensity of the fluctuations varies with k, a relation which is parameterized by the scalar spectral index $n_s(k)$, defined by

$$n_s - 1 = \frac{d\ln \mathcal{P}_\mathcal{R}(k)}{d\ln k} \; . \tag{39}$$

To evaluate this expression for slow-roll inflation, we use the last expression in Eq. (23) for $P_\mathcal{R}(k)$; we recall that the expression is to be evaluated at t_{ex} (or some fixed number of Hubble times later), and find that Eq. (21) leads to $d\ln k/dt_{\mathrm{ex}} = H$. Then using Eq. (9) to write $\dot\varphi_0 = -V'/(3H)$, these equations can be combined to give[30]

$$n_s - 1 = -\frac{V'}{8\pi G V}\frac{d\ln(V^3/V'^2)}{d\varphi} = -6\epsilon + 2\eta \approx 4\frac{\dot H}{H^2} - \frac{\ddot H}{H\dot H} \; . \tag{40}$$

The WMAP seven-year paper[29] quotes $n_s - 1 = -0.032 \pm 0.012$. This result suggests that the slow roll parameters are indeed quite small, and furthermore they have very plausible values. The time of Hubble exit is typically of order 60 Hubble times before the end of inflation, depending mainly on the reheat temperature, which means that the natural time scale of variation is of order $60H^{-1}$. If each time derivative in the right-hand expressions of Eqs. (14) and (15) is replaced by a factor of $H/60$, one sees that the slow roll parameters are plausibly of order $1/60$.

The case $n_s = 1$ is called scale-invariant, because it means that $\mathcal{P}_\mathcal{R}(k)$ is independent of k; that is, each mode has the same strength, at the time of Hubble exit, as any other mode. Since $\mathcal{P}_\mathcal{R}(k)$ is constant while the wavelength is long compared to the Hubble length, all modes also have the same strength at the time of Hubble reentry. Single-field inflation produces density fluctuations that are approximately scale-invariant, because all the modes that are visible today passed through Hubble exit during a small interval of time during inflation, so the conditions under which they were generated were very similar.

In addition to the nearly scale-invariant spectrum that we just calculated, there are two other key features of the density fluctuations that follow as a consequence of slow-roll single-field inflation. The first is that the fluctuations are adiabatic, which means that every component of the matter in the universe — the photons, the baryons, and the dark matter — fluctuate together. The temperature can be related to the density of photons, so it also fluctuates with the density baryons or dark matter. The reason for this feature is clear, because the time delay affects all properties of the matter in the universe the same way. Until the perturbations reenter the Hubble length (after which complicated things can happen), every region of space behaves just like any other region of space, except for a time offset. Thus the matter content of any one region can differ from that of some other region by at most an adiabatic compression or expansion. The WMAP team[29] has tested this relation for the possibility of non-adiabatic fluctuations between photons and cold dark matter. By combining WMAP data with other data, they find at the 95% confidence level that the non-adiabatic component is at most 6% of the total in the case of "axion-type" perturbations, or 0.4% in the case of "curvaton-type" perturbations.

The other key predication of slow-roll single-field inflation is that the perturbations should be Gaussian. Why are they Gaussian? They are Gaussian because $\delta\tilde{\varphi}(\vec{k}, t)$ is calculated in a quantum field theory. The perturbations are small so we expect accurate results at lowest order, which means that we are only calculating free-field-theory expectation values, and they are Gaussian. There are of course higher order corrections, which in a given model can also be calculated, but they are generically very small. So, to first approximation, we expect the answers to be Gaussian, which means in particular that the three-point correlation function should vanish. There has been a lot of effort to look for non-Gaussianity, but so far no convincing evidence for non-Gaussianity has been found.

The calculations shown here stop just after the end of inflation, but with a lot of work by many astrophysicists the calculations have been extended to make detailed predictions for the fluctuations that can be detected today in the cosmic microwave background. The success is beautiful. To process the data, the temperature pattern observed in the CMB is expanded in spherical harmonics, which is the spherical equivalent of Fourier transforming, providing information about how the intensity of the fluctuations varies with angular wavelength. Figure 3[31] shows the observed temperature fluctuations as a function of the multipole number ℓ, using the 7-year WMAP data[29] for $\ell < 800$, and ACBAR data[32] for higher ℓ. The red line is the theoretical curve that comes about by extending the inflationary predictions to the present day in a model with dark energy (Λ) and cold dark matter, using the best-fit parameters found by the WMAP team:[29] $\mathcal{P}_{\mathcal{R}}(0.002 \text{ Mpc}^{-1}) = 2.42 \times 10^{-9}$, $n_s = 0.966$, $\Omega_\Lambda = 0.729$, $\Omega_{\text{dark matter}} = 0.226$, $\Omega_{\text{baryon}} = 0.045$, and $\tau = 0.085$, where τ is the optical depth experienced by the photons since the "recombination" of the primordial plasma at about 380,000 years after the big bang. While there are 6 free parameters, 4 of them have values that are expected on the basis of theory ($n_s \approx 1$) or other observations (Ω_Λ, $\Omega_{\text{dark matter}}$, and Ω_{baryon}), and they agree well. One of the free parameters determines the overall height, so one should not be impressed that the height of the primary peak matches so well. But the location, shape, and relative heights of the peaks are really being predicted by the theory, so I consider it a spectacular success.

For comparison, the graph also shows predictions for several alternative theories,

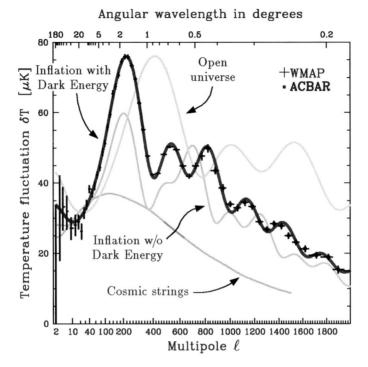

Fig. 3. Comparison of the latest observational measurements of the temperature fluctuations in the CMB with several theoretical models, as described in the text. The temperature pattern on the sky is expanded in multipoles (i.e. spherical harmonics), and the intensity is plotted as a function of the multipole number ℓ. Roughly speaking, each multipole ℓ corresponds to ripples with an angular wavelength of $360°/\ell$.

all of which are now ruled out by this data. The yellow line shows the expected curve for an open universe, with $\Omega_{\text{total}} = 0.30$. The green line shows an inflationary model with $\Omega_{\text{total}} = 1$, but with $\Omega_{\text{dark matter}} = 0.95$ and no dark energy. The magenta line shows the expectations for fluctuations generated by the formation of cosmic strings in the early universe, taken from ref. 33. Structure formation caused by cosmic strings or other "defects" was considered a viable possibility before this data existed, but now cosmic strings are completely ruled out as a major source of density fluctuations.

(There are possibly alternative ways to generate density perturbations with the same properties as those of inflation, but there is not yet a consensus about how easy it is to construct a plausible model. The cyclic ekpyrotic model[34–38] was claimed to naturally produce such fluctuations, but these claims were disputed by a number of authors.[39–42] Now at least some of the founders of ekpyrosis[43,44] agree that the original models do not give a nearly-scale invariant spectrum, as had been claimed. But these papers and others have proposed newer, more sophisticated versions of bouncing universes, generally involving either multiple fields, or settling for scale invariance for only a limited range of scales. Baumann, Senatore, and Zaldarriaga[45] have argued that any single-field model with attractor behavior has to be very close to de Sitter space to remain weakly coupled for at least the required ∼10 e-folds needed to account for observations.)

6. Outstanding Questions About Density Perturbations

There are still a number of important, outstanding questions concerning density perturbations:

(1) Will B-modes be found? Experiments are starting to measure the polarization of the CMB, for which the spherical harmonic expansion for the temperature pattern is replaced by an expansion in E-modes and B-modes.[46] The E-modes are those that can be expressed as gradients of scalar harmonic functions, and they are produced as a by-product of the density perturbations that we have been discussing. The B-modes are orthogonal to the E-modes; they cannot be expressed as gradients of scalar modes, and they cannot be produced by density perturbations. There can be foreground contamination, but the only known primordial source of B-modes is a background of gravitational waves. Thus, gravity waves might be discovered in the CMB before they can be seen directly. The discovery of a primordial gravity wave background would be very exciting, because it is the only thing that will give us a clue about the energy scale at which inflation happened. As far as we know now inflation might have happened anywhere from the electroweak scale up to the grand unified theory (GUT) scale, or a little beyond. The discovery of gravity waves would end the uncertainty, and would also give strong evidence for the inflationary picture. There are, however, many inflationary models for which the energy scale would be too low for the gravitational waves to be visible.

(2) Can sub-Planckian physics influence the calculation of inflationary density perturbations? A typical GUT-scale inflationary model would include about 60 e-folds of inflation, expanding by a factor of $e^{60} \approx 10^{26}$. From the end of inflation to today the universe would expand by another factor of $\sim 10^{15}$ GeV$/3$ K $\approx 10^{27}$. This means that a distance scale of 1 m today corresponds to a length of only about 10^{-53} m at the start of inflation, 18 orders of magnitude smaller than the Planck length ($\sim 10^{-35}$ m). With a little more than the minimal amount of inflation — which would be a certainty in the eternal inflation picture to be discussed below — even the largest scales of the visible universe would have been sub-Planckian at the start of inflation. So, it is relevant to ask whether inflation can possibly offer us a glimpse of sub-Planckian physics. There is of course no solid answer to this question, since there is no real understanding of how this process should be described. Kaloper, Kleban, Lawrence, and Shenker[47] have argued that the perturbations are determined primarily by local effective field theory on the scale of order H, so that sub-Planckian effects would be invisible except possibly in unconventional models for which the fun-

damental string scale is many orders of magnitude below the four-dimensional Planck mass, $\sim 10^{19}$ GeV. Some authors[48,49] have reached similar conclusions, but other authors[50–53] have concluded that the effects might be much easier to see. The conclusions of ref. 47 seem plausible to me, but certainly the role of sub-Planckian physics is not yet fully understood.

(3) Will effects beyond the single-field slow-roll approximation be found? With multiple fields, or with unusual features in the potential for a single field, models can be constructed that predict significant non-Gaussianity, non-adiabaticity, or spectral distortions. There is an active industry engaged in studying models of this sort, and in looking for these nonstandard features in the data. The WMAP seven-year analysis[29] reports "no convincing deviations from the minimal model," but we all await the data from the Planck mission, expected in less than a year, and the data from a variety of ground-based experiments.

7. Fluctuations on Larger Scales: Eternal Inflation?

Since the density perturbation calculations have been incredibly successful, it seems to make sense to take seriously the assumptions behind these calculations, and follow them where they lead. I have to admit that there is no clear consensus among cosmologists, but to many of us the assumptions seem to be pointing to eternal inflation, and the multiverse.

The mechanism for eternal inflation is described most efficiently by separating the cases of the two types of potential functions shown in Fig. 1. For the new inflation case, that state for which the scalar field is poised on the top of the potential hill is a metastable state, often called a "false vacuum," which decays by the scalar field rolling down the hill. This state decays exponentially, but in any working model of inflation the half-life of the decay is much longer than the doubling time associated with the exponential expansion. Thus, if we follow a region for a period of one half-life, at the end of the period only half of the original region would be still be inflating. However, the half that is still inflating will have a volume vastly larger than the volume of the entire region at the start, so the process will go on forever. Each decay will lead to the production of a "pocket" universe, and the creation of pocket universes will go on forever, as pieces of the ever-growing false vacuum region undergo decays. Once inflation starts, it never stops.[j]

For the case of a chaotic-type potential, as in Fig. 1(b), naively one would think that the field would inexorably roll down the hill in some finite amount of time. However, Linde[57] discovered that when quantum fluctuations are taken into account, this need not be the case. To understand this, consider an inflating region of space of size H^{-1}, with the inflaton field φ approximately uniform over this region, at some value φ_0. After one Hubble time (H^{-1}) the region will have expanded by

---

[j]The first models of eternal new inflation were proposed by Steinhardt[54] and Linde.[55] Vilenkin[56] was the first to describe eternal inflation as a generic feature of new inflation.

$e^3 \approx 20$, and can be viewed as 20 Hubble-sized regions which will start to evolve independently. The average field φ in any one of these regions will usually be lower than φ_0, due to the classical rolling down the hill, but the classical evolution will be modified by random quantum jumps, which can be estimated as $\sim H/(2\pi)$. It is therefore possible that in one or more of these 20 regions, φ can equal or exceed φ_0. A back-of-the-envelope calculation shows that if

$$\frac{H^2}{|\dot{\varphi}|} \gtrsim 5 \,, \tag{41}$$

then the expectation value for the number of regions with $\varphi > \varphi_0$ is greater than one. That implies that the number of Hubble-sized regions with $\varphi > \varphi_0$ will grow exponentially with time, and the inflation becomes eternal. Note that $H^2/|\dot{\varphi}| \approx \sqrt{\mathcal{P_R}} \approx (GV)^{3/2}/|V'|$, so the eternally inflating behavior is really the large-φ, long-wavelength, tail of the density perturbation spectrum. Since $V^{3/2}/|V'|$ grows without bound as $\varphi \to \infty$ for most potentials under consideration, almost all models allow for eternal inflation.

There is certainly no proof that we live in a multiverse, but I will argue that there are three winds — that is, three independent scientific developments, arising from three different branches of science — which seem to be leading to the multiverse picture.

(1) *Theoretical Cosmology: Eternal Inflation.* As I just described, almost all inflationary models are eternal into the future.

(2) *String Theory: The Landscape.* String theory predicts that there is not just one kind of vacuum, but instead there are a colossal number of them: 10^{500} or maybe more.[58,59] The underlying laws of physics would be the same everywhere, but nonetheless each type of vacuum would create an environment in which the low-energy laws of physics would be different. Thus, if there is a multiverse, it would be a varied multiverse, in which the different pocket universes would each appear to have their own laws of physics.

(3) *Observational Astronomy: the Cosmological Constant.* The third "wind" has its roots in the fine-tuning that our universe appears to exhibit. In the past a minority of physicists argued that things such as the properties of ice or the energy levels of carbon-12 appeared to be fine-tuned for the existence of life, but not very many scientists found this convincing. If these properties were different, then maybe life would form some other way. However, a form of fine-tuning that many of us find much more convincing became evident starting in 1998, when two groups of astronomers[60,61] announced that the expansion of the universe is not slowing down due to gravity, but is in fact accelerating. The simplest explanation is that the acceleration is caused by a nonzero energy density of the vacuum, also known as a cosmological constant. But that would mean that the vacuum energy density is nonzero, yet a full 120 orders of magnitude smaller than the Planck scale (M_{Pl}^4, where $M_{\mathrm{Pl}} = 1/\sqrt{G}$), the scale that most theoretical physicists would consider natural. Physicists have struggled to find a physical

explanation for this small vacuum energy density, but no generally accepted solution has been found. But if the multiverse is real, the problem could go away. With 10^{500} different types of vacuum, a small fraction, but nonetheless a large number of them, would be expected to have an energy density as small as what we observe. The smallness of the vacuum energy density would be explained, therefore, if we could explain why we should find ourselves in such an unusual part of the multiverse. But as pointed out by Weinberg and collaborators[62,63] some time ago, there is a selection effect. If we assume that life requires the formation of galaxies, then one can argue that life in the multiverse would be concentrated in those pocket universes with vacuum energy densities in a narrow band about zero. Thus, while a typical vacuum energy density in the multiverse would be on the order of the Planck scale, almost all life in the multiverse would find a small value, comparable to what we see.

While the multiverse picture looks very plausible in the context of inflationary cosmology — at least to me — it raises a thorny and unsolved problem, known as the "measure problem." Specifically, we do not know how to define probabilities in the multiverse. If the multiverse picture is right, then anything that can happen will happen an infinite number of times, so any distinction between common and rare events requires the comparison of infinities. Such comparisons are not mathematically well-defined, so we must adopt a recipe, or "measure," to define them. Since the advent of quantum theory essentially all physical predictions have been probabilistic, so probability is not a concept that we can dispense with. To date we do not understand the underlying physical basis for such a measure, but much progress has been made in examining proposals and ruling out many of them.[k]

One might guess that this problem is easily handled by choosing a finite sample spacetime region in the multiverse, calculating the relative frequencies of different types of occurrences in the sample region, and then taking the limit as the region becomes infinite. This seems like a very reasonable approach, and in fact most of the measure proposals that have been discussed are formulated in this way. The problem is that the answers one obtains are found to depend sensitively on the method that is used to choose the sample region and to allow it to grow. The dependence on the method of sampling seems surprising, but it sounds plausible if we remember that the volume of the multiverse grows exponentially with time. A sample spacetime region will generally have some final time cutoff, and the spacetime volume will generally grow exponentially with the cutoff. But then, no matter how large the cutoff is taken, the volume will always be dominated by a region that is within a few time constants of the final time cutoff hypersurface. No matter how large the final time cutoff is taken, the statistics will never be dominated by the interior of the sample region, but instead instead will be dominated by the final time cutoff

---

[k]For a recent summary of a field that is in a state of flux, see ref. 64. For a new proposal that was advanced since this summary, see ref. 65.

surface. For that reason, it is not surprising that the method of choosing this surface will always affect the answers.

There are a number of important questions, in the multiverse picture, that depend very crucially on the choice of measure. How likely is it that we observe a vacuum energy density as low as what we see? How likely is it that our universe has a mass density parameter Ω sufficiently different from 1 that we can hope to measure the difference? How likely is it that we might find evidence that our pocket universe collided with another sometime in its history? And, if there are many vacua in the landscape of string theory with low energy physics consistent with what we have measured so far, how likely is it that we will find ourselves in any particular one of them? If we live in a multiverse, then in principle all probabilities would have to be understood in the context of the multiverse, but it seems reasonable to expect that any acceptable measure would have to agree to good accuracy with calculations that we already know to be successful.

As discussed in ref. 64, we can identify a class of measures that give reasonable answers. It seems plausible that the ultimate solution to this problem will give similar answers, but the underlying principles that might determine the right answer to this question remain very mysterious. Nonetheless, the success of inflation in explaining the observed properties of the universe, including the density perturbation predictions discussed here, provides strong motivation to expect that some solution to the measure problem will be found.

References

1. A. D. Sakharov, The initial stage of an expanding universe and the appearance of a nonuniform distribution of matter, *Zh. Eksp. Teor. Fiz.* **49**, 345 (1965) [*JETP Lett.* **22**, 241 (1966)].
2. A. H. Guth and S.-Y. Pi, Fluctuations in the new inflationary universe, *Phys. Rev. Lett.* **49**, 1110 (1982).
3. A. A. Starobinsky, Dynamics of phase transition in the new inflationary universe scenario and generation of perturbations, *Phys. Lett. B* **117**, 175 (1982).
4. S. W. Hawking, The development of irregularities in a single bubble inflationary universe, *Phys. Lett. B* **115**, 295 (1982).
5. J. M. Bardeen, P. J. Steinhardt, and M. S. Turner, Spontaneous creation of almost scale-free density perturbations in an inflationary universe, *Phys. Rev. D* **28**, 679 (1983).
6. V. F. Mukhanov and G. V. Chibisov, Quantum fluctuations and a nonsingular universe, *Pis'ma Zh. Eksp. Teor. Fiz.* **33**, 549 (1981) [*JETP Lett.* **33**, 532 (1981)]; Vacuum energy and large-scale structure of the universe, *Zh. Eksp. Teor. Fiz.* **83**, 475 (1982) [*JETP* **56**, 258 (1982)].
7. G. Hinshaw et al., *Nine-Year Wilkinson Microwave Anisotropy Probe (WMAP) Observations: Cosmological Parameter Results,* arXiv:1212.5226 [astro-ph.CO].
8. A. A. Starobinsky, A new type of isotropic cosmological models without singularity, *Phys. Lett. B* **91**, 99 (1980).
9. A. A. Starobinsky, The perturbation spectrum evolving from a nonsingular, initially de Sitter cosmology, and the microwave background anisotropy, *Pis'ma Astron. Zh.* **9**, 579 (1983) [*Sov. Astron. Lett.* **9**, 302 (1983)].

10. V. F. Mukhanov, Quantum theory of cosmological perturbations in R^2 gravity, *Phys. Lett. B* **218**, 17 (1989).

11. M. Sasaki and E. D. Stewart, A general analytic formula for the spectral index of the density perturbations produced during inflation, *Prog. Theor. Phys.* **95**, 71 (1996), arXiv:astro-ph/9507001v2.

12. B. A. Bassett, S. Tsjikawa, and D. Wands, Inflation dynamics and reheating, *Rev. Mod. Phys.* **78**, 537 (2006), arXiv:astro-ph/0507632v2.

13. K. A. Malik and D. Wands, Cosmological perturbations, *Phys. Rept.* **475**, 1 (2009), arXiv:0809.4944 [astro-ph].

14. V. F. Mukhanov, H. A. Feldman, and R. H. Brandenberger, Theory of cosmological perturbations, *Phys. Rept.* **215**, 203 (1992).

15. D. H. Lyth and A. R. Liddle, *The Primordial Density Fluctuation: Cosmology, Inflation, and the Origin of Structure* (Cambridge University Press, Cambridge, 2009).

16. S. Weinberg, *Cosmology* (Oxford University Press, Oxford, 2008).

17. V. Mukhanov, *Physical Foundations of Cosmology* (Cambridge University Press, Cambridge, 2005).

18. S. Dodelson, *Modern Cosmology* (Academic Press, San Diego, CA, 2003).

19. E. Noether, *Invariante Variationsprobleme (Invariant Variation Problems), Nachr. d. König. Gesellsch. d. Wiss. zu Göttingen, Math-phys. Klasse,* **235** (1918); For an English translation by M. A. Tavel, see arXiv:physics/0503066v1 [physics.hist-ph], reproduced by F. Wang.

20. A. H. Guth, The inflationary universe: A possible solution to the horizon and flatness problems, *Phys. Rev. D* **23**, 347 (1981).

21. A. D. Linde A new inflationary universe scenario: a possible solution of the horizon, flatness, homogeneity, isotropy and primordial monopole problems, *Phys. Lett. B* **108**, 389 (1982).

22. A. Albrecht and P. J. Steinhardt, Cosmology for grand unified theories with radiatively induced symmetry breaking, *Phys. Rev. Lett.* **48**, 1220 (1982).

23. A. D. Linde, Chaotic inflation, *Phys. Lett. B* **129**, 177 (1983).

24. L. -M. Wang, V. F. Mukhanov and P. J. Steinhardt, On the problem of predicting inflationary perturbations, *Phys. Lett. B* **414**, 18 (1997) [astro-ph/9709032].

25. A. H. Guth, to appear.

26. T. S. Bunch and P. C. W. Davies, Quantum field theory in de Sitter space: Renormalization by point splitting, *Proc. Roy. Soc. Lond. A* **360**, 117 (1978).

27. G. W. Gibbons and S. W. Hawking, Cosmological event horizons, thermodynamics, and particle creation, *Phys. Rev. D* **15**, 2738 (1977).

28. A. R. Liddle and D. H. Lyth, *Cosmological Inflation and Large-Scale Structure* (Cambridge University Press, Cambridge, 2000).

29. E. Komatsu *et al.* [WMAP Collaboration], Seven-Year Wilkinson Microwave Anisotropy Probe (WMAP) Observations: Cosmological interpretation, *Astrophys. J. Suppl.* **192**, 18 (2011) [arXiv:1001.4538 [astro-ph.CO]].

30. A. R. Liddle and D. H. Lyth, COBE, gravitational waves, inflation and extended inflation, *Phys. Lett. B* **291**, 391 (1992) [astro-ph/9208007].

31. This graph was prepared by Max Tegmark. An earlier version of it appeared in A. H. Guth and D. I. Kaiser, Inflationary cosmology: Exploring the universe from the smallest to the largest scales, *Science* **307**, 884 (2005) [astro-ph/0502328].

32. C. L. Reichardt *et al.* [ACBAR Collaboration], High resolution CMB power spectrum from the complete ACBAR data set, *Astrophys. J.* **694**, 1200 (2009) [arXiv:0801.1491 [astro-ph]].

33. U. -L. Pen, U. Seljak and N. Turok, Power spectra in global defect theories of cosmic structure formation, *Phys. Rev. Lett.* **79**, 1611 (1997) [astro-ph/9704165].

34. J. Khoury, B. A. Ovrut, P. J. Steinhardt and N. Turok, The Ekpyrotic universe: Colliding branes and the origin of the hot big bang, *Phys. Rev. D* **64**, 123522 (2001) [hep-th/0103239].

35. J. Khoury, B. A. Ovrut, P. J. Steinhardt and N. Turok, Density perturbations in the Ekpyrotic scenario, *Phys. Rev. D* **66**, 046005 (2002) [hep-th/0109050].

36. A. J. Tolley, N. Turok and P. J. Steinhardt, Cosmological perturbations in a big crunch / big bang space-time, *Phys. Rev. D* **69**, 106005 (2004) [hep-th/0306109].

37. J. Khoury, P. J. Steinhardt and N. Turok, Inflation versus cyclic predictions for spectral tilt, *Phys. Rev. Lett.* **91**, 161301 (2003) [astro-ph/0302012].

38. J. Khoury, P. J. Steinhardt and N. Turok, Designing cyclic universe models, *Phys. Rev. Lett.* **92**, 031302 (2004) [hep-th/0307132].

39. R. Brandenberger and F. Finelli, On the spectrum of fluctuations in an effective field theory of the Ekpyrotic universe, *JHEP* **0111**, 056 (2001) [hep-th/0109004].

40. D. H. Lyth, The Primordial curvature perturbation in the Ekpyrotic universe, *Phys. Lett. B* **524**, 1 (2002) [hep-ph/0106153].

41. S. Tsujikawa, R. Brandenberger and F. Finelli, On the construction of nonsingular pre - big bang and ekpyrotic cosmologies and the resulting density perturbations, *Phys. Rev. D* **66**, 083513 (2002) [hep-th/0207228].

42. P. Creminelli, A. Nicolis and M. Zaldarriaga, Perturbations in bouncing cosmologies: Dynamical attractor versus scale invariance, *Phys. Rev. D* **71**, 063505 (2005) [hep-th/0411270].

43. J. Khoury and G. E. J. Miller, Towards a cosmological dual to inflation, *Phys. Rev. D* **84**, 023511 (2011) [arXiv:1012.0846 [hep-th]].

44. J. Khoury and P. J. Steinhardt, Generating scale-invariant perturbations from rapidly-evolving equation of state, *Phys. Rev. D* **83**, 123502 (2011) [arXiv:1101.3548 [hep-th]].

45. D. Baumann, L. Senatore and M. Zaldarriaga, Scale-invariance and the strong coupling problem, *JCAP* **1105**, 004 (2011) [arXiv:1101.3320 [hep-th]].

46. M. Zaldarriaga and U. Seljak, An all sky analysis of polarization in the microwave background, *Phys. Rev. D* **55**, 1830 (1997) [astro-ph/9609170].

47. N. Kaloper, M. Kleban, A. E. Lawrence and S. Shenker, Signatures of short distance physics in the cosmic microwave background, *Phys. Rev. D* **66**, 123510 (2002) [hep-th/0201158].

48. N. E. Groeneboom and O. Elgaroy, Detection of transplanckian effects in the cosmic microwave background, *Phys. Rev. D* **77**, 043522 (2008) [arXiv:0711.1793 [astro-ph]].

49. A. Avgoustidis, S. Cremonini, A. -C. Davis, R. H. Ribeiro, K. Turzynski and S. Watson, Decoupling survives inflation: A critical look at effective field theory violations during inflation, arXiv:1203.0016 [hep-th].

50. J. Martin and R. H. Brandenberger, The transplanckian problem of inflationary cosmology, *Phys. Rev. D* **63**, 123501 (2001) [hep-th/0005209]; R. H. Branbenberger and J. Martin, The robustness of inflation to changes in superplanck scale physics, *Mod. Phys. Lett.* **A16**, 999 (2001) [astro-ph/0005432]; R. Brandenberger and X. -m. Zhang, The Trans-Planckian problem for inflationary cosmology revisited, arXiv:0903.2065 [hep-th].

51. A. Kempf and J. C. Niemeyer, Perturbation spectrum in inflation with cutoff, *Phys. Rev. D* **64**, 103501 (2001) [astro-ph/0103225].

52. R. Easther, B. R. Greene, W. H. Kinney and G. Shiu, Inflation as a probe of short distance physics, *Phys. Rev. D* **64**, 103502 (2001) [hep-th/01044102; *Ibid.*, Imprints of short distance physics on inflationary cosmology, *Phys. Rev. D* **67**, 063508 (2003)

[hep-th/0110226]; G. Shiu, Inflation as a probe of trans-Planckian physics: A brief review and progress report, *J. Phys. Conf. Ser.* **18**, 188 (2005).

53. H. Collins and R. Holman, Trans-Planckian enhancements of the primordial non-Gaussianities, *Phys. Rev. D* **80**, 043524 (2009) [arXiv:0905.4925 [hep-ph]].
54. P. J. Steinhardt, Natural inflation, in *The Very Early Universe,* Proceedings of the Nuffield Workshop, Cambridge, 21 June – 9 July, 1982, eds. G. W. Gibbons, S. W. Hawking, and S. T. C. Siklos (Cambridge University Press, Cambridge, 1983), pp. 251–66.
55. A. D. Linde, Nonsingular regenerating inflationary universe, unpublished preprint Print-82-0554 (Cambridge).
56. A. Vilenkin, The birth of inflationary universes, *Phys Rev D* **27**, 2848 (1983).
57. A. D. Linde, Eternally existing selfreproducing chaotic inflationary universe, *Phys Lett B* **175**, 395 (1986).
58. R. Bousso and J. Polchinski, Quantization of four form fluxes and dynamical neutralization of the cosmological constant, *JHEP* **0006**, 006 (2000) [hep-th/0004134].
59. L. Susskind, The anthropic landscape of string theory, in *Universe or Multiverse?*, ed. B. Carr (Cambridge University Press, 2007), pp. 247–266 [hep-th/0302219].
60. A. G. Riess *et al.* [Supernova Search Team Collaboration], Observational evidence from supernovae for an accelerating universe and a cosmological constant, *Astron. J.* **116**, 1009 (1998) [astro-ph/9805201].
61. S. Perlmutter *et al.* [Supernova Cosmology Project Collaboration], Measurements of Omega and Lambda from 42 high redshift supernovae, *Astrophys. J.* **517**, 565 (1999) [astro-ph/9812133].
62. S. Weinberg, Anthropic bound on the cosmological constant, *Phys. Rev. Lett.* **59**, 2607 (1987).
63. H. Martel, P. R. Shapiro and S. Weinberg, Likely values of the cosmological constant, *Astrophys. J.* **492**, 29 (1998) [astro-ph/9701099].
64. B. Freivogel, Making predictions in the multiverse, *Class. Quant. Grav.* **28**, 204007 (2011) [arXiv:1105.0244 [hep-th]].
65. Y. Nomura, Physical theories, eternal inflation, and quantum universe, *JHEP* **1111**, 063 (2011) [arXiv:1104.2324 [hep-th]].

Prepared comments

V. Mukhanov: Quantum Fluctuations in Cosmology

I would like to address the question what the theory of quantum origin of the universe structure really predicts and how these predictions come in agreement with the most recent observations of the Cosmic Microwave Background (CMB) fluctuations. In 1980 we have found that the quantum metric fluctuations can explain the observable structure of the universe if and only if the *expanding* universe went through the stage of cosmic acceleration. This stage is now called cosmic inflation. At present there are hundreds of different inflationary scenarios and to understand what the theory really predicts it is convenient to describe inflation using the effective hydrodynamical approach, when the state of the matter is entirely characterized by its energy density ε and the pressure p. We assume that in the past the universe went through the stage when dark energy with equation of state $p \approx -\varepsilon$ was dominating and, hence, the universe was accelerating. The cosmological constant corresponding to the equation of state $p = -\varepsilon$ cannot serve our purpose because finally one has to have graceful exit from inflation. Therefore, from the very beginning there should be small deviations of the equation of state from the cosmological constant, that it, $(\varepsilon + p)/\varepsilon \ll 1$, but non-vanishing. This ratio, which should be smaller that about 10^{-2} to provide us necessary duration of inflation, grows and finally when it becomes of order unity inflation ends and the decelerated expansion begins. One can realize the needed equation of state using the condensates of the scalar fields, the R^2 gravity and in some other ways. The key point is that the microscopic origin of the dark energy does not play crucial role for the predictions. Everything what we need is the "decaying cosmological constant". Then irrespective of the initial conditions one can make concrete robust predictions for observations. What are these predictions?

- First of all, inflation predicts that the cosmological parameter Ω should be equal to unity within an accuracy about 10^{-5}, which means that at present the universe has a flat Euclidean geometry. This prediction was first confirmed only at the end of the nineties and the experimental/observational data were in strong disagreement with it before that.

The other set of the robust predictions concerns the amplified quantum fluctuations:[a]

- The produced inhomogeneities should be adiabatic. I would like to stress that about thirty years ago the observational data were more supportive for the entropy perturbations. However, nowadays they are ruled out by the precision CMB measurements, which confirmed adiabatic nature of the primordial inhomogeneities.

---

[a]V. Mukhanov and G. Chibisov, *JETP Lett.* **33**, No.10, 532 (1981).

- The primordial inhomogeneities are Gaussian. This is related with the fact that these inhomogeneities were originated as a result of amplification of the initial Gaussian quantum fluctuations by the external gravitational field. The expected corrections to the Gaussian gravitational potential Φ_g, due to nonlinear corrections to the linearized Einstein equations are of order $O\left(1\right)\Phi_g^2$, that is, $\Phi = \Phi_g + f_{NL}\Phi_g^2$. The present experimental bound $-10 < f_{NL} < 70$ is in agreement with the prediction of the theory, according to which f_{NL} is expected to be about ten or so. The expected accuracy of the Planck mission $\Delta f_{NL} \simeq 5$ will allow us to improve further the bound on the non-gaussianity.

- The most nontrivial prediction for the perturbations is the weak scale dependence of the amplitude of the gravitational potential of the generated perturbations. Namely, the value of the potential Φ should logarithmically depend on the scale λ, that is, $\Phi\left(\lambda\right) \propto \ln\left(\lambda/\lambda_\gamma\right)$. Within the observable range of the scales this logarithm can be approximated as $\Phi \propto \lambda^{1-n_s}$, where $n_s \simeq 0.96 \div 0.97$. The exact value for n_s depends on the unknown particle physics at the energy scales above TeV. However, this uncertainty is not more than about one percent. The logarithmic dependence of the spectrum has deep physical origin and is due to a small deviation of the equation of state from cosmological constant of order few percents during the last, relevant for observations, 50-70 e-folds. From the most recent WMAP, ACT and SPT measurements it follows that $n_s = 0.966 \pm 0.011$ in an excellent agreement with the prediction of the theory. The accuracy of the determination of spectral index from the Planck mission is expected to be at least twice better. However already now the logarithmic dependence of the gravitational potential is confirmed at the level of 3σ.

If any of the above predictions would contradict to the observations then inflation as a predictive theory (sometimes called simple inflation) would be ruled out. Of course there are more complicated (multi-parameter) scenarios of inflations which could "explain" nearly any outcome of the measurements. However, these scenarios are not experimentally falsifiable and hence cannot be ruled out or confirmed.

- One more robust prediction of inflation is the existence of the long-wave gravitational waves.[b] Their amplitude is not predicted by inflation. For instance, R^2 and Higgs inflation give us much less amount of the gravity waves than, for instance, the $m^2\phi^2$ inflation. Hence non-detecting these waves would not rule out the predictive inflationary theory. On the other hand their detection will provide an additional evidence that the universe went through the stage of cosmic acceleration.

I would like to stress that the predictions mentioned above are extremely non-

---

[b]A. Starobinsky, *JETP Lett.* **30**, No.11, 682 (1979).

trivial and were for a long time in conflict with observations. For example, in the 80th, along with the theory of quantum initial perturbations there were competing theories of cosmic strings, textures and entropy perturbations. Now all of them are ruled out by observations and only the theory of quantum cosmological perturbations with all its nontrivial predictions is brilliantly confirmed by observations. Moreover, although there are still the claims in the literature that there are alternatives to inflation, there is no any alternative to the quantum origin of the universe structure.

E. Silverstein: Inflationary Theory and the Quantum Gravity World

Inflationary cosmology is UV-sensitive, requiring small coefficients for higher dimension Planck-suppressed operators. The Lyth bound relates detectable tensor modes to a super-Planckian field excursion. A complete treatment of inflation requires control of these effects descending from the UV completion of gravity and particle physics.

There are several UV-complete inflationary mechanisms developed thus far in string theory, with different properties. The process can be controlled in a Wilsonian-natural way by a shift symmetry. Axionic fields (one or multiple-field) in string theory realize this idea, with a potential which is generically unwound by monodromy and flatter than $m^2\phi^2$ inflation. This leads to distinctive predictions for the tilt and the tensor to scalar ratio. Various earlier models with small field range make different predictions, some leading to cosmic string production and others falsifiable via their level and shape of non-Gaussianity. Additional progress in organizing the perturbations more systematically has come from low energy effective field theory. Of course the inverse problem is nontrivial. But this interface will get very interesting in the near future as observations will cross important thresholds, distinguishing between very different mechanisms for inflation.

E. Rabinovici: Singularities

My intervention is based on work with Jose Barbon over the last years.

During the first Solvay meetings the participants were challenged by the appearance of various singularities in the classical theory of light and matter. They also struggled to clarify in their own minds the precise meaning and significance of the new particle/wave duality. They struggled to become fluent in both narratives as well as to establish and master the dictionary/complementarity between them.

I would like to discuss today how merging different narratives/symmetries of string theory can shed a new light on some singularity issues in Gravity. The provocative take away slogan is: *String theory can live with some big*

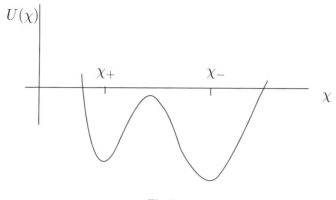

Fig. 1.

bang/crunch singularities!
I start on the symmetry side. When one probes the properties of a system with extended objects the different mathematical characterizations of the system become ambiguous. There are examples where the absolute meaning of each of the following notions crumbles. Length, topology, number of dimensions (for both large and small spaces), commutativity of the manifold and time like singularities. One case is when a large system can be narrated as if it was a small system, this is but the tip of an iceberg. I find it unsatisfying that this magic needs to be dug out on a case by case basis. A lack of a description in which all these symmetries and more are explicitly evident is lacking. This call for more "transparency" may be related to that expressed by Nati Seiberg yesterday. We will add a new pair of narratives to address a big crunch space singularity.

The singularity side: the various singularity issues in gravity were very nicely described by Gary Gibbons in the 2005 Solvay meeting. Usually the detection of singularities is a reflection of the breakdown of the physical picture used: one can add QM, one needs to add forgotten massless states etc. In gravity there may be new aspects for example one may hide behind the skirt of a horizon. Consider a class of big crunches. This is done in an AdS set up so that Holography can provide a guide. Two dualities will be used: one relating the bulk crunch to a boundary theory where there is no explicit evidence for the crunch and the other, a complementarity conformal mapping between that boundary theory and one with an explicit big crunch. In figure 1 one sees an effective field theory potential with a metastable vacuum and a stable one. Both with negative energy. In the absence of gravity in flat space a metastable state will always decay into the true vacuum. In the presence of gravity the negative cosmological constant would require a AdS background and the "metastable" state would either remain stable or generically crunch. This depends on the parameters of the system. It is the crunch case that we consider. It can be mapped (Thanks

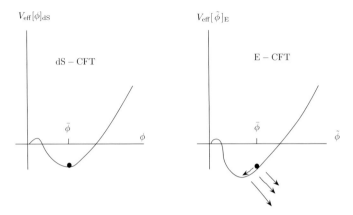

Fig. 2.

Juan) holographically to a boundary theory which (for a small expansion rate) is generically a well defined theory on an expending dS world volume with time independent couplings. There is nothing singular in that calm narrative. Moreover the big crunch features are recovered by a conformal complementarity map for which the world volume is Einstein and time independent. The couplings are time dependent and diverge at a fixed time beyond which the system ceases to exist. A crunch worthy of it name. In more detail the boundary theory reflects the big crunch when it exhibits a bread and butter spontaneous symmetry breaking associated with a large number of massless degrees of freedom. This is seen in the left of figure 2. This is a well defined potential in the dS narrative. In the complementary system, the right of figure 2, the potential is bounded from below for any time before the crunch and an infinite energy gap forms at the moment of the crunch. From AdS/CFT we have already gained confidence that if a finite amount of entropy is behind a BH horizon one can recover it on the boundary even though the explicit complementarity map is not yet known. Here we suggest an explicit complementarity map for a case where an infinite amount of entropy is at stake.

This leads us to suggest that string theory can take care of at least some space like singularities without needing to wait for a new knight on a white horse. A more profound understanding of systems which are defined only for finite times should be addressed, on the route of understanding this one may need to confront the possibility that time can both emerge and dissolve. It may turn out that because of yet to be discovered properties these singularities will be regulated, this will require among other things that dS would be unstable and that on $O(d, 1)$ symmetry present here be broken to $O(d) \times U(1)$. At this stage we see no need for that to happen. Dualities can take the sting out of these singularities. Thank you.

E. Verlinde: Quantum Gravity and String Theory

I start by presenting a formula that empirically relates dark matter, dark energy, and ordinary matter. It follows from the observed rotations curves and velocity dispersion profiles in galaxies and clusters of galaxies. As is well known, these profiles do not obey the laws of Newton if one just includes baryonic matter. One has to add a dark component Φ_D to the Newton potential.

The energy in the gravitational field up to a radius R is given by

$$E = \frac{1}{8\pi G} \int_{r \leq R} |\nabla \Phi_D|^2$$

One finds from the data, that this gravitational energy due to the dark matter obeys a relation which rather magically, involves the present day Hubble constant H_0. It takes the suggestive form

$$E = \frac{1}{2} N_B kT,$$

where

$$kT = \frac{\hbar H_0}{2\pi}, \quad N_B = \frac{M_B c R}{\hbar}$$

Note that T is exactly the Gibbons Hawking temperature. This formula not only describes galaxies, but also clusters of galaxies, even the universe: i4 percent baryonic matter leads to precisely 22.5 percent dark matter, as observed.

Any explanation of this relation, even a tentative one, should be taken seriously. I will argue that this relation can not be explained using our conventional framework, but is a reflection of the emergent nature of space-time, and provides strong support for ideas that come from string theory.

What is string theory? String theory is an effective framework, just like quantum field theory. I believe there exists a refinement of the Wilsonian renormalization group in which coupling constants become dynamical fields. While integrating out degrees of freedom, one integrates in the coupling fields. This is the generalized RG step. At a fixed point, this step becomes an invariance, and gives the familiar UV/IR correspondence of string theory which makes it look UV complete. These properties are emergent, however, as is the apparent duality between gravity and gauge theory.

What does this imply for gravity in our physical world? In string theory we can only count black hole states microscopically in a regime that does not involve space time coordinates, called the Higgs branch. It represented by variables that live in an abstract space, and schematically can be thought of as bi-fundamentals characterized by two very large integers N_A and N_B. Its phase space volume is given by a universal expression in terms of these integers: $S = 2\pi\sqrt{N_A N_B}$, hence it is a dynamical system with a very large Hilbert space.

Space time emerges in the transition from the Higgs branch to the so-called Coulomb branch. This transition can be described with the help of matrix valued coordinates, whose eigenvalues describe the positions of the smallest bits of matter. The size of these matrices is given by N_B, the smallest of the two number N_A and N_B. I will argue that this number of bits depends on the size of the system, and is precisely given the product of the mass and the radius. In fact, yesterday, Gia Dvali came to a similar conclusion from a very different starting point.

The forces of Nature emerge together with space time in a very natural way. In the past ordinary matter particles entered the Coulomb branch, which is represented by space time. But a big part of the full wave-function still lives on the Higgs branch. The fast dynamics on the Higgs branch is driven by the slow motion on the Coulomb branch. This leads to a reaction force. As I learned from the papers of Michael Berry, these reaction forces can be studied in a systematic way. The leading reaction force is the adiabatic Born-Oppenheimer force. In the next order a force emerges due to the Berry phase, and there can even be situations with non-abelian Berry phases. This is the way that gauge forces emerge. The adiabatic force, I claim, is gravity, or actually inertia. It is determined by the principle that the volume of the underlying phase space remains constant. This naturally explains why gravity knows about the entropy of the microscopic phase space, and why the gravitational action is equal to the entropy.

Now we are in business, because we can use gravity to determine the phase space volume and read off the numbers N_A and N_B. One of the requirements is that when N_A and N_B are equal that we obtain the Bekenstein Hawking entropy. This fixes them uniquely: N_A is precisely equal $1/2\pi$ times the Bekenstein-Hawking entropy and N_B is as given above. If we give the system a finite temperature, say $\frac{H_0}{2\pi}$, the energy contained in it is equal to the total energy of a universe with H_0 as Hubble constant. Hence, one reaches the conclusion that the Dark energy is the energy contained in the Higgs branch.

It is standard lore in string theory that Newtonian gravity arises by integrating out the off diagonal modes from the matrices X that represent the coordinates. This system still interacts with the degrees of freedom on the Higgs branch. In most situations the wave function can be treated as separable, but at very long time cosmological scales they do mix. This leads to decoherence of the wave functions on the Coulomb branch, which may have implications for the quantum measurement problem. For us the main point is that these interactions can be calculated using standard statistical physics, and lead to the presented formula.

This framework leads to a very different scenario for the birth of our universe than the usual Big Bang scenario, in my view one that it is much more logical and consistent with basic principles, like the preservation of phase space. In fact, my intuitive picture is the evolution of the universe is like the aging of a glassy system. Matter and space time emerged in the past from a glassy transition from the Higgs branch to the Coulomb branch. The matter we see are the degrees of freedom that got stuck on the Coulomb branch. Ordinary general relativity describes the short time scale behavior of gravity, but at long time scales on sees a difference. Dark matter was formed in this glassy transition, and is real localized energy. Its amount is determined by the present baryonic matter, but it is not stuck to it. So the famous bullet cluster is not a problem for this framework.

I realize that this description is radical change from our conventional way of viewing the universe. But his idea leads to one simple formula that is in accordance with many observed data. This is called "a smoking gun".

Discussion

J. Polchinski The chair has asked that I stimulate a comment from an astrophysicist.

R. Blandford I have a question actually for Guth. You characterize the transition from the inflationary epoch to the heated epoch as a wiggly line, whereas there is dynamics associated with it. I guess the question to Guth is: what do you see as the major unsolved issues as to describing that transition and the possible observable implications?

A. Guth In terms of the calculations of the density perturbations on the wavelengths that I am talking about, I think the details do not matter. What I did not mention is the length scales involved. The typical perturbations that we are interested in — that we see today in the CMB — had a wavelength at that time that was maybe something like 10^{20} times the Hubble length, so even though those lines have a thickness of some number of Hubble times, the width of the wavelength we are interested in is vastly larger, so it really is, on the appropriate scale, an incredibly thin line where you go from inflation on one side to ultimately having reheated matter on the other. On the other hand, there are other scales involved and there could be very interesting things going on. On smaller length scales than the ones that we observe in the CMB, things can get very complicated and it is hard to know exactly what that might lead to. People are trying to figure out what the observable consequences might be. So I don't have a good answer to your question but I do think there might be something there and people are certainly continuing with research on it. [Note added in proof: If there are multiple fields involved in the transition, which is quite possible, then the dynamics can be much more complicated. The fields at different locations can then differ in more ways than just a time offset, so the argument I gave above about wavelengths no longer implies that the details of the transition are unimportant. Nonadiabatic perturbations and significant non-Gaussianities can then be generated, so it is important to look for these effects.]

J. Hartle Guth mentioned the difficulty of the measure problem and we lack any sort of fundamental definition of the measure, but one source of a fundamental measure is certainly the universe's quantum state. If the universe is a quantum system, it has such a state. Not as an initial condition in some fixed spacetime, but as a theory of the four-dimensional histories which would include the universe from the beginning to the end. And so theories of the initial state will supply some kind of measure and we should investigate these along with the others that have been proposed.

M. Shifman It is a question. I think I sense some kind of tension, if not contradiction, between Guth's talk and Mukhanov's talk concerning eternal inflation, but I do not quite understand what is the essence of this. Maybe somebody will explain to me.

V. Mukhanov I was not speaking about eternal inflation, I was speaking about the real world. Eternal inflation is a different story. As you understand it is some kind of speculation which perhaps we will never be able to verify. It can be logical or not logical but nevertheless not verifiable. But the main thing about inflation is not what was before it, but if you would postulate that there was this stage of the expansion then it has robust predictions. These robust predictions right now are in excellent agreement with all observations, unexpectedly excellent, not only CMB, by the way, also large scale structure. And Guth was exploiting more philosophical things in the second part.

M. Shifman Ok.

G. Dvali Maybe we should add to these assumptions about inflation. It is absolutely true that the inflationary predictions are incredibly robust. I think this was reflected in both talks, actually in three. But there is an extra assumption which goes there, that there was a quasi de Sitter state of expansion, but then there was a clock which was counting time somehow, because inflation finished, but also that clock was weakly coupled. In other words, that you can trust your computations. These three assumptions are enough to derive to leading order, with an incredibly accuracy, these density perturbations. The models that affect this weak coupling of the clock, in other words if they introduce a new cutoff in the theory, which is lower than Planckian, or as Silverstein was saying, fields change on trans-Planckian distances, of course those models would change slightly subleading predictions. But I don't know any model which changes the leading predictions and is not strongly coupled. Maybe there are such models, but it is hard to imagine.

F. Englert What we hear about eternal inflation, or the string landscape, seems somehow unavoidably to lead to some kind of multiverse. However, it seems to me there is a fundamental problem there. Once of course you have the multiverse, then you can start playing around and try to find probability or getting to the anthropic principle, or whatever. But the point is that the picture is essentially a classical one, and it is difficult to see that if you have many universes, coming essentially with an inflationary state, that there would not be plenty of horizons in this. Now the quantum mechanics of horizons is, I think, perfectly not understood. The simplest example is the black hole, where after all nobody knows really if the problem lies in the singularity or if it lies really already in the horizon. Therefore one seems to be faced with a fundamental problem if one tries to play with the multiverse. Either it is quantum mechanically inconsistent, in which case probably it is just wrong, or you have to go to a form of quantum mechanics which is not known and which has to come at that level as a different way of understanding things. Of course, I have no answer for this, but I think that problem, conceptually, seems to be in my opinion fundamental.

V. Rubakov This is kind of an astrophysical comment. I think we should not overestimate what we actually know about the primordial cosmological perturbations, because what we know are very basic things: they are Gaussian, they have flat or nearly flat spectrum, they are adiabatic and they is very few, if any, admixture of gravity waves. Gaussianity means that we are dealing with vacuum fluctuations – originally, vacuum fluctuations – of some free fields; maybe even that is not necessary. Flatness of the spectrum in inflationary theory is a consequence of de Sitter symmetry. You might think of different symmetries; for example, you might think of spontaneously broken conformal symmetry as the reason behind the flatness of the spectrum. Adiabaticity is kind of a triviality. Small admixture of gravity waves – ok, we have to wait until this admixture is found. So, what I am saying is that we are at a very interesting time when we know basic things about the primordial perturbations and hopefully there will be a lot more to learn in the future. When the details come, including gravity waves, non-Gaussianity, statistical anisotropy, then we will know what was the quantum mechanics in the early universe.

L. Randall Just one brief comment and one brief question. One comment is that there are some model building assumptions that have been sort of swept under the rug. The fact is we can have multifield inflation, the scale of inflation is not known and the decay time of the inflaton was sort of briefly mentioned, and these can all affect predictions and we have not distinguished among these possibilities yet, so there is some interesting stuff to do. I actually had a question for Verlinde: one of the most notable things about all these energy densities is that they are sort of comparable in size, that the dark energy, dark matter and baryonic matter, while not exactly the same, are more comparable than we really have a right to expect. Is there any understanding of that in what you are saying?

E. Verlinde I have at least one relation among them, in the sense that I can calculate from the baryonic matter density what is the dark matter density, and as I said, if you take 4 percent of baryonic then I get 22 percent dark matter. So from the three, I can already relate two. This is. . .

L. Randall That was based on the observations of the rotation curves if I understand, so it is sort of a phenomenological. . .

E. Verlinde No, but I also derived the same formula.

L. Randall But it is a phenomenological fit if I understood what you are saying.

E. Verlinde It fits it and it can be derived, so I first presented the empirical formula and then I explained it, so it is a prediction.

D. Gross As chair I would like to ask Blandford, the only astrophysicist in the room, whether he would comment on this empirical relation between dark energy and dark matter, which suggests – Verlinde did not stress this but predicts this – that dark matter will not be found at the LHC.

R. Blandford I have to confess I do not understand the theory that Verlinde just outlined. The astrophysicists I think are at least working on the assumption that they have ruled out all of their known "suspects" of regular astronomical sources of dark matter and have turned to the two best particle physics candidates, which are some supersymmetric particle perhaps or an axion. As everyone in this room knows, there is a large search, not just from the LHC, which one can characterize as being on-ground, but also above-ground with satellites looking for indirect signatures of dark matter, and below-ground looking for direct detection. All three of course are working with upper limits and rumors at the moment, but I think there is an optimism in all three communities that the observation space will be opening up. Speaking personally, and I think for many of my colleagues, I am hopeful that one of these three – still for me most plausibly the LHC – will actually find a candidate for dark matter. But it would be, again, equally exciting if the upper limits ruled out a lot of the conventional supersymmetric theories. I would throw that coin back at everyone else in this room: what are the implications of the LHC failing to find any signatures of supersymmetry?

D. Gross But the empirical relation that he showed, which came from astrophysics, of rotation curves up to the edge of the galaxy, is that correct?

R. Blandford The fact that the mass rises more than the mass associated with the dark matter is concerned, yes, I believe that is correct. One of course sees evidence for dark matter in the universe at large and in clusters of galaxies too.

E. Verlinde May I comment on this? The relation is known as the baryonic Tully-Fisher relation and has been checked quite carefully. There are similar relations – things that people try to fit for clusters – that did not work so well, but I indeed recently realized that the same formula actually does fit clusters and the other thing I mentioned about the baryonic and dark matter ratio is just a back-of-the-envelope calculation. So in that sense I am quite convinced about these relations and I have a feeling that a lot of the astronomical community who are looking at this and try to use ordinary particle dark matter are ignoring a lot of data and a lot of signs for all kinds of scaling relations that are there and have no explanation whatsoever from a particle picture and particles also give all kinds of problems if you want to fit the actual density profiles like having much higher density in the core of a galaxy. It is clear to me, I think there is a challenge to the particle dark matter paradigm.

S. Kachru To change the subject a little, I have a question for Hartle or maybe Guth or both. Guth summarized the confusions about measures and Hartle, I think, in his talk stated rather strongly that the no-boundary proposal for the wavefunction predicts a measure which volume-weights, and my question is – I do not follow this measure business too carefully – but I thought that that kind of measure was highly disfavored by something called the youngness paradox, if not for other reasons. Is that a strike against the no-boundary proposal or did I misinterpret your comments?

J. Hartle I presented a sort of simplified story in order to save time. The more exact statement is: comparing eternally inflating universes and universes that have finite volumes, the measure suppresses the ones with finite volumes, so I do not think there is this problem, but we could look at it a little bit more.

L. Randall I just wanted to comment about dark matter models. I that in our community a lot of the stuff we have heard is stuff that is pretty old, but I think dark matter model building is actually a field that has evolved a lot in the last even year or so. With respect to first of all just WIMP models which you are referring to – we always knew that it could go beyond a supersymmetric partner. Any weak scale particle had the potential to be a candidate. But there are a lot of models that we now know of – and I and many others are working on – that show a relationship directly between dark matter density and matter density and energy density, and that has to do with some interaction between the two that could have been present in the past or could be present today. Something that could violate both symmetries, for example, that could connect them. Those models unfortunately would be much harder to test. For example, some of these interactions could turn off. I think a lot of the dark matter searches today very much rely on the WIMP paradigm. There are particular cases that you can test but it really becomes much more model-dependent.

R. Blandford Sometimes that is called the WIMP miracle, which means the fluxes of dark matter particles are within the realm of being measurable today based on interactions in the early universe.

L. Randall No, I am telling about something entirely different. The WIMP miracle relies on the relic abundant spacetime in being a WIMP mass particle. I am telling about models where for example you can have an asymmetry in dark matter related to an asymmetry in ordinary matter or some other interaction between WIMPs and ordinary baryons or leptons. So there are many other types of interactions that could be present that in some sense much more naturally give rise to a relationship between energy density in dark matter and energy density in ordinary matter and many of those have very different testable consequences.

G. Gibbons I would like to aks a question that has always troubled me and I think many people, which is that there seems to be no prospect but maybe there is not. Can Guth comment on what prospects there are – if any – for identifying the inflaton by conventional terrestrial means. Will we ever be able to find out what the inflaton is, or just invoke some unknown object or entity called the inflaton?

A. Guth I think it is hard to predict that far into the future – it certainly will be in the future. There are a number of papers that claim that the ordinary Higgs field could be the inflaton. In which case, we have already found it. My guess is that is not true – but I believe that it is conceivable. I think the question really is linked heavily to progress on string theory and progress on beyond-the-standard-model particle physics. I think we do not really know where either of those fields are going to be ten years from now. So I would say it is an open question.

V. Mukhanov I just want to add something to Guth's comment. Take three inflationary models. One is Higgs inflation, the other is R^2 gravity, and the third $m^2\phi^2$ but change the kinetic term, so make non-trivial speed of sound. All these three theories will have identical predictions for what we see, or what we will measure. After that, what kind of principle you could use to distinguish them? For the third okay, perhaps you can say that you could use fluctuation non-Gaussianity etc., but the previous two are just indistinguishable. Therefore I just do not believe – it does not matter how accurate measurements will be – that we will ever identify a fundamental field which is responsible for inflation. For that, there is not enough information. But it is enough to have just $p = -\epsilon$, that is all.

G. Gibbons What about details of reheating, for example?

V. Mukhanov For that, you have to know particle physics. How can you get particle physics at energy density GeV? No way, practically, maybe cosmic rays but they are very dirty. About accelerators, we can forget of course. The details of reheating depend on the concrete Lagrangian and there are thousands of them which I can write. And all of them will be successful.

E. Silverstein Maybe it is worth briefly commenting on the non-Gaussianity. At the single-field level this effective field theory technique has shown that precisely in the case with small sound speed that Mukhanov alluded to there is non-Gaussianity of a very special shape that cannot be produced by slow-roll inflation and that is also not the kind that is produced by multiple-field slow-roll versions of inflation. So there is some hope there – it may not be too likely – that those are the examples of inflation that exist in the real world, but one thing to say about them is: rather than having inflation proceed because the potential is very flat, it is possible to have a steep potential, but interactions of the inflaton slow it down on a steep potential, and it is that latter class of models that naturally gives larger non-Gaussianity. So it is a reasonable possibility – I do not know how to weigh the probability, but it is interesting.

N. Arkani-Hamed I would like to re-ask a question that was already asked at the last Solvay meeting here in 2005, which is I think the very hard question at the heart of trying to make sense of physics in our accelerating universe. It has to do with the basic fact that precise quantum mechanical predictions force us to drag infinitely large classical apparatuses around with us everywhere we go and that is difficult with gravity. That makes local observables impossible, it forces us to make sharp observables associated with boundaries of spacetimes. That ideology has served us very well in the idealized situations that we have learned to deal with, but we cannot deal with it in de Sitter space with access to a finite number of degrees of freedom, so we cannot make the needed separation between an infinitely large apparatus and a small system to even talk about precise quantum mechanical observables. That is a question that was already raised six years ago and I think it is an important question to raise again and continue to think about: What could we possibly mean by a precise theory for anything we might see in our universe? Now something which changed – at least a development in the last six years – has been taking more seriously the picture of the spacetime offered by eternal inflation that suggests there are regions in spacetime where we tunnel out of the de Sitter phases and go out into flat space, and there have been a number of suggestions that it might be possible that we find precise observables with those guys – with things that survive into these asymptotic flat space regions. This seems like a very interesting general idea to me, but a general concern is that the observables associated with those guys seem to have absolutely nothing whatsoever to do in any direct continuously related way to anything we might observe. So I think this to me is the most interesting basic conceptual problem, which is that the ideology of "stick with precise observables, go out to boundaries where they are defined" has served us so well, and yet we seem to have lost it entirely in dealing with these cosmological questions. What I think is fascinating and important is that it goes back to very basic things about quantum mechanics, and about what it means to have precise observables. I have no concrete proposal, but I think it is a question that needs to be talked about in any such discussion.

D. Gross A comment. As I think Wilczek suggested in his talk, and I think I in my opening introduction, perhaps we need to specify boundary conditions at the end in the future. I mean, what is special about the past? Physics is time-reversal invariant. We could get rid of a lot of problems if we specify what the future was – the future will be.

F. Wilczek I have a question for everybody and an answer to part of it. In my presentation I mentioned what I think are two of the outstanding problems of fitting gravity into our description of the world, which are these two hierarchy problems: Why is gravity so much weaker than other forces? Why is the vacuum density so much smaller than other particle physics densities? There has been kind of deafening silence on those two questions in this description of gravity, but they are the most striking questions I think. There is a partial answer, I think, that is quite reliable to one, which is sort of to answer it in a backwards way: Part of why gravity appears so weak, is why protons are so light compared to the Planck scale, and that we can understand because of the evolution of couplings being logarithmic. You start with a moderately small coupling, it takes a long long run in energy until it becomes truly strong and makes the proton mass. But that is from the particle physics end. I am wondering if there are addition insights from the gravity end – other than the anthropic arguments, which I think we all are familiar with.

D. Gross But Wilczek, if all we had was gravity, it would be both strong, weak, and...

F. Wilczek But if we do not have only gravity,...

J. Polchinski It looks like Kachru wants to give you an answer.

S. Kachru This is a comment related to what Gross said. There is one sense in which it is very clear that the initial state is very odd and we do not yet know if that is so clear about the final state, and that is that we have an arrow of time. So it is very clear that at least at some stage we had an initial state with very low entropy. We have not heard anything about the emergence of the arrow of time here, but I think that does point to a very special initial state at least for the evolution that we see.

D. Gross But there might be a special final state as well. We do not have much observations.

A. Guth Concerning the special initial state, I have never been so persuaded about that statement as other people seem to be. All we really know about the history of our universe is that we came from a patch that was very hot. The total entropy we do not know anything about, we only know something about entropy within our patch. As far as achieving an arrow of time, I have lately become somewhat enthralled with an idea which may have begun by Sean Carroll — I am not sure really where it began. If you imagine starting — I think you need an infinite phase space for this to be possible — if you have a system that has an infinite available phase space, I would presume that it is at least plausible that no matter how you started it, if you followed it very far into the future, it would start spreading out into that phase space and entropy would be increasing as it spreads further and further out. If you followed it very far in the opposite time direction, you would see the same thing, but with entropy increasing toward the past. But that would mean that the infinite future and the infinite past would each have an arrow of time, and it would come about quite naturally without any particular fine-tuning of an initial state. No matter how you started, if it was a normalizable probability distribution, it could always spread out into this infinite available phase space. That seems to be very natural.

L. Randall I just want to briefly comment in response to Wilczek's question. That is exactly what warped geometry is, it is a gravitational way of explaining the weakness of gravity that is dual to the kind of scaling that you talked about. Whether or not you believe it, is a question that you can leave open, but it is actually a gravitational explanation that is dual to the kind of thing you are talking about.

G. 't Hooft I also want to respond to Wilczek's question. I have no answer to the hierarchy problem today, and I had no time for this subject in my talk, but I could imagine what an answer could possibly look like in a more sophisticated theory. In my theory you still had to identify the algebra: You have to fix the algebra – that is the gauge group and the representations – and then everything else should follow. Now if you just fix the algebra to be something like say $SO(10)$ – name an example – then the details of the theory can be arranged such that the coupling constant would naturally become of the order $1/10^2$. Now how do we get mass parameters much smaller than that? Imagine a theory where some of those parameters are due to instantons. Then you have a tunneling effect depending exponentially on the couplings, so you get e^{-100} and that sounds a lot better. So perhaps we can have theories where some of the masses are typically due to instantons and take the values of the order e^{-100}, I could imagine a theory where the hierarchy problem could be solved along such lines.

J. Hartle To return to the arrow of time, Gell-Mann and I, building on work of very long ago by Aharonov, Bergmann and Lebowitz, formulated quantum mechanics in a time-neutral way, so there is no quantum mechanical arrow of time, with initial and final conditions. In that kind of framework, every time-asymmetry comes from asymmetries between the initial and final conditions. And it is not correct to say that there is no observable effect: If you have a very special final condition, you will find observable effects today. Therefore the question is not only what is our initial state, but what is our final state and what are the observable predictions of both? It is exactly the same as the empirical question we face just ordinarily in cosmology, except that it has this additional degree of freedom, and it has this beautiful symmetry that there is nothing special about quantum mechanics: as far as the arrows of time go, it has the same status as all the other models.

E. Silverstein I just wanted to comment briefly on Arkani-Hamed's point, which I completely agree with, and say one more thing about it. As you go into what Arkani-Hamed was calling the flat space region, which is really a decelerating FRW expansion, two things happen. The covariant entropy bound goes off to infinity, so there is a chance of a precise description, and secondly it turns out that anyway anybody has proposed to holographically dualize de Sitter space by itself involves fluctuating gravity. This is true of the dS/CFT correspondence, when you compute general observables and also what we call dS/dS and so on. But what happens in this FRW phase is that the lower-dimensional Planck mass goes off to infinity at late times, so that is another hopeful sign that there is a possibility of a precise dual description. Of course, all these things raise more questions than answers at this point.

Session 7

General Discussion and Conclusions

Chair: *David Gross*, Kavli Institute, Santa Barbara, USA
Rapporteurs: *Alain Aspect*, Institut d'Optique, France; *Peter Zoller*, University of Innsbruck, Austria; *Bertrand Halperin*, Harvard University, USA; *Howard Georgi*, Harvard University, USA; *Joseph Polchinski*, University of California at Santa Barbara, USA
Scientific secretaries: *Glenn Barnich* and *Pierre Gaspard*, Université Libre de Bruxelles, Belgium

Alain Aspect: Foundations of Quantum Mechanics and Quantum Computation

A. Aspect Thank you, David. Tony Leggett had to leave this morning and he gave me the opportunity to clarify some points at breakfast. I will not waste time in thanking all the rapporteurs and speakers of the session for keeping strictly the time. I would like now to make a few remarks and then have a discussion, but a discussion which would go beyond usual suspects, I mean people who have liked arguing about these points forever. It would also be nice if everybody around the table would feel free to say something even if he, or she, thinks he or she is totally naive in the subject. So to start I will not make a summary. I was advised not to make a summary, but rather extract a few points which have struck me. Thereafter, I will make some personal remarks and I will finish with provocative remarks to start the discussion. Before starting you must be aware that I am totally biased because I am an experimentalist, moreover, working on small scale experiments of the kind you can control with one or two persons. I work in AMO physics and the problem that I have is that I am not only biased but also schizophrenic because, on the one hand, I am fascinated by these questions of interpretation but, on the other hand, when I am in my lab or when I write a paper or analyze data, I am also an acolyte of the church of "shut up and calculate" .

W. D. Phillips Just a member.

A. Aspect OK, just a member, not an acolyte, a temporary member, when I do the experiments and write a paper.

D. Gross A deacon.

A. Aspect Yes, a deacon only on Sunday. This being said, I will start by extracting some points which, in my opinion, are not controversial. I have been struck by the fact – well I am biased again – that in this question of foundations, more than anywhere else, the connection and the interplay between theory and experiment is crucial. Until there were these Bell's inequalities allowing us to do a test, the discussion between Einstein and Bohr was just a matter of taste, an epistemological or philosophical discussion, and then there were these Bell's inequalities allowing for a test. I think that, at this point, I will follow Tony Leggett in his conclusion that it is safe to claim that Bell's inequalities have been violated. All loopholes have been closed separately. There is what Tony calls the locality-collapse loophole and I clarified that with him this morning. It is nothing else than the question whether we are sure that, when a photomultiplier clicks, the photomultiplier itself is no longer in a superposition. Well, I must admit that, as a naive experimentalist, I have no doubt that when my photomultiplier makes a click and there is something registered in the memory of the computer I think the measurement is made, but it is a point which is raised.

The second point which is fundamental in this discussion is the question about linear superposition in macroscopic systems, for which we have some experiment although further experiments are strongly demanded. We have heard of the experiments advocated by Tony Leggett for at least a decade, probably more, on superposition of currents in Josephson junctions. Anton Zeilinger refused to talk about the beautiful experiments with fullerene for very good reasons. He explained we miss a criterion, something equivalent to Bell's inequality, to decide if it is really macroscopic or not. Of course, we have an idea but there is nothing as clear as Bell's inequality. Anyway, having experiments is absolutely important. At this point, it is important to stress the remark of Murray Gell-Mann about connecting a superposition of a microscopic system to a macroscopic system, like in the Schrödinger cat story. Although it is trivial to make a superposition between the microscopic and the macroscopic objects, what is not trivial is to demonstrate that the cat is still in a superposition. So we must keep that in mind.

Now, there is another question raised by John Preskill about quantum computing that would demand experiments and for which we have no clear idea of what it would be, up to now. The question is the following: When we have many entangled particles, we have a fantastically large Hilbert space. Are we sure that we have access to the whole Hilbert space, or are there states which are excluded by some kind of selection rules? We do not know yet and it would be very important to find conclusive experiments in order to answer this question. These experiments have been and are still important for other reasons that have been clearly advocated and expressed by Anton Zeilinger. It is concerned with the cross fertilization between experimental tests of foundations and the development of quantum information. I will allow myself to give a personal testimony of this link. In the early eighties, when I was doing experiments on single photons or entangled pairs of photons, these were experiments only to test the foundations of quantum mechanics. It was only later that the development of quantum information really started. So I fully endorse what Anton Zeilinger said about that.

Now after, I hope, the consensus on the interest for experiments, what are the fundamental questions at stake and what are the open questions? In the discussion on the Einstein-Podolsky-Rosen paradox and Bell's inequality, the key points are realism and locality. I have checked with Tony Leggett this morning that what he calls the counterfactual definiteness is some more sophisticated and more precise way of expressing what I name just realism, that is to say the fact that you can attribute given properties to a certain system. From this point of view, I would like to insist that the local realistic world-view of Einstein is very appealing. The idea that you have a physical reality attached to a finite volume in space-time seems to me strongly linked to the idea that this volume of space-time cannot be influenced instantaneously by something happening in a space-like separated region of space-time. Otherwise, what would it mean to say that you have a full set of properties for the first volume of space-time? Therefore, I think that Einstein's point of view lead to a a beautiful, appealing and consistent model of the universe except but we have to renounce it. Now, when we renounce it, can we keep only a part of it? Well, I think at this point it is a matter of taste. What is important is to realize that we cannot stick to that view of the universe because, for instance, if some days I belong to the church renouncing to the concept of locality, then of course I am fully aware of the problems with the arrow of time or whatever you call the fact that you cannot keep the ordering between events, and things like that. Hence, the important point is that we have to renounce a beautiful and consistent world-view.

The other point which is clearly at stake is the boundary between the quantum and the classical worlds. It can be raised in many different forms. For instance, we have already mentioned the question of superpositions for large objects. The most difficult point for me is the question of decoherence. First the non-provocative statement. I have been impressed by the remark of Michael Berry that we should better have decoherence otherwise classical chaos would be strongly suppressed. I do not know what I draw from this remark, but at least it is an important point, I think. Now, I come to the provocation, my friends. I feel uncomfortable with the words decoherence, pointer basis, or decoherent consistent histories, because it seems to me to be magic formulae to sweep under the carpet the difficulty that you have in understanding what happens. At the end of the day, is this very different from using the postulate of projection of the wave packet? I have the feeling that these big words are related with this postulate in the same manner as the so called demonstration fo Boltzmann's H theorem is related to the second principle of thermodynamics. I would be glad to hear the reaction of the tenants of decoherent consistent histories, but of course everybody else would be welcome to react.

W. H. Zurek The key to what I wanted to do in my five minute presentation and what I am going to restate now is that, if you look at textbook postulates, admittedly they are confusing because you start with some very quantum postulates – states which live in Hilbert spaces, unitary evolutions – and then you throw in postulates which have explicitly to do with measurements, for example Born's rule, and these do not really fit consistently within a single list. Bohr got away from the problem by saying that part of the world is classical, but we would like to do better. We have the instinct for unification. The aim of what I was saying, and I am just going to restate it, was that you can use purely quantum postulates of quantum mechanics – states living in Hilbert spaces, unitary evolutions, Hilbert space of a composite tensor product of Hilbert spaces of the constituents, plus repeatability, which is an important ingredient of our classical intuition about the universe because this is what gives persistence – to claim existence and to derive the business-end of the collapse postulate, i.e. the preferred set of states. This follows simply from unitarity and essentially from something that is a version of the no-cloning theorem and you can derive Born's postulate, a result which I have not presented. Thus, it is possible to show that probability is equal to the amplitude square. So with this, you cut the confusing long list to a short list, which has the appeal of being completely consistent. This was the main point that I wanted to make in response to Alain Aspect and this gives you the preferred basis on a platter.

Now, there is one more point that I would like to make and that is loosely related only to what you have raised. I wanted to say a word or two about the relation between entanglement and quantum computing.

The discussion on Wednesday could have left one with the impressions that, whenever there is a quantum ingredient in correlation, it is entanglement and, whenever you have a gain in computing or information processing power, it is also due to entanglement. Neither of these impressions are true. It is now clear that entanglement measures quantum ingredient and how you put a pair of systems or several systems together. There is another quantity which has to do with how you pull them apart using classical means. These two quantities do not coincide. The other quantity is known as quantum discord. You can have states which have non trivial large quantum discord and zero entanglement and you can use such states to speed up information processing.

E. Witten I just wanted to make a small comment about Bell's inequality. It is marvelous that quantum mechanics has been tested in the context of Bell's inequality but we really should not lose perspective on the whole process of testing quantum mechanics.

The main reason to have confidence in quantum mechanics is not just that it was confirmed in that situation but that a thousand and one predictions of quantum mechanics have been tested.

A. Aspect This is true, but it was remarkable to discover, when Bell's inequalities were looked at, that most of the situations which are successfully described by quantum mechanics could as well be described by a classical looking model and so it was very interesting to have a place to look specifically.

E. Witten I would restate that a little bit differently. The way I would say it is that the test of the situation with Bell's inequality showed that it was actually impossible, while previously had it been implausible, to replace quantum mechanics by a classical hidden variable model.

A. Aspect Yes, OK.

S. Das Sarma I just want to reiterate what Ed Witten just said. Let us take examples of things that I understand, such as superconductivity. I do not know of any classical or classical looking model that can explain superconductivity. It is a very simple example. I know that Tony Leggett is very, very intrigued by superposition of Josephson junction flux, but the theory of superconductivity works.

Now, I want to go back to the issue of decoherence, which you brought up as a very confusing thing. I think it is confusing because it is made very mystical, at least in the condensed-matter and atomic physics context. Again, I do not want to comment on the wave function of the universe because I do not understand this concept, but I do understand decoherence very clearly in condensed-matter and atomic physics. If we have a system, what precisely the system is defines what you mean by decoherence. For instance, if I am talking about a few electron spins and I am doing some manipulations on them – let us say q-bit manipulations or something like that – I am leaving out all the nuclear spins surrounding it, so that I am looking at only part of the Hamiltonian. If I looked at the whole Hamiltonian in this particular context, including nuclear spins and so on, there is no decoherence, there is just an eigenstate which evolves unitarily. The concept of decoherence arises in the experiment, e.g. which is looking at just electron spin, because whatever I am leaving out would have an effect on the system. If we want to call that decoherence, that is fine since it is very well defined. In any particular experimental context, what is meant by decoherence is never confusing or confused. At least, in condensed-matter physics, I mean it is very well defined. It is only when we start talking about reality in a very broad way that I get confused. I have never come across an experimental situation where decoherence is a problem because I can calculate what it is by using master equation, density matrix, or self-energy diagram. Thus, I think that this problem has come because people have harped on the word realism as Einstein defined it and insist on some classical definition of realism, which as Murray Gell-Mann said, is simply not there. We should probably get away from this realism business.

M. Berry I wanted just to endorse what has just been said. I do not think of decoherence in terms of these fundamental interpretational issues about what is real and what is not real. Indeed, as I have said to some of you here privately, I do not think we would lose very much if we banned the word real from our discussions. This is a separate matter. I think of decoherence as an extension to quantum mechanics of what we learned to do already in classical physics. Namely, if you want to treat a system and you know that there are influences from outside of which you are largely ignorant, then you deal with this by averaging in some way or other, better defined in some context than in another, but you do average over what you do not know. Now in the case of quantum mechanics, the particular consequence of this kind of averaging is that it very quickly, and more quickly as you go more classical as I said, removes interference effects, suppresses them, so that decoherence is a very natural term. I think of it as a practical consequence of the fact that we usually do not apply quantum mechanics to the whole universe in practical situations. We apply it to part of the universe. Sometimes, we can get away with the old fiction of the isolated system which served us so well for many decades, but sometimes you cannot. I think of it in this very practical way.

J. Hartle Well, I agree with what the previous two speakers have said. Decoherence is not some mysterious addition. It is a fact. For the variables that we are interested in and the superpositions, the phases are carried away by interactions with other systems. That is true in a measurement situation and it is also true in the early universe. You can have an approximate formulation of quantum mechanics which describes measurements, but if you seek to ask a question about the early universe, where no experiments were being done, then the alternatives which you can describe are ones which decohere, because they become entangled with other systems. It is just a very natural extension of quantum mechanics, decoherent histories.

M. Gell-Mann I do not have much to add to what my colleague Jim Hartle just said. It is in that direction that we have to proceed, I think, and not to worry much about realism which, if you take it in Einstein's sense, is ruled out, but just simply to worry about how to get a formalism that provides sensible answers. One problem is that the probability of A or B, when these are exclusive alternatives, has got to be equal to the probability of A plus the probability of B, and that is not true if you have interference. If you use the linear definition of the probability instead of the quadratic one, then that is automatic, but then the problem is to save probabilities lying between 0 and 1. In each case, this is done by coarse graining. An intervention occurs in the otherwise unitary evolution.

A. Aspect So you agree that there is a relation to the H-theorem and this kind of points.

M. Gell-Mann I guess so.

Y. Aharonov When we speak about non-locality in the case of the EPR and Bell's inequalities, we should realize that this is a kind of soft non-locality, because certainly what we do in one place does not affect the probability of what happens in other places. It is a kind of a passion-at-a-distance but not truly non local, unless you want to assume that there are some hidden variables. On the other hand, there is another non-locality in quantum mechanics which is of the form of a strong non-locality in the sense that what happens in one place affects non-locally the probability in another place. The simplest example of that is the AB effect when you think about a magnetic field here that affects the behavior of an electron at some place else. Now, it turned out that, if you look at the Heisenberg equations of motion, you see that the non-locality is not in the classical theory. The difference between the Heisenberg equations of motion and classical theory is the difference between commutators and Poisson brackets. When you look at simple functions like momentum, the commutator of the momentum with the Hamiltonian and the Poisson bracket are the same. But if you look at the quantities that are relevant for interference, which are periodic functions of momentum, i.e. the modular variables, then their equations of motion are classically local and quantum-mechanically non-local. That is extremely important in order to understand interference phenomena. If you think of a particle moving through one slit, you can say that the reason why the other slit being open or closed is relevant for it is because of this non local equations of motion.

W. D. Phillips I have been sitting here listening to everyone saying that there is no problem with decoherence and I agree. However, I think that we have perhaps lost sight of why we started to discuss it in the first place and there are two reasons.

One is that some people, apparently no one here, but some people have claimed that understanding decoherence is a way to solve the quantum measurement problem. I believe it does no such thing and I think Wojciech Zurek would agree with that, that it does not solve the philosophical problem of how you go from a superposition state to an actual result. It may explain a little bit about how the apparatus works to do that, but it does not relieve the angst that people feel, but that is their problem.

The other issue is that some people, notably Tony Leggett, have suggested that there may be something beyond the quantum mechanics that we understand. There may be something else that specifically prevents us from having macroscopic superpositions to explain why we never observe them and Tony has suggested that we might experimentally look for these things sometimes called fundamental or intrinsic sources of decoherence. This is something entirely different from what we have been discussing here. This would be something that would be an addition to quantum mechanics as we know it and that is something that could be measured. So it is worthwhile thinking about it because you could think of an experiment that could test whether that was in fact the case.

G. 't Hooft I am still a bit puzzled why people deduce immediately from the violation of Bell's inequalities that there is no underlying realistic theory. I think that is wrong because Bell's theorem is based on nothing more than that, if you have non-commuting operators, you cannot measure them both simultaneously. But in a classical system, you can also introduce non-commuting operators, which means that there still could be an underlying theory where the original operators are the commuting ones, but where you can add non-commuting operators that survive the renormalization group all the way to the standard model scale. That could be an explanation as to why we see non-commuting operators today. Another way of saying things is that the mystery of quantum mechanics today is the dynamics, not the statics, whereas Bell's inequalities are clearly statements about the states that we are looking at.

W. H. Zurek Just to define precisely what I believe decoherence does solve and what it does not. I think what it does solve and what was not known before, is where a small corner of the Hilbert space (which is huge) that corresponds to classical states comes from. Decoherence does this. It was not known how to do it before. Now, you can change it by changing interactions, by changing systems, or by changing Hamiltonians. Whether the other problems which are very personal are addressed by decoherence is in the eye and the soul of the beholder and I am a member of the church of larger Hilbert space.

A. Zeilinger Just one small remark. Besides the question of the possible fundamental importance or not of decoherence, decoherence costs a lot of money. It is one of the big challenges for quantum computation and many people are worrying about that. So independently of whether you believe that this solves the measurement problem or not, or whatever it is, it is relevant.

A small question I want to throw out. There are some people who worry when we talk about the wave function of the universe that we might run into a self-referential problem of the Gödel type because we have to represent ourselves also in this wave function. Do people believe this or not? This is just a question.

B. Altshuler What I want to tell is that there is a lot of mystic talk about decoherence. But actually there are many examples how you can actually learn a lot about the environment by studying decoherence. For instance, when you study weak localization, you measure this decoherence time with pretty good accuracy and I always tell that a bad q-bit can be an excellent probe of the environment. You do not need to philosophize in order to learn what is actually going on in the environment when you look on how decoherence depends on certain parameters. If I may, I would like also to ask a question to everybody. Sometime ago, there was a question that John Bell asked and later Alain Aspect answered. That answer is that, yes indeed, the Bell inequality is violated experimentally. Do you have anything like this now? Is there any question that theorists can ask to experimentalists or that experimentalists have, or are we only discussing philosophy?

A. Aspect Thank you, Boris. It is exactly the point that I was raising in saying that it would be wonderful to have such a clear-cut criterion for macroscopic superposition or things like, but I do not know that we are at this stage. So, I close the discussion.

Peter Zoller: Control of Quantum Systems

P. Zoller Rather than summarizing the session, I would like to use my time to make a few comments because I think that some things were left out from the session that could be discussed but time was actually too short.

Let me sort of summarize as follows. The basic notions are to have control of quantum systems, to connect these things to quantum information or condensed-matter physics, of course, in the context of quantum computing, quantum simulation, and quantum communication. What was left out was the discussion about using certain quantum states to make for example better measurements like in the context of say squeezing. It is something which is of course based on foundations, but we talk really about applications here. Thus, the question is the physical realization, bringing these things to the lab. The basis of all of that is of course the control of Hamiltonians, but we should also emphasize the notions of preparation of an initial state, doing measurements in the end of also a single quantum system. The theory of continuous measurement and of quantum noise is very close to all of these things. If you ask me whether there has been a theoretical contribution of this whole subject to, for example, quantum physics, I would say it is really the theory of quantum noise, probably. Let me make a few remarks in the context of quantum computing. Fifteen years ago, the ideas of building a quantum computer, I should say a small quantum computer, were purely theoretical. As we saw in our discussion, this field is nowadays firmly in the hands of the experimentalists. In small model systems which demonstrate the basic ingredients in the lab, reported progress is truly impressive both on the AMO side and also of course mesoscopic physics. But the general purpose quantum computer, as the computer scientists would like to have it, is of course far in the distant future. My guess is that these things will happen, but we should maybe not hold our breath or invest in stock of quantum computing companies yet. Given the fact that all these things are far in the future, the question is what can we do now with the limited resources that are available. I think the session has given a pretty clear answer to that. There was a lot of discussion focused on quantum simulation. So I would like to say a little bit about quantum simulation now and then also make some comments. Before, I want to show one slide that shows another system that was left out from the discussion and maybe coming less from condensed-matter physics than from quantum computing. This is an optical lattice. What you can see on the left-hand side is sort of the dream of every theorist. You have single q-bits that are stored on sites of a lattice. You would like to have a mechanism to entangle them with hopefully a very high fidelity, to address single atoms, and to do single shot measurements of all of that. On the right-hand side, there is the optical lattice that is the physical implementation. During the last two years, one has gotten really a significant step closer to these things and I think that this is one of the most promising systems to do general purpose quantum computing, besides the ions and the mesoscopic systems that we have heard about. On the right-hand side, you can see photographs that show the single addressing of q-bits or of atoms in a lattice. One is actually quite close to have almost the full control over them in a quantum sense and to be able to perform measurements. I think that is a very important ingredient in all of this engineering of quantum systems.

Now, let me come back to this question of quantum simulation and all of that. It is very clear that quantum information is no longer a new field, but has evolved into a quite mature field. In light of the promises that were made concerning large entanglement in the context of quantum computing and quantum information, I think the challenge is now to built a quantum device involving some proven large-scale entanglement which cannot be represented classically and which is hopefully also useful in some context. As I said, I personally believe that this will happen in the context of quantum simulation. I guess the session very clearly illustrated that. Let me just remind you that, in quantum simulation, we talk about two possible ways of doing that. One is the emulator which is obtained if you try to build an analogue system that you can control fully. But there is also a second version which was not really mentioned in detail in this session and which is closer to what a quantum computer is, but requires maybe less control. It is a digital quantum simulator. For example, you use the time evolution of the system that you can represent in terms of gates in order to mimic an effective Hamiltonian. Now, until very recently, all of these things were pure theory. I just want to point out this recent experiment in the context of ion traps that have demonstrated the basic ingredients. I find truly remarkable when these experimentalists now do up to 150 gates in systems of about six spins and I think this is opening a door to a new direction of these quantum simulations. But what can we do with them when we talk about quantum simulation? Well, the emphasis in the session has been very much on condensed-matter physics with cold atoms realizing interesting quantum phases, maybe exotic quantum phases where nature is not so kind to give to you the systems that we write down in our Hamiltonians. But I would say that there are also important new frontiers that go beyond the standard paradigms. Nowadays, if we have very well isolated systems, we can look at time evolution of a closed many-body system. I guess these things are still largely unexplored from a theoretical point of view in the context of many-body physics. We can talk about thermalization and, of course, there is this whole world of non-equilibrium physics and non-equilibrium phases. These systems are extremely stable and promise all of that. Besides engineering Hamiltonians to do interesting quantum dynamics and states, you can also for example couple them to reservoirs in a controlled way. We have just heard about decoherence. You can also think about controlling the coupling to an environment. Of course, this opens the way to carry out the quantum simulation of open systems with all of this non-equilibrium physics that maybe we do not really know so much about at the moment. So I guess that this is the door opened to some really new physics. The claim is of course that these things are simpler than quantum computing. There are some open problems here. We still have to understand better the role of errors in quantum simulation. Maybe you do not have to do quantum error corrections in the full blown form in these kind of systems that I am just showing on the slides. Also things like verification of a quantum simulator are somehow open.

I would like to continue then with some comments on quantum simulation *per se*, because today it seems fashionable to talk about quantum simulation, but one should clearly distinguish what we actually mean by this, what is the holy grail, what is interesting, and what is the ultimate goal in this context. Now, when I listen to discussions about the Feynman quantum simulator with an exponential large Hilbert space, I hear about superpositions. Of course, all of us know quite well that, when we talk about ground states of many systems such as one-dimensional systems, we can compress them and write them in the form of matrix-product states, maybe PEPS or MERA, i.e. the kind of things we have learned from the quantum information side. Very clearly, for a quantum simulator to be really useful, you should talk about systems where we have some dynamics of states evolving from initial conditions with a Hamiltonian that leads at the very end to something that involves really large-scale entanglement. Thus, a quantum simulator has its proper meaning when it really involves large-scale entanglement beyond what we can represent classically, so that all of these DMRG ideas fail.

At the moment, there is actually quite an interesting situation in experiments. Sankar Das Varma commented on this in a way that I found quite interesting. For example, you can talk about quantum simulation in a system with a few degrees of freedom, but of course you can write yourself a Mathematica program to do the same. If so, why do I need quantum simulation? A little bit more delicate is the situation in the case of these very beautiful phase diagrams that the experimentalists are able to measure at the moment up to a few percent for some very complicated strongly interacting many-body systems, like for example the unitary Fermi gas, where the phase diagrams can be measured very accurately. But if you take bosonic systems and we talk about equilibrium condensed-matter physics, then we have the Monte Carlo method, and these things agree extremely well. I think that the first glimpse at what quantum simulation will be or should be is shown by experiments in optical lattices doing a quench. You take an initial condition that corresponds to a very highly excited state and wave packet. You let these things propagate. For example, when Immanuel Bloch does an experiment on the quench and looks at density fluctuations, you can see as a function of time that, when these things progress, the DMRG stops after a short time because entanglement gets too large and, of course, the experiment goes on much longer. That is an indication that we have large-scale entanglement in these systems. It is necessary to describe these things if you want to have a theory able to predict and control these phenomena. So much about quantum simulation.

Let me also make a few comments about other things before we can start our discussion. I found particularly interesting the notion that was brought up as to what the connection between gravity and AMO physics might be. I think that there is a lot of experiments to be done in this context. The secret hope is of course to see fundamental decoherence creeping in somehow from gravity in some of these experiments, but maybe that is just a dream. There is also a lot to be said about the things that Anton Zeilinger mentioned in the context of quantum communication, in particular, the idea of building a repeater and all of that.

Let me just conclude with one remark. I have mentioned open systems before as one of the new frontiers in this whole field, I think. Maybe you can control the coupling to the environment and all of that, and talk about non-equilibrium. To be complete, I mention that some of my colleagues are very excited about quantum biology, but myself I am not entirely sure that I know what this really is.

Here, I would like to conclude. I am sure there are some questions and comments and the session is now open for discussion.

W. D. Phillips I have a question that, in a sense, overlaps both this and the last session that has to do with the accessibility of Hilbert space. It is really just a clarification. I think it is directed to John Preskill. The issue is whether there are inaccessible regions of Hilbert space given the fact that we have all these unitary operators that we can perform on our q-bits and that we can entangle them at will. What I am wondering is the following. In order to have a completely general quantum computer, is it necessary to be able to access all of Hilbert space or can you do everything that you want without getting to every spot in Hilbert space?

J. Preskill I would put it a little differently: A quantum computer, as we normally define and envision it, will be capable of starting with some simple prepared state, like a product state that may be prepared by doing a measurement, and then we perform a sequence of gates each of which will act on some small number of q-bits, like two or three. Then we can ask how many states can we reach, how many unitarities can we reach if we have circuits of reasonable size. The essential point is one of scaling. The Hilbert space is exponential. It has a volume which is exponential in its dimension. Its dimension is exponential in the number of q-bits, so that the size of Hilbert space in that sense is doubly exponential in the number of q-bits. But if you ask how many circuits can you perform, how many unitarities can you reach, with a good approximation with a circuit that has a size that scales reasonably with the number of q-bits, like polynomially, it is an exponential of a polynomial, far, far, far smaller than the double exponential. Thus, in practice, most of Hilbert space will be inaccessible and that is what I meant. But what we can reach with the circuits of reasonable size is still a hell of a lot. As Peter Zoller has emphasized, the type of entanglement we really want to explore is not the sort of bipartite entanglement that is tested in Bell inequalities but what Wen called long-range entanglement, i.e. many-body entanglement. We have understood some ways of characterizing it. It can be used to define different phases of matter but there is a much richer as yet largely unexplored classification of types of many-body entanglement. That is what we hope to explore with quantum simulators, studies of highly correlated matter, and quantum computers.

G. 't Hooft I want to just make a brief remark, a challenge if you will. Imagine a classical computer where I could scale up its performance such that its memory would be as small as the Planck size and its processing speed as fast as the Planck time. We cannot make such a classical computer but if one would be allowed to scale up the properties of a classical computer in that sense then my claim, conjecture, or theory, is that that will always out-perform a quantum computer. This is of course a non-trivial prediction because quantum computers are supposed to solve non-polynomial problems. So I would be very eager to see if there is any possibility to make a quantum computer which would be better than such a classical computer in any of its tasks.

W. H. Zurek There is a number of connections that were raised, including those raised by Peter Zoller now, between this session and the next session on quantum condensed matter. I would like to raise one more connection which comes up in several contexts. For example, one of the schemes of doing quantum computing is the so-called adiabatic quantum computing. You start with a very simple initial state in a finite system and then, slowly enough not to excite the system out of the ground state, you drive it into a Hamiltonian whose ground state supplies you with the answer to the question you want.

If you go from one state to the other, you encounter what essentially looks like a quantum phase transition in a finite system. Understanding the dynamics of such phase transitions is important for the purpose of adiabatic quantum computing. Similarly, if you implement in real life the sort of systems that we have heard about in the session on quantum condensed matter, you do a quench produced in a finite time. As a result of both of these situations, you end up for example with topological defects. Systems cannot smooth out their order parameter enough to exclude them. I thus think that there is a very interesting area, which involves dynamics of quantum phase transitions, where the overlap between atomic physics and condensed-matter physics can be further explored.

S. Das Sarma I wanted to comment on the issue brought up by Phillips, Preskill, and 't Hooft. What has not been emphasized very much is that a quantum computer does not do everything well. In fact, there are very few tasks that a quantum computer does well. When Bill Phillips asked his question, it was not clear to me what he was asking. If we want to do Shor's algorithm, Shor showed that you do not need to access the whole Hilbert space, so that the question of accessing the whole Hilbert space is remote. If you have to access the whole Hilbert space, it is not going to be a fast computer. Actually, Shor showed how you can solve a problem in polynomial time that classically takes exponentially many steps by concentrating yourself on selected parts of the Hilbert space. This is the key issue, which concerns only the problems that those who advocate a quantum computer are interested in solving. Now, suppose that you want to solve an arbitrary quantum problem, which is what I think Bill was asking, then of course you have to explore the whole Hilbert space and that is going to be very difficult because it depends on the arbitrariness of the problem. Or suppose you want to solve problems that classical computers can solve. That goes back to 't Hooft's question and it may very well be that a quantum computer has to be as large as a classical computer to solve that particular problem. In fact, there has been very little progress in the last ten or twelve years, on the classes of problems that quantum computers can solve efficiently. We are only interested in the problems that it can solve efficiently. Factorization is one for which we know that you do not need to explore the whole Hilbert space. We know what parts of the Hilbert space have to be explored and that a quantum computer can do very efficiently.

G. 't Hooft My claim is that any of the tasks of a quantum computer would be out-performed and that includes in particular of course the Shor problem.

G. Dvali I got intrigued by this idea to have a quantum or classical computer of the Planck size, but what I do not understand is that Planck length is an intrinsically quantum length. Actually, that is an absolute bound on the number of information bits that gravity puts on us. Therefore, any computer of the Planck size cannot even process one bit of information.

G. 't Hooft No, the number of memory sites per square centimeter should be the square centimeter divided by the Planck length square, so that is an enormously big classical computer.

G. Dvali Yes, of course, then it is a normal thing. This brings me to this question of connection between what you were discussing in gravity. The idea of quantum mechanics and quantum measurement is based on the fact that, if you have a lot of money and a lot of time, you can do arbitrarily precise measurements. However, gravity fundamentally limits this because there is an absolute bound on the information storage. Information gravitates and any measurement is based on extracting or putting information.

G. 't Hooft Gravity might also bring about decoherence.

F. Wilczek I would like to make a remark which is probably whimsical but maybe not. The three degree microwave background radiation was discovered by mistake, or by accident, when people turned very sensitive antennas to the sky. Quantum computers are very sensitive, very delicate as we have learned. Delicate in ways in that previous objects have not been delicate. Delicate against weak perturbations that upset coherence over many, many degrees of freedom. So it could be that there are things out there in the sky that will foul up quantum computers, concretely, maybe the neutrino background, maybe an axion background, or maybe something else. It might be worthwhile thinking the other way round, whether you could use this technologies to try to make antennas that detect cosmic noise.

P. Zoller Are you saying that instead of giving money to the LHC, you want to give it to quantum computing?

F. Wilczek Other people's money, well yes.

S. Sachdev I want to come back to a comment that Peter Zoller made when he was discussing quantum simulation. You mentioned some experiments of Immanuel Bloch, I think, where you said that the time evolution was already well beyond the range that DMRG could simulate, suggesting that there was complex entanglement developing in the system. Don't you expect most quantum systems at long times to essentially reduce to some effective classical model and that there is a classical description which, in fact, would be out of range for DMRG? It is not immediately implied that you are going to get entanglement, I guess, because DMRG has not been able to describe it successfully.

P. Zoller No, but if you do a DMRG calculation, you are essentially expanding in terms of the entanglement. If you fail after a certain time because the entanglement gets large in the system, it is a self-consistent statement about the corresponding wave function. What you point out is that, for some low lying or low order observables, it may not be necessary to have this complete information and I entirely agree with that point, of course.

S. Sachdev I mean there could be classical variables which you only see if you properly "decohere" the rest of the Hilbert space and DMRG may not do that efficiently.

J. Preskill Just two quick points as far as quantum computers not being able to do much. It is true they have limitations but what they can do is simulate unitary evolution governed by any local Hamiltonian, any Hamiltonian that can be written as a sum of terms, each of which acts on some constant number of q-bits. That is a lot because, as far as we know, they can efficiently simulate any physical process in nature and classical computers cannot.

As far as quantum computers being sensitive to influences like neutrinos, error correction works in a way that is effective against weak noise that is not correlated, or only weakly correlated. Therefore, the danger would be very intense neutrino bursts or a rare event like an earthquake that shakes up the lab, but probably not normal neutrinos.

J. Maldacena When we think of finding the ground state of an Hamiltonian, it is useful as a theorist to think about the evolution in imaginary time that projects onto the ground state. Is there any way to do that in a quantum computer, to choose your Hamiltonian and evolve in imaginary time?

I. Cirac Unfortunately not. We know that it will not be possible because there are many Hamiltonians for which finding the ground state is a QMA-complete problem, meaning that not even a quantum computer could solve it efficiently.

E. Witten On this last point of finding the ground state of a quantum system, is it clear you cannot find the ground state by cooling down your quantum computer? You put it in contact with the heat bath and lower the temperature.

B. Altshuler The point is that finding the ground state of a complex system is an NP complete problem and the answer to Ed Witten's question is basically a one-million-dollar answer because the Clay Institute will give a million dollars for proving that P is equal to NP, or disproving it. The answer to Ed's question is whether you can find in polynomial time a solution of an NP-complete problem.

Bertrand Halperin: Quantum Condensed Matter

B. Halperin Thank you very much. I will try to give a very brief summary of the highlights of the discussions we had in our session, but I would like to first emphasize, as Peter Zoller did in his summary, that many topics were left out. The subjects we covered in the condensed-matter session are really only part of condensed matter physics. Arguably, they form the most exciting part. I would certainly say it is the part that is most relevant for the interests of many of the other participants at this conference, particularly those in field theory, string theory, and cosmology, but it is not the whole story and I do want to emphasize that condensed matter is a much broader field.

What did we talk about? We emphasized here what we would call universal properties of systems. We discussed almost entirely systems in or near equilibrium, in the limit of low temperatures, primarily homogeneous systems, systems infinite in extent, homogeneous on a macroscopic scale, translationally invariant or spatially periodic on a microscopic scale. We also did talk a bit about behavior at surfaces which we would think of as a boundary of a say semi-infinite system, or an interface between two of them, but still translationally invariant in the direction parallel to the surface. This is obviously quite a bit of restriction.

What I meant by universal properties is that we are mostly focusing on low-energy properties such as for example the spectrum of elementary excitations at low energies. We talked about the response of the system to various perturbations at low frequencies and long wavelengths, and there is much more.

But much of condensed-matter physics is also concerned with particular properties as opposed to universal, the sort of quantitative details that distinguish one material from another. At the basic level for example, we would like to be able to calculate ground state energies of a collection of electrons in the presence of some atomic configuration, to find out what favors one crystal structure over another, what are the energies that cause reconstruction at surfaces or interfaces, or figuring out what the energies for formation of defects are, or we might look at kinetic problems of crystal growth at surfaces, how fast they happen, chemical reactions, catalysis, all these are quantum problems and they are important. There are also finite energy problems such as trying to calculate the electronic band structure. Particularly, you might be interested in the values of band gaps. Even in insulators, these are difficult problems. We look at optical absorption, calculating the dielectric constant at finite frequencies. We might for example be interested in what happens when you photoexcite some kind of electrocarrier above a band gap, and then how it moves. This would be things that are very important for photoelectric devices. Typically, these are very difficult problems and the condensed-matter people often approach them by making drastic approximations. So we start with a Hartree approximation or various approximations based on density functional theory, and then the problems involve a lot of fairly massive computations, while the approximations are not very well controlled from a theoretical point of view. But there has been a lot of progress and it is important. By fine tuning these things and using empirical data, people have gotten a lot of insight. There have been improvements over the years in the algorithms, extensions to things like dynamical mean-field theory and other more sophisticated ways of trying to incorporate moderately strong electron-electron interactions in some useful way. More accurate calculations on simpler models also are major parts of what we are interested in. I wanted to mention this as one of the things that was left out.

Another thing that was clearly left out was the effects of disorder. They were mentioned by Boris Altshuler and Alain Aspect in Session 3 on control of quantum systems. They talked about the localization problem which has a lot of very important theoretical aspects, but disorder is very important in many other ways. It will affect for example the classification of phases that you can you get. It has of course effects on properties of all different kinds. Transport properties are often dominated by disorder particularly at low temperatures, either external disorder or disorder due to frustration. This can lead to glassy behavior in which you do not ever reach equilibrium. You have various modes with very slow relaxation times. There is a spread of relaxation times, so that things are relaxing on any laboratory scale. These certainly affect experiments at low temperatures and they lead to decoherence. They may make it just very difficult to cool, because you have reservoirs of heat that keep releasing energy. Of course, there are very important $1/f$ noises. So this is another area that is very active, and we just did not have much time to say anything about it.

I think very little was said about mesoscopic systems and devices, except perhaps a little bit, again, in the context of Session 3, since solid-state approaches to quantum computers for example are based on mesoscopic devices. There are a lot of very interesting problems in them that are different from macroscopic systems, particularly, when you get to systems whose size is small compared to the mean free path for an electron crossing the sample, to undergo inelastic scattering, or other mechanisms for decoherence. So again issues of quantum coherence play a very crucial role in understanding these kinds of systems.

There are many experiments that are typically done out of equilibrium. Often you have a mesoscopic system weakly coupled to the external world. An electron may tunnel into it, spend a long time there, long compared to the decoherence time, or maybe short compared to the decoherence time but long compared to the transit time, and we want to understand what happens. There are interesting very important effects of interaction of the electron spins with say nuclear spins that have been touched upon, but not discussed here.

I am not going to say much more about this. I will just say a couple of words about what was discussed to remind you of some of the highlights. A lot of what we are interested in and what was talked about was the classification of different kinds of systems, a universal classification of their low energy behavior. Subir Sachdev introduced three broad categories: first of all systems with an energy gap in the bulk; secondly systems which do have zero energy excitations but only at isolated points in the Brillouin zone, so that there is only a small density of states for those low energy excitations and in many cases they have a linear dispersion, they look kind of relativistic; and the third category was the systems which have zero energy excitations along a whole curve or surface in the Brillouin zone. These include normal Fermi liquids but interestingly also other types of systems, as was emphasized for example by Matthew Fisher and Leon Balents as well. There seem to be systems, made out of fermions, that have Fermi surfaces but are not conventional Fermi liquids. Their Fermi surface may not be in the place you would normally expect. There are Bose systems, Bose metals, which seem to have something like a Fermi surface. These are clearly a frontier for continued research, and perhaps one of the most exciting frontiers, so we may expect to hear more about that in the near future.

For the classification of different types of systems, of course the low-energy behavior in the bulk is crucial, but in addition, as you have heard, there is a lot of excitement about the ideas of topological classification. That is dividing up some of these systems further by so-called topological distinctions which may be manifest particularly if there is a surface. So two systems with properties that look more or less identical in the bulk may have very different properties at the surface. Some of them may have low-energy excitations, some not. The first example of course are quantum Hall states, which are two dimensional systems with a bulk gap that have always propagating low-energy modes at the edges. Another early system which I might mention is the Haldane phase of the one-dimensional spin-one antiferromagnetic chain, which Haldane and others demonstrated very beautifully in the early nineteen eighties to have a gap in the bulk, but necessarily has to have at each end a low-energy spin one-half degree of freedom which just comes out of nowhere. That is the beginning of this whole subject, and clearly there is much more going on as Xiao-Gong Wen explained. The classification of topological orders is not necessarily completely understood at present and there are many links to the concepts of quantum entanglement, quantum entropy, and many other open questions.

I would like to use these remarks as a stimulus for people to talk about where we might go from here.

S. Das Sarma We heard about topological classification and Subir Sachdev of course talked about AdS/CMT at first in condensed matter, and we heard about AdS/CMT in the string-theory session also. What I was wondering is something that I am very interested in: Is there a stage where these two paradigms, these two concepts, can come together? They have been discussed separately here, but they are both high powered field-theory techniques that exist both in particle physics and in condensed matter. Could the AdS/CMT approach shed any light on the kind of condensed-matter problems, mostly two-dimensional problems, which are described by these boundary conformal field theory and topological quantum field theory? Is there a connection somewhere? It is beyond me. This is a comment, but also a question. I did not see any connection in the discussion, but if there is, then the subject certainly expands enormously.

G. Horowitz One comment. I have very limited understanding of these topological insulators, but one thing I think is true is that typically they are described by some sort of topological field theory in the bulk and you have sort of degrees of freedom on the boundary, but really no local degrees of freedom in the interior. AdS/CFT is really very different because there we have a bulk theory with lots of degrees of freedom. All of general relativity and everything operates in the bulk, and nevertheless there is a duality with a theory with lower dimension.

H. Ooguri I am not sure whether I am answering that question, but certainly there are attempts to understand the classification of Fermi surfaces from the point of view of string theory.

B. Halperin Again my understanding was that one might have more hope of learning something from the AdS/CFT connections by applying them to systems which did not have a gap in the bulk, or else they were perhaps at a phase transition, but in any case, would be richer.

L. Balents I just want to make a comment, but not about AdS/CFT. We heard a lot of emphasis about classification of different types of phases, especially ones that have some kind of subtle topological or quantum order, exotic phases we would call them. That is an important problem. I would say probably the more pressing problem for us in condensed-matter physics is actually figuring out how to identify them in actual experiments, especially, given that these phases often are characterized by the lack of conventional orders, which is usually what we look for. So, I think one of the pressing problems for theory is to figure out how it is we could actually look for some of these things experimentally. That is just a comment.

B. Halperin Let me make an additional comment which I think is pretty relevant. A lot of progress that may occur is not going to be necessarily driven entirely by theoretical ideas but by improvements in materials, improvements in experiments. To return to this issue very close to what Leon Balents was saying, obviously, we need ideas for good experiments to do when we talk about these topological insulators and some of the ideas that maybe you can create very exciting new phases. If you look at the topological insulator which has electrons on the surface in contact with a superconductor, in contact with magnetic fields, we can produce non-abelian anyons. All this depends on being able to create the materials. Somehow, we know which materials we would like to have, but how do you control the defects? The materials that we have, they are not insulating enough for example. Well, there have been big improvements, but I think it is hard to predict what will be successful and what will not. There are a lot of exciting new things to come, for example interfaces between different oxides. It is another problem that I think poses a lot of very interesting theoretical problems. We know that it looks like you can get superconductivity and you can get various effects there. These are not necessarily universal effects but they are effects of great interest to us and I think those are the kind of directions we could go. Maybe other people will suggest very specific things that they expect to happen.

S. Das Sarma I want to give a partial answer to Leon Balents' question, since this is very big on my mind. I am working closely with the experimentalists precisely to answer this question. I do not see any generic answer as Bert just alluded to, of how to look for topological effects in a given system experimentally. But at least for fractional gap topological insulators, fractional quantum Hall effects, Majorana systems, Kitaev lattice, and related systems, I think the technique is to look theoretically or numerically for topological degeneracy. That of course cannot be seen experimentally. Well, there are some ideas of looking for entropy. Bert has some work on it, but I think those are very difficult experiments. Therefore, I think one has to do non-generic experiments looking for the fractional excitations, the anyons themselves. For fractional quantum Hall effects, you look for the fractionally charged excitations, since the Majorana mode is not neutral but becomes charged there for a technical reason. For the superconducting system, you look for the zero-energy Majorana mode. Again you do a fractional Josephson effect experiment. Thus in my view, in the end you have to look for the excitations because that is what distinguishes these systems. They are connected deeply with this topological degeneracy. It is not a very satisfying answer but right now I do not see any generic answer, any signature. I agree with you, this is the thing. Maybe you are surrounded by topological systems. We just do not know how to identify them.

S. Davis I would like to bring up a point that Frank Wilczek often emphasizes. In the history of condensed-matter physics, there are very many important ideas and observations which led eventually to contributions in the high-energy physics sector, and perhaps even in the cosmological sector, that remains to be determined. Breaking of $U(1)$ gauge symmetry, superconductivity, superfluidity, actually you break $SU(3)$ in superfluid helium 3, condensates, etc. From the point of view of our high-energy and astrophysical colleagues, do they have any suggestions for us for new condensed-matter table-top experiments which could be germane to the high-energy and cosmological issues which are so important today?

B. Halperin That is a challenging question.

D. Gross Well, there were non-equilibrium phenomena in the early universe, cosmic strings, or things like that. What happens with defects, relaxation, those are all things which were asked twenty years ago and never delivered. I think what Gary Horowitz was alluding to was that you could imagine posing problems if this connection via AdS/CFT becomes better understood, but probably we are not there yet.

F. Wilczek The community working on field theory and particle physics brings a different culture to the subject which may be helpful. In a way, I already touched on this in fact. We are very comfortable working with effective field theories in kind of eyeballing their consequences where a condensed-matter theorist would be calculating band structure and maybe after a lot effort extract similar consequences. For instance in the topological insulators, some of us think of them as $\theta = \pi$ electrodynamics and that suggests a certain kind of question about what happens at the surfaces, how you can relate different realizations to each other, and what happens when they go superconducting. That kind of different perspective is there, I think. I can talk about particular suggestions too, but maybe this is not the time.

W. H. Zurek Just to amplify what David Gross already said on the business of phase transitions. Normally in the lab they are seen as equilibrium phenomena, investigated through a sequence of equilibria as you go through the critical point. Now, if you are dealing with the early universe, you do not have the luxury of waiting for the system to come to equilibrium. So I think the business of what happens if you go through the phase transition at a rate given by whatever you are going through is very interesting, and topological defects are part of that business but there are many more questions that are related.

S. Davis Let me just respond to that point for a second. Tony Leggett always emphasizes that we cannot actually address that question because when we go through a phase transition, we take away the energy through a refrigerator to another location, whereas in the beginning of the universe you did not have another refrigerator.

W. H. Zurek You do not need to refrigerate to take energy out. You can drive the system through the phase transition by changing some other parameter, and that is possible. In fact, that brings back some of the issues that have been raised earlier. In adiabatic quantum computing, you start with one Hamiltonian and you end up in the ground state of another Hamiltonian. That is the way of getting ground states of arbitrary Hamiltonians without cooling, but you need to know the behavior of the system right as you cross the critical point.

R. Blandford I just make one more comment on a slightly different topic, in terms of Seamus Davis, for the connection to astrophysics. The big example apart from cosmology is the neutron star and there has been a large interplay backwards and forwards between understanding the physics of the interiors and the surfaces of the neutron stars and experiments in condensed-matter physics, and I think that remains a rich field.

I. Klebanov Since there was some discussion of classifying three-dimensional critical points, I think what AdS/CFT can tell us reliably is that, for certain infinite classes of three-dimensional conformal field theories with very large numbers of degrees of freedom, their classification is essentially mapped to classifying seven-dimensional Einstein spaces. The reliable low-curvature gravity duals are of the form of four-dimensional anti-de Sitter times seven-dimensional Einstein space. We actually know various infinite classes. As you go down in the number of supersymmetries, the wealth certainly increases. For example in the case of extended supersymmetry, we have various infinite classes. I guess the big question is whether somewhere in this class there is a non supersymmetric theory that is possibly realizable.

F. Wilczek One more thing I think we might have to add is questions and challenges. Particularly, I think very concretely that we desperately want to see measurements of fundamental electric-dipole moments. It would be a wonderful thing if somehow we could find experimental correlates of the entanglement entropy that we love to calculate. Sensitive detectors for cosmic backgrounds of various kinds. We have an agenda that I think would not be answered by the LHC but might be answered by atomic, molecular, or condensed-matter physics.

––––––––

Howard Georgi: Particles and Fields

H. Georgi I am not planning to sum up because I was really enjoying the lively discussion that we were having at the end of the session and I would like to see it continue. To my mind, one of the most interesting issues was the tension between the enormous success of – in Frank Wilczek's words – the radically conservative quantum field theory, which I interpret to mean effective field theory as nicely discussed by Erik Verlinde and others, and the indications from various sources noted by Nati Seiberg, Misha Shifman, Nikita Nekrasov, David Gross and others that there is much more structure to it. So, I have suggested to some of the speakers that it might be useful to think about this a little more concretely and ask what we might see at the LHC that would make it seem more urgent to think about quantum field theory in a different way. With that I am going to throw the floor open and hope that there is some interesting discussion.

M. Shifman Frankly speaking, I do not know exactly what to say about LHC, but I want to make a brief comment regarding Nati's presentation yesterday and a little bit about Frank's beautiful talk. It goes without saying, in areas with curiosity-driven research, it is more than natural to look beyond the horizons of knowledge in every possible direction. However, without hints from nature, it is very hard to decide which direction is more important. There are infinitely many directions. For instance, one might ask oneself whether it is possible that quantum mechanics completely fails at distances such as 10^{-25} centimeters, long before gravity becomes of order one. In principle, I do not see why it cannot. It could be the case but should we now start research in this direction? Practically speaking, I would say no. I suggest that we adopt a kind of practical attitude. When it is needed and very useful, we can give up the Lagrangian approach and there are many examples, for instance in two-dimensional conformal field theory, where that is a routine procedure, because the algebraic construction is more general than the Lagrangian construction and people do that all the time. And there is nothing bad or revolutionary in doing that. On the other hand, in three or four dimensions, I see in fact no urges, no hints from nature that effective field theory as formulated in the Wilsonian formulation should fail, or not be useful.

Maybe I did not find good words during my presentation, but what I wanted to emphasize is that Yang-Mills theories are unique in the sense that they are consistent, they are closed and, in principle, could be considered in isolation. They present a huge variety of various dynamical scenarios, like Coulomb phase, Higgs phase, confinement phase, conformal phase, and there is a number of exotic phases. I want to mention one of those that Frank somehow did not manage to, I think. There is an exotic phase at high-temperature/high-density QCD, which might even be found in neutron stars. It happens when color and flavor gets locked. In this color-flavor locked phase, there is color superconductivity. Personally, I think that this is a very interesting phase that may even be observable. By the way, this phase supports the type of non-Abelian strings which I mentioned in my presentation.

H. Georgi Very good. We have to go on and let other people contribute. The reason I asked the question the way I did, is that we may only get one shot at

LHC energies and we do not want to miss anything by not being imaginative enough in thinking about what we are looking for. So let me go on to Ignatios.

I. Antoniadis I would like to mention one concrete example of a possible realization of a theory that has a non-Lagrangian description, as mentioned in Nati's comment yesterday: it is the so-called little string theory. This theory can be obtained as a weak-coupling limit when g-string goes to zero of some particular configuration of string theory with Neveu-Schwarz five-branes. It is a theory with no gravity. The idea is to put back gravity, but very weakly coupled, in other words to consider a configuration in which the string coupling is very small, so that gravity appears, but very weakly. One can use this setup to address the hierarchy problem. This idea we had ten years ago with Savas Dimopolous and Amit Giveon, but it was very difficult to say something more, but last year we had a way to make a concrete prediction by using duality and find a dual description in terms of a gravity dual, which as usual has one extra dimension, the holographic dimension, in which the background is not AdS but it is flat and there is a dilaton (a scalar field) which varies linearly in the extra dimension. This is the so-called linear dilaton background. In this background, one is able to compute the gravity spectrum and one finds a very peculiar spectrum with a mass gap which could for instance be at the TeV scale and then a series of very dense excitations, gravity extensions. Every excitation couples with inverse string-coupling squared, which means inverse TeV. As Savas says this would be like a jackpot because one does not find just one excitation, but several of them.

C. Bunster I would like to make an ultra-conservative remark. It is the following: If we are going to execute field theories or action principles, we should execute them for crimes that they have committed and not for crimes that they have not. So, in the context of Nati Seiberg's remarks of yesterday, I want to put on record that, contrary to what he stated, there is an action for self-dual p-forms. The simplest case is that of a chiral boson in 1+1 spacetime dimensions. The action is local and Lorentz invariant, but the Lorentz invariance is not manifest . Moreover, for the cases in which one does not have self-dual fields but one can formulate the idea of an electric-magnetic duality rotation, there exists an action which is manifestly duality invariant. Again, the action is local but not manifestly Lorentz invariant. This addresses the point of hidden symmetries in a very elementary way, and it provides an example of a more general fact; namely, that even hidden symmetries can be searched for within existing action principles.

H. Georgi Thank you. Gian.

G. F. Giudice I just wanted to reply to your question. A big hope for the LHC is based on the naturalness idea, right? And that is based on effective theory, on the separation of scales on which we can separate infrared from ultraviolet. Maybe we have an indication that that does not work and that is the cosmological constant. So the best indication to see a failure of effective theories at the LHC will be to see the Higgs and nothing else. Of course that is a pretty dismal possibility but I should say that recently people have thought of other motivations for physics at the LHC and not based on just naturalness, for instance split supersymmetry is one case in which you could see something but not (naturalness) necessarily.

E. Verlinde The question of naturalness is usually associated with trying to extend the paradigm of quantum field theory to higher and higher energies and eventually it will break down. Actually, I had a discussion yesterday with Gia Dvali about the possibility that effective field theory would indeed break down at a much lower energy scale. Maybe people are too much looking for signatures at LHC that only could come from some sort of field-theory thinking. Maybe we should start thinking about what would indeed be the sign not just in terms of particles but anyhow in certain phenomena that are not based field theory. An example could be say the black-hole formation that was somehow not there, but maybe people did not really look at things in the right way because it was maybe done in a too classical form.

N. Arkani-Hamed The most obvious opportunity for something that would force an interesting extension of the way we think about field theory would be exotic strong dynamics that we cannot think about in normal ways. The problem for twenty years has been that this wild and wonderful possibility has been in a big straight jacket since the early nineties when we had very good evidence for a perturbative picture of at least electro-weak symmetry breaking. Having said that it also gives me an opportunity to mention not just the LHC but another class of experiments that might give us access to interesting sectors that might have strong dynamics which, amongst other things, might force us to use some of these alternative descriptions of field theories. It could be for one thing that there are sectors that have really nothing to do with electro-weak symmetry breaking, they are just lying around for no particular reason. That is the sort of lucky accident that has happened to us before. Such a thing might happen at the LHC and of course we might be able to see that stuff. There is no reason, it is not particularly constrained from anything we know so far and that is one possibility. Another one along the same lines, is that we could also have hidden sectors that are much much lighter that the weak scale. We could be weakly coupled to them, not horrendously weakly coupled to them, but weakly coupled to them enough to be interesting. One could have a whole slew of new particles at the GeV scale for example with the coupling just suppressed to us by a part in a thousand. That possibility is not remotely excluded. The way to look for sectors like that is not through high-energy experiments like the LHC, but through very-low-energy experiments with very very high intensity beams. A variety of experiments like that are actually going on right now, in Mainz in Germany and at the Jefferson lab in the US. That is just a general thing. There are places where exotic dynamics could be lurking which are not strongly limited by what we learned already from experiments in the early nineties.

D. Gross I wanted to address Howard's question which was what observable effects and surprises at the LHC might force us to consider the kind of theory, or worse, that Nati challenged us with. To some extent, Nima answered that in the first part of his three-part answer, namely, the discovery of particles plus couplings that could not be accommodated by field theory, or by effective field theory. Now, that is hard to imagine because we have never seen anything like that. That is the power of quantum field theory. Since we know nothing or very little, as you say, about the dynamics of such theories, we would be surprised. We would try to fit them to anything and it would not work but, finally you, who will accept the following challenge, will know enough about say little string theory or two-zero theory to say that this particle phenomena fit into the phenomenology of such theories. In order to answer Howard's question, it would help a lot if one looked at these unnatural quantum field theories that we do not know how to describe easily in a Lagrangian framework and try to predict whether they would have strange observable consequences we could look for. It is very hard to discover surprises because we do not know how to look for surprises.

H. Georgi That was the reason for my question.

N. Seiberg I made a short list here of phenomena which are extremely unlikely, but we should keep an open mind. I think the first thing, which again I emphasize I bet money against it, of no Higgs. This would clearly be a breakdown of our understanding of quantum field theory. Next unlikely thing: there is a Higgs, but nothing else. That is a hierarchy problem, naturalness and so forth. I can go down the list, there can be unparticles, as our chair recommended, a new phase of quantum field theory. It is not going beyond quantum field theory but we need deep knowledge of quantum field theory. I sympathize with Misha that there could be all sorts of other phases of quantum field theory that we never thought about. They might materialize at the LHC. Then we can go further. Things would go beyond quantum field theory. Antoniadis mentioned the little string theory. It is just by accident that we stumbled on it. For all we know, there are trillions of other such things. So I think David is right. We should keep our eyes open and try to stick with standard quantum field theory but if not, be happy.

S. Das Sarma I also want to address the LHC question that Howard raised. Nati to some extent answered my question, but I want to ask it anyway. This is a question I asked Ed Witten I guess six or seven years ago in a conference in Santa Barbara as an outsider and he gave what I consider to be a compelling answer. The question is what if the LHC does not see any sign of SUSY, none whatsoever, you know. What is the wiggle room? I mean there is this huge edifice that has been built which is very beautiful but it is not seen at all. Is the wiggle room huge? What Ed Witten told me at that time: Life would be difficult. That is what he said. That was seven years before it was built. You were pretty sure it would be seen but I said that is not the question "what if it is not seen?" and you answered then, after thinking, that life would be difficult.

So first I want to know what you say now, but then I want to know what others say.

E. Witten Let us hope we will not have to cross that bridge.

S. Dimopoulos This is just an elaboration on what Ignatios started. We studied in detail experimental consequences. The amazing thing would be you would have a mass gap in the Kaluza-Klein graviton and then effectively a quasi-continuum of states, so you produce tons of them. They couple strongly with an inverse TeV strength, so they are easy to produce and they decay inside the detector. Furthermore, you produce a slew of them. So many of them give you displaced vertices and, because they are graviton-like, they decay to leptons. So they have very clean experimental signatures. What I learned from Nati while we were doing a Tintin walk this afternoon, is that in the full theory you not only have the graviton excitations, but you have many brothers of the graviton which have again the feature of a mass gap and continua with different spins and similar couplings. Therefore, seeing two or three such towers of mass gaps followed by continua would be very exciting.

N. Arkani-Hamed A very very brief comment, not to rain on the parade of the excitement of two-zero and little string theories at the LHC, but these are examples of theories where the difficulty is giving them a Lorentz invariant Lagrangian description in higher dimensions. Both of these theories, both the two-zero theory and the little string theory can perfectly be described in terms of stringing together chains of four-dimensional gauge groups coupled to each other. In that sense if you discover them, you are not forced to abandon the realm of four-dimensional quantum field theory at all. In fact, they have a beautiful description in terms of four-dimensional field theory. Of course, one would discover that it has some higher dimensional Lorentz invariance and that would be the mystery that one would try to explain. This is just to stress that it would not take us out of four-dimensional field theory language yet.

M. Douglas Just a short comment. We are talking about exotic possibilities. As Nima said, sometimes what looks exotic might also have a four-dimensional description, but we can ask the following question. Suppose the underlying theory is string theory. Should we expect to see any of these things or not? We were expecting supersymmetry. The easy loophole out is to push the masses of the superpartners up by a factor of ten or a hundred, which is consistent within string theory. My comment is just that most string vacua, by any way we know to count, measure or construct them, have many more degrees of freedom than the standard model and the ones we have seen. They have tens or even hundreds of gauge groups and they have charged matter under these things. We do not know how we should expect them to couple to the standard model, but that is the sort of thing we can try to find out by continued study of the string-theory landscape. This might be reason for optimism that there should be really much more out there to some day discover.

E. Silverstein Let me add a brief comment to that. When it is said that string theory expects supersymmetry, that is probably true as a cultural statement but I do not think that it is at all clear from the top down that supersymmetry is expected at low energies, where I include an order of magnitude or more that Mike just alluded to. There are these extra dimensions. They can take many many different shapes and most shapes they can take are very curved and will break supersymmetry a priori. It is hard to do all the calculations required to count the relative distributions of these things and answer this question, but to me it is not obvious at all that low-energy supersymmetry is preferred by string theory. So when Sankar asks about wiggle room, I mean for string theory there is, unfortunately perhaps, there is plenty of wiggle room, as far as I understand it.

G. 't Hooft Let me briefly mention that naturalness was a beautiful guideline, but imagine that it is wrong because there are some problems, the hierarchy problem for instance. What if we try to attach a more important significance to conformal symmetry in theory. Maybe we are entering a conformal phase, where basically no structure is left. That would be the complete opposite of what many people are expecting: perhaps LHC will see very little because we are entering into a conformal regime.

A. Polyakov I will not speculate about what will be discovered or not ... or maybe I will actually? What I am going to say is that we are used to describe QCD by the Lagrangian, and we have gluons, etc. Then we hope that we have the gauge-string correspondence, in which case we describe everything in terms of flux lines and open strings with their ends at infinity corresponding to gluons. These strings do not have any higher excited states. Now, let us imagine that the ends of the strings are not fixed precisely at infinity but somewhere else. Then we will have a whole spectrum of the open strings. What I am saying is that it is not impossible, I think, that the primary description of QCD is not in terms of fields and Lagrangians, but in terms of strings. If we have conditions which fix the ends of these open strings, what we will see is an infinite number of excited gluons. So it will be a modification of QCD, where apart from normal gluons and normal quarks, you will also have an infinite sequence of excited states. That is a speculation, but I think that such a picture could be made consistent.

I. Klebanov Yesterday, when I was giving a talk, it was cruelly cut short by the alarm clock just a second before I was about to say a magic word. This magic word was not LHC, it was actually entanglement. So one thing that I did not have time to mention is that this calculation for a CFT on the three sphere actually computes a certain kind of quantum entanglement entropy that came up in Subir Sachdev's talk for example. Namely, if you consider the entanglement between inside and outside of a circle for a theory in two spatial and one time dimension, the universal term is negative and it is precisely minus this F that I defined in my talk. This interesting result was established recently by Casini, Huerta and Myers. So it gives you a kind of tool for computing entanglement and perhaps explains better why that quantity has something to do with the number of degrees of freedom in a CFT.

H. Nicolai I wanted to come back to something Eva Silverstein said. Many are under the impression that somehow superstrings predict low-energy supersymmetry and you said that this is by no means the case. But is there any viable model of a string compactification that gives the right massless states and does not produce any tachyons or pathologies? Is there any known example of such a compactification?

E. Silverstein I mean there is no example known with or without supersymmetry that gets the entire spectrum right, the cosmological constant right and so on.

H. Nicolai My understanding was that when you do not have supersymmetry there are always instabilities.

E. Silverstein No. The methods that you use to control the calculations are very basic. It is just perturbation theory and you insist that there be a local minimum of the potential. These techniques do not depend on supersymmetry. You may be thinking of certain ten-dimensional string theories which happen to have tachyons and also happen not to have supersymmetry but, even there, projections can remove the tachyon from the spectrum. So there is nothing close to a theorem that string theory predicts low-energy supersymmetry and I think it could well go the other way. I may be wrong though.

H. Georgi Let Mike give a brief answer and then maybe Gabriele.

M. Douglas Just to elaborate on Eva's point. One can debate what seems very very likely from what we know as you could push the masses of the superpartners in some constructions way up, maybe up even to the GUT scale. I am not here talking about experimental constraints, but just in terms of the internal consistency of the compactifications, which of course from our experimental point of view is as if they are not there. Now, naturalness was always the answer to why the superpartners should be light but what we start to realize in string theory is that the traditional idea of naturalness is just one component in a more complicated calculation that has to involve details of early cosmology and how different vacua are populated. That calculation can easily have many factors that go the other way weighing against low-energy supervacua. It is that type of consideration that Eva is alluding to. String theory might not predict it, or might even predict the opposite, if we could understand it.

G. Veneziano My comment is also related to the question of Hermann Nicolai. If indeed supersymmetry is not found at the LHC and is pushed very high, then there are two mysteries about quantum corrections, the absence of a large cosmological constant and the big renormalization of the Higgs mass, which in my opinion call for an ultraviolet completion of effective field theory. Since precisely these quantities depend on high momenta in loops, I mean they may be sensitive to what is the completion of say, some effective field theory. Then to come to a more specific question, which was maybe answered actually by Michael, can superstring theory push the scale of supersymmetry breaking very high and yet protect the mass of the Higgs?

H. Georgi OK, I think we will postpone that for a later discussion. Wen, did you still have an important comment?

X-G. Wen Actually, this is a question. There are so many experts here I could not help ask the question whether the fundamental theory for our universe is simpler at the cut-off Planck scale or is very ugly? If we have a very ugly theory at the Planck scale which produces an effective standard model at a low energy, do we accept this as the theory of our world? You know, basically from condensed matter experience when you go up in energy, the theory is getting uglier and uglier, rather than simpler and simpler.

D. Gross If it works, it works. Usually, if it works, it is simple.

F. Wilczek There is also a selection effect: if it is ugly at low energies, you call it chemistry.

H. Georgi So we have to close the session, but let me do so by giving a slightly different answer to the question of what happens if supersymmetry does not show up at the LHC. So we have a slide here which is from a conference in 1994 in Erice and this is a bet between our chair and Ken Lane and the moral of this bet is that one needs to be patient, because you will notice that they have decided the bet only after 50 inverse femtobarns of integrated luminosity have been delivered to the detectors. So David or Ken have a little time before they have to pay up on this bet. With that, let me close the session.

D. Gross Let us thank Howard.

David GROSS'
Ken LANE (Outsiders)

BET:

1 - DINNER AT GIRANDET
FOR THE ABOVE AND THEIR
SPOUSES (AT THE TIME) AND

THE FOLLOWING
AFTER LHC STARTING TIME (DEFINED
AS $5 \cdot 10^{40}$ 1/cm² INTEGRATED

LUMINOSITY) SUPERSYMMETRY WILL
HAVE BEEN DISCOVERED
31 JULY 1994   David Gross   Ken Lane

Joseph Polchinski: Quantum Gravity and String Theory

J. Polchinski We shall start with one more short talk by Stephen Hawking.

S. Hawking The no-boundary wavefunction gives the probability for the entire universe. However, we do not observe the entire universe. Our observations are limited to a small patch mostly along a part of our past light cone. Probability for local observations involves a sum over free metrics and fields on a surface of constant density. The sum is weighted by the volume of the surface that take into account the different possible locations of our past light cone. Volume weighting has a significant effect on the probability distribution for the amount of inflation when the potential has a regime of eternal inflation and leads to the prediction of a long period of inflation in our past. The usual Euclidean path integral for the volume weighted no-boundary wavefunction is difficult to define when the potential has a regime of eternal inflation. Eternal inflation occurs where epsilon, the slow roll parameter of the gradient of the potential, is less than the potential V. This is because the scalar will be effectively massless in this region and will have large fluctuations. It will not be a good coordinate. We shall replace the eternal inflation region with anti-de Sitter space. The fluctuations on this background should be a good approximation to the fluctuations on the boundary of the region of eternal inflation. This provides a framework in which to understand the creation of the universe and the fluctuations that cause all the structures in it, including ourselves. Thank you for listening.

J. Polchinski If there are any immediate comments or questions on Stephen Hawking's talk and also Jim Hartle's related talk, perhaps Jim can respond to those.

V. Mukhanov How should we see this wavefunction of the universe if for instance the universe is infinite, like an eternal universe? This should be a completely classical object. The action is infinite. Everything is infinite.

J. Hartle I am not sure I understood your question.

V. Mukhanov If you take a very very big system, for which the action goes to infinity, then this system is classical. For instance, it does not make sense to quantize an open or flat universe. You can only quantize a closed universe. If the universe is eternal then it is an analogy to an open or flat universe.

J. Hartle We are assuming that the universe is closed where the target for prediction are alternative classical histories. We are assuming the cosmological constant is positive that the universe may exhibit. So they have basically de Sitter behavior.

V. Mukhanov So, in this closed universe, there was a beginning, then there were quantum fluctuations, and after that ...

J. Hartle That is a very spacetime picture. We only get a reasonable spacetime picture when we have classical histories which we calculate from the wavefunction itself. We calculate those by looking at the saddle points of the action that satisfies the no-boundary condition and we find that the saddle points can be represented in different ways, some of which correspond to de Sitter behavior. They can also be represented by a behavior which is like anti-de Sitter space, and in that representation you can use a dual field theory to replace the regime of eternal inflation.

D. Gross Jim, in your presentation of this in your talk, there was one term I did not understand. You added "assuming that we are rare".

J. Hartle That was for expositional purposes. We do not have to assume that we are rare. If we are not rare, then we have to calculate based on the probabilities that our data exist in a given Hubble volume. That in fact regulates the volume divergence, the divergences you get for large volumes. So, assuming the sum is actually over locations including, in the vast universes that are considered in this type of cosmologies, the possibility that our data replicated it in different places. That was too complicated a thing to present in five minutes, as is evident from the fact that you are shaking your head, so I opted for making the assumption and then we get these results.

J. Polchinski I will give not so much a summary but a perspective, not on the whole earlier session but on the part of it concerning the question of "what is quantum gravity". Afterwards, people can give their perspective on this question or ask their own questions.

So the answer to this question is certainly constrained in many ways by observation. But observation probably cannot get to the heart of the question because the Planck length is so small. We have known that since Max Planck calculated it twelve years before the first Solvay meeting. Fortunately, it is also constrained by consistency. A number of people remarked that quantum mechanics is hard to modify. For a variety of reasons, it is also true that it is very hard to combine quantum mechanics with gravity. If you try, you can easily run into infinities, negative probabilities, large violations of Lorentz invariance, or violent violations of energy conservation. In exploring these clues, we have been led to find this very connected structure that we call string theory although, as we learn more about it, maybe the strings have a less and less prominent role and we still have not fully discerned its nature. The most remarkable and deep thing we have learned about it is the thing that Juan Maldacena focused on, this duality between gauge theories and gravity. Therefore, the thing we have been looking for, a theory of quantum gravity, has all along been hidden in the exact framework of gauge field theory that we were using to build the standard model. It was not only hidden there, but hidden with it were the strings, the branes and the full structure of string theory. Now, as Gary Horowitz has mentioned, this allows one to find gravitational models over a wide range of real-world physics, which is a remarkable connection among different parts of physics. But I want to focus on the other direction. That is, this gives us a construction of quantum gravity in terms of something we sort of understand, quantum field theory. Of course, we give this construction in a very special space, a very special box, anti de Sitter space. Anti de Sitter space is sort of where particle physicists would study gravity if they could. You control the boundary conditions, you control the initial conditions. You throw stuff in. You see what comes out. A lot can happen in the box. You can form black holes, they evaporate. You can have Planckian scattering. You can have singularities as Eliezer Rabinovici discussed. You can have topology change and all these things are described. But of course, there is a lot that does not happen in the box and we want to get out of the box and I will get to that in a second.

Before, I want to mention a couple of lessons from this, a couple of words. The first word is holography. One of the lessons of black hole quantum mechanics is that there is a tension between quantum mechanics and locality. Now, I do not thinking that this has anything to do with the reality and EPR discussion, which also is a tension between quantum mechanics and locality, but there is a tension, and thus far, quantum mechanics has won completely. That is, the framework for this gauge-gravity duality is just the framework of quantum mechanics with states and superpositions and all that stuff, whereas locality has given way not just in sort of a string-scale, Planck scale fuzzing out of spacetime, but much more radically, the basic variables, the things that appear in the equations or the Lagrangian do not live in any sense near spacetime points but very non-locally projected on the sphere at infinity. Perhaps this is the best way to think about it, but we do not even know in general how to think about it. So the first comment is that locality loses, quantum mechanics wins. Another quantum comment is that this gauge and gravity duality is essentially quantum. It is when the gauge fields, the Yang-Mills fields are maximally quantum fluctuating that the classical gravity description becomes the effective one. Really, what you have is one quantum theory with one limit in terms of classical gauge fields and one limit in terms of classical gravity. So in fact, the quantum world is more connected than the classical world. Now, it is also true that we cannot derive this, and maybe this is a footnote, but it is striking that this duality makes many sharp mathematical statements, and yet we have no derivation. The straightforward quantum field theorist's way to do this would be to define the right change of variables in the path integral. This has failed for thirty years. Maybe we have not been clever enough, but it is one reason for thinking along Nathan Seiberg's lines that there is a deeper way think about quantum theories.

So, how do we get out of the box? It is very very hard to say. As you know, we are used to formulating physical systems in boxes and then taking the boxes away. But for a holographic theory, the theory is the box and so, when you take the box away, nothing is left. Therefore, it is really very hard to interpret. Whatever is the theory of gravity, we do not live in a box: we live in this cosmology. All the stuff that Alan Guth talked about cannot happen inside this box. So we need to understand quantum gravity in this bigger space and certainly, locally, it is the same theory that we understood in the box. But globally and conceptually, we know nothing. Does it have a wavefunction? If it is a wavefunction, what is that a function of? How is that wavefunction related to observables? None of these things seem to extrapolate directly from understanding the theory in the box. We have gotten a lot of clues from paradoxes and problems: the short-distance infinities, black hole puzzles, and we need more puzzles. One puzzle is what is the initial condition or what replaces it. There seems to be interesting resonance between the Hartle-Hawking wavefunction and AdS/CFT. One is the measure problem that Alan Guth mentioned. One is the problem that the dynamics seems to go to some multiverse. The holographic principle tells us in some sense that we should not look beyond any single horizon, as I think Englert mentioned, and so there is a tension there. So maybe these are the paradoxes that we need to move forward. Ideas from quantum information may help. There have been some interesting feedbacks on the black hole information problem. For example, the work of Hayden and Preskill on the rapid re-emission of information from black holes, a surprising result. Ideas of entanglement and information seem to arise in trying to understand how spacetime emerges. My time is short so I shall say just one more thing which is surprises, as various people like Nima Arkani-Hamed mentioned. One example of surprise was the landscape. There was a time when string theorists thought that our theory had some magic symmetry that would set the cosmological constant to zero and maybe give us the unique vacuum. And partly due to our own internal understanding and partly due to observation, we now know that our vacuum is no better than any ground state of condensed-matter physics. That is, there are many of them, they are metastable, they have a wide range of energies and effective properties. One surprising effect of this is that it has redefined naturalness. It has changed the way we think about the physics that we might see at the LHC. Surely, answering the question "What is quantum gravity?" will transmute the way we think about other questions as well. Now, let me stop and open the floor to comments.

X-G. Wen We know that in topological quantum field theories, all the low energy degrees of freedom are localized at the boundary. I wonder whether gauge-gravity duality also implies that quantum gravity itself is like topological quantum field theory, where all the degrees of freedom is actually at the boundary.

J. Polchinski The analogy has been made many times but there is a dynamics in the bulk. So perhaps someone else has an insight here but, because there is a dynamics in the bulk, it seems hard to draw this parallel. But maybe we have not thought about it right. Gary.

G. Horowitz I wanted to make a comment about the black hole information puzzle and, particularly the question of what is wrong with Hawking's original argument which seemed to imply information loss. Many people seem to have the idea that you can start with Hawking's calculation and make small corrections. I think Juan said that one has to include non-perturbative effects and everything will be fine. But there is actually a theorem that Sumir Mathur has proven which says that that will not work. Basically, you cannot start with Hawking's calculation and make small modifications to get the information out. One has to think much more radically and I believe what Joe Polchinski was saying about locality breaking down is going to be a key part.

N. Seiberg I have a question for the experts in the audience. What is the status of getting inflation out of string theory?

E. Silverstein I gave a talk about that. If that was not sufficient, maybe you should ask more specific questions. Can I comment on the question about the boundaries? It is not true that people say that the theory lives at the boundary. What is true is that the field theory is dual to the whole gravitational system and I think that is hopeful for the cosmological case because there, we do not have an extreme decoupled time-like boundary. But we do have low-energy regions from redshift effects that we might use to approach this problem.

A. Polyakov It is curious I think that one of the most important problems have almost not been mentioned so far. This is the problem of the cosmological constant. I want to make a comment about some approach which I believe is correct and will lead to the solution of this problem, which remains to be seen of course. Namely, the general idea is that you can have infrared corrections, infrared screening, the kind of phenomenon we are used to have in quantum electrodynamics when the charge is screened and leads to an almost zero value for the charge. A similar thing can happen with the cosmological constant. There is a huge amount of technical details which would take a couple of hours to present. Anyway, it turns out that it is very useful to consider several dynamical systems simultaneously. There is an analogy between a supercritical item, a back hole, constant electric field and Schwinger pair production, which is analogous in turn to de Sitter space. By analytic continuation, the electric field goes to the magnetic field and this is analogous to anti de Sitter space. On a technical level, all these problems have a lot in common. For example, both the electric field and de Sitter space have fantastic mixing between UV and IR. The renormalizability breaks down in the external field to some extent and you get large infrared corrections, which are not summed so far, but which presumably will lead to the screening of the cosmological constant. I will end with a very simple physical picture which I expect to appear but which is not proved so far. Imagine yourself on a balloon accelerating with a constant acceleration and imagine that there is a mechanism for populating the surface of the balloon with particles in such a way that the particles do not dilute as the balloon expands. Then imagine that there is a gravitational attraction between the particles. What will happen? Obviously, this gravitational attraction will put a break on the acceleration so your car will start to slow down. What can be seen so far are just the first logarithmic corrections to this, but hopefully it will lead to the full solution of the cosmological problem. The only bad feeling I have about this balloon analogy is that it very much resembles the theory that worms are generated by wet soil, but otherwise I think it is nice. Thank you.

J. Malcadena What I had said is not that you take Hawking's calculation and you correct the answers slightly, but you need a framework that allows you to compute the matrix elements exactly. Hawking's calculation does not allow you to calculate a single amplitude because it only gives you the density matrix for example. If you formulate a very clear problem where you cover interference between black holes and so on, you would see these small corrections. Now, I would like to make another comment that is related to the fundamental meaning of symmetries. In quantum field theory, we are used to the idea that if we go to higher energies, we can have a theory that has more and more symmetries and then they are spontaneously broken at low energies. Now, in a theory of gravity, like in our universe perhaps, we can have supersymmetries at some higher scales. But in what sense do we have supersymmetry in the full theory given that we do not have supersymmetry at long distances? If we try to explore the theory at very high energies, we produce black holes which are so big that they also sense the supersymmetry breaking. So this is something intermediate of some part of the theory. This shows that supersymmetry cannot be very fundamental to the theory.

J. Polchinski I would like to comment on that more generally on that symmetry breaking because Frank Wilczek described this very successful paradigm in which things get more and more symmetric as you look at them more deeply. But we know that in quantum gravity, there are no global symmetries and we also know that local symmetries are not symmetries. They are there basically because they are infrared stable, they are there because there are phases that are stable against large changes, and so it is the opposite paradigm. So, as you are saying, it seems that there is some middle ground where the maximum symmetry including supersymmetry rules. Maybe that is the right place to think, but past this point, well who knows?

S. Kachru A general comment which is one of the reasons we are so quick in connecting general relativity to the real world is that the four-dimensional cosmology that we inhabit happens to be one of the maximally symmetric solutions of relativity. It is one of the first solutions you would write down, and it was written down by 1922. In string theory, we have notoriously been much less successful in connecting to the world and I think this is related to what Joe Polchinski and Juan Maldacena were saying in response to Frank Wilczek. You could say that the story of the standard model is a triumph of symmetry. From the point of view of string theory, the four-dimensional world we inhabit and the standard model we see, is really awfully unsymmetric. We know much more symmetric states of the theory. Part of the problem is that, if string theory does describe our world, we live in a sort of provincial asymmetric solution of the theory, which makes it very hard to identify.

F. Wilczek Thank goodness there still is quite a bit of symmetry. Sort of along the same lines actually, I seem to be a sponsor of failed paradigms. But one thing that has led to the standard model, and led to so much progress in physics, is precisely locality. So why does it look so good if it is fundamentally flawed. And secondly, what would be the experimental manifestation of its flaws? It is a question. I do not have any answer.

J. Polchinski In some sense, it is the same answer. It is emergent. In AdS/CFT somehow it is a stable property of phases in a broad range of systems. It is a very subtle non-locality. It shows up in response to the information problem which in the end is sort of the question of what the realistic description of the black hole is, the inside and the outside.

F. Wilczek Are you proposing something for LHC?

J. Polchinski No, but I would love to.

N. Arkani-Hamed A very brief comment. Lorentz invariance is also broken by cosmology, so that we really have nothing as far as symmetry goes. Of course, I agree with what everyone else has said, but even good old Lorentz invariance is concerned. I also wanted to make a brief comment about the way I think a lot of people have thought about AdS/CFT in terms of the CFT being king so to speak, being the well-defined object that we know how to control and understand, and put on the lattice and simulate to arbitrary precision, and so on. As somehow as the coupling is cranked up, some strong coupling miracle occurs and all these incredible phenomena become apparent, we grow in dimension, etc, etc. As Joe Polchinski said, that attempt of deriving AdS/CFT has gone nowhere. The idea that you try some kind of change of variables in the path integral to see how this phenomenon happens has not been particularly fruitful. I just want to point something out which has not been mentioned in this meeting so far but is, to my mind, one of the most spectacular things that happened in the field in the past five years or so. It is the complete solution of the spectrum of anomalous dimensions in $N = 4$ super Yang-Mills. It is work that was pioneered by Juan Maldacena and friends, and completed by Niklas Beisert, who is here, and Mathias Staudacher. This is the first set of exact quantities we have computed in a four-dimensional gauge theory of any sort that extrapolate from weak coupling to strong coupling. To me, the most interesting qualitative thing about it is that it does not look like that paradigm that you start from the definition of the theory at weak coupling and somehow some strong coupling magic happens. The central objects that make an appearance there are some third thing. There are some very beautiful set of integral equations. When you solve those beautiful equations, at weak coupling they reproduce the answers of quantum field theory and at strong coupling, they reproduce what you can recognize as bulk AdS physics. So I think that is some indication already from a very very successful solution that we might be looking for something which is a third thing. It does not have any of the redundancies that we have been used to for eighty years. It does not have gauge redundancy. It does not have diffeomorphism redundancy but there is some third object and I think the other attempts to reformulate these theories that we know and love in different ways are at least hunting around for different formulations of this sort.

R. Blandford I was encouraged by David Gross to say just a couple of words about Erik Verlinde's ideas that he presented. I had a nice chat with him at lunch time and I was very impressed by the coincidence, numerically, of the scale of acceleration in Milgrom's theory and H over 2 pi, and that is certainly encouragement. If you think about the three arenas where his ideas could be applicable, the first is galaxies which is what he paid most attention to. A brief comment, as one goes down to low luminosities, the so-called Tully-Fisher law changes slope and that might be a clue or might be a signature of your ideas. The second is in the clusters of galaxies, which I think you describe as a work in progress thinking about how your ideas might apply there, the new observation is the outer parts of clusters of galaxies, the gas entropy is measured and is very high, indicative of a change of the ratio of dark matter to baryonic matter. The third is cosmology itself, specifically, the microwave background. As we have already seen at this meeting, the exquisite coincidence of the standard model of cosmology, including of course regular dark matter whatever it may be, and the microwave background observations, and this is obviously a major challenge to theories like yours, and Bekenstein's and Milgrom's to accommodate that. It is also an opportunity because in two years Planck will have even more precise observations, and it is an opportunity to maybe make some different predictions. I would like to make one final comment. I also had a nice chat with Lisa Randall about the possibility of high mass WIMPS and so on, since I knew about her work on this. But I just would like to advertise that there is a proposed Cherenkov telescope array which does TeV gamma ray astronomy which might be a very good facility, if this turns out to be what there is out there, of finding indirect signatures of these high mass WIMPS.

David Gross: Conclusions

D. Gross According to the rules that I established I was supposed to end this session by summarizing the summaries, which have summarized the various sessions. Also, according to the schedule, I have exactly minus ten minutes left to talk. So I will try to go rapidly through whatever comments I had which have not already been discussed by others. I must say that among the few messages that we can all take away from this meeting is that after a full century quantum mechanics is alive and well. Physics is alive and well and making incredible progress. There are wonderful problems that are being explored, from single-degree-of-freedom quantum mechanics to the wave function of the universe. Anyone who will read the conference volume, which should be out in one year we hope, will be impressed with the vitality of physics as a whole and, in particular, quantum physics, a hundred years since the first Solvay conference. This conference was conceived, rather ambitiously, to cover all of quantum physics. No mean task, a bit of a gamble. Can we still talk to each other across the enormous divide from nano-physics to quantum cosmology? I think that we have proved that we can. We can talk. We can argue. We can debate. I am not sure that we can agree but we can certainly have fun arguing. This success is a great tribute to physics as a culture that has still has retained its unity, and to all of you. I only have a few comments and I will try to go through them quickly, I probably will not succeed but this is the end of the conference and we have still have left some negative time. Not imaginary, negative.

During the second session, we got into a debate that would have been familiar at the first Solvay Conference on the nature of reality. I think we all came away more convinced of our own notions of reality. What I found truly amazing were the tests that realized the textbook examples of quantum mechanics in the laboratory with such great precision. Much has been said about the implications of these tests, whose results have surprised no one. There was one point that I found very interesting, the whole question of entanglement. John Preskill made a provocative remark about the nature of Hilbert space, where I think we are still learning to live. Normally, we start describing the physical world with a ground state, with a vacuum. Igor Klebanov reminded us that the vacuum is usually highly entangled but John Preskill told us it that most of this entanglement cannot be used, is not a resource for, say, quantum computing. Much like the fact that the vacuum has a lot of energy but this energy is untappable. Hilbert space is very large, but the parts that we can use or excite are only those portions of the Hilbert space that are accessible by the operation of multiple local unitary operators. This is true for both quantum computers and for describing the real world. That is interesting, it is a kind of redundancy. Hilbert space is enormous, yet we cannot access most of it. Is this redundancy like some of the other redundancies, such as gauge symmetries, that are endemic in our descriptions of nature? Do we really need all the Hilbert space and what is the non-redundant way of characterizing what we need? Is there a less redundant description of some quantum mechanical situations?

Atomic physics and quantum control were the subjects of the third session. I have a few disconnected remarks. First, I am envious of the condensed-matter theorists who can now convince their amazing experimental colleagues to construct analogs of model Hamiltonians in the laboratory and explore physics they cannot yet solve. The big breakthroughs, such as solving the Hubbard model, have not yet happened, but are on the way. It would be lovely if one could construct analog models of supersymmetric field theories, non-Abelian gauge theories, and maybe even duals of string theory. Peter Zoller said, in the summary of the session, that a place where new physics might emerge from such optical lattice simulations is the dynamics and the nonequilibrium behavior of closed quantum systems, something about which theory has had, so far, very little to say. Theorists have not said much over the last century because the problem is so hard and because there has been so little data. Now, with the possibility of experimentally exploring dynamical behavior, relaxation to equilibrium and other nonequilibrium behavior, I think that theorists are going to be stimulated and might begin to understand these phenomena. Of course, the experimentalists might beat them to the punch.

There is another enormously exciting area of atomic physics and optics that was not discussed at all in this meeting, the use of very short pulses of radiation, femtosecond and attosecond pulses, to probe and control atoms and molecules. This is a totally amazing development that can be used for chemistry, quantum control of chemical reactions and to explore atoms. I dream that one day, maybe by 2111, there will be pulses that would not only be able to pull electrons out of atoms and watch them snap back, but pull quarks out of nuclei and let them snap back. We also had marvelous discussions of quantum control, driven by dreams of technological applications: certainly, quantum computers but many other applications. There have not been many speculations as to the nature of the technology that will emerge since it is much harder to predict technology than it is to predict advances in basic science. In basic science, nature poses the problems; in the case of technology often the market controls the developments. So it is hard to say what the future of technology is but, listening to the talks on quantum control, mesoscopics and spintronics, it is clear that in 2111 technology will be unimaginable. Finally I also want to add my plea to the atomic physicists to continue with precision tests of fundamental physics, that they survived on for many years before they discovered quantum simulation. Please do not stop. Most likely the result of these precision tests will get a null result, but these tests are very important. They challenge our fundamental concepts and assumptions, they limit our speculations and sometimes they point the way to the future.

The next session was on condensed matter physics. We heard from Subir Sachdev much about AdS/CMT; the duality between string theory in AdS space and conformal field theory that describes quantum critical points. The string theory community is incredibly happy about this development for many reasons that are obvious, but also because such connections between quantum field theory and condensed matter physics have always been a source of enormous stimulation for both high energy and condensed matter physics. We currently have a program on this development at the KITP where it is amusing to hear condensed matter physicists, like Subir Sachdev and others, talk about string solutions, Einstein's equations, and black holes, and similarly string theorists talk about Fermi surfaces and Lutinger liquids. It is a wonderful development and I have no doubt that it will produce results. Somebody said, I think it was Nima Arkani-Hamed, that no one has derived AdS/CFT; no one has proved that it is correct. In fact I think we are well on the way to a proof, by following something like a better version of the Wilsonian renormalization group and extending quantum field theory from the boundary into the bulk, the extra dimension being the scaling variable. Surely this will be done by 2111. Interestingly enough, the renormalization group from the field theory point of view is a sophisticated block-spin technique wherein you lose information by constructing an effective Lagrangian or Hamiltonian. You lose observables and, in a sense, information and entropy increases, although it is not stated that way. I found it quite remarkable that a similar approach was discovered by condensed matter physicists who were trying to construct better wave functions for complicated condensed-matter systems by using tensor products as trial wave functions, and by increasing entanglement step by step they, in effect, discovered an extra dimension, which could be interpreted geometrically much as in AdS. So there is something really interesting going on with the connection between geometry and information, which I think we are just beginning to understand. We have many examples of this connection using AdS/CFT, but it seems to me much broader and much deeper. Coming back to the session, Subir Sachdev raised the question: what are the quantum phases of matter? It seems kind of amazing that after a hundred years of quantum mechanics, that this is an open question. Maybe all quantum phases are dual to the possible string solutions, maybe not. But it is a fascinating question that might be answered by 2111.

The fourth session was devoted to particle physics. Frank Wilczek, in his talk, said that there were two small number problems, or large number problems, that he thinks are critical. I think we understand one of them. I mean we do understand why gravity is weak, if you give us our understanding of the standard model, especially asymptotic freedom, and you give us string theory. Even if you start with closed strings and pure gravity, string theory invariably has open strings, gauge interactions, and standard model-like matter. The converse is also true – if one starts with standard model-like matter and gauge interactions – one discovers gravity. The two are connected. If you start with gravity, you will have the rest, matter and gauge interactions, and then, given our understanding of the standard model, you can calculate the 10^{-19} ratio of the proton mass to the Planck mass. So I do think that we understand one out of these two small numbers. The other small number, which is related to the so called the cosmological constant problem, is also about equal to 10^{-19}, if you take the fourth root, which is correct since you want to make a mass scale out the cosmological constant and not an energy density. That one we have not solved. As you know, there are solutions that I do not like but with the experience of having solved one of those problems, I think we can have faith that we can solve the other by 2111.

Howard Georgi asked a wonderful question: What discoveries at the LHC would shake our faith in quantum field theory? Wouldn't it be nice if that turned out to be what we discovered at the LHC. I think that Ed Witten's response to Sankar Das Sarma's statement that, if we do not find supersymmetry at the LHC life would be tough is correct. If we do not discover anything new at the LHC – or just the Higgs which is the real nightmare – then life will really be tough for us, because the only clues we have had will not have been realized at this accessible energy and we will have no guide as to what the next threshold is, except for the Planck scale or the GUT scale, and no way of convincing society to pay for bigger accelerators.

Finally, we have just concluded the session on string theory and quantum gravity. I have a few comments. We discussed the question of observables in quantum gravity. This is a perennial and difficult problem in quantum gravity and I have not heard any good answers. What you do, especially in closed universes of the type that Jim Hartle discussed where it is pretty hard to think of any observables at all, or in cases like de Sitter space where every degree of freedom separates eventually from every other degree of freedom outside one's horizon. We talk about the wave function of the universe, but I am not sure what the observables are. This question really hits us when we try to discuss quantum cosmology. Then there is the issue that I brought up in the introduction, and was also brought up by Frank Wilczek and many others: Is the separation between dynamics and kinematics natural? And what picks out of the framework that we have, say string theory, which allows for many so-called vacua (a term I object to because most string "vacua" are metastable and have singularities in their past), the actual observable universe. What picks the initial and the final condition? We have some interesting suggestions but we have to invent the rules as we go along.

Juan Maldacena said that there is very strong evidence that there is an exact string theory. I do not know what he meant by an exact string theory. My view in recent years is that string theory is a framework, which is not continuously disconnected from quantum field theory. We have learned that quantum field theory is actually a much bigger framework, containing what we call string theory whatever that is. But I never thought of quantum field theory as a precise theory; rather it is a framework in which we specify other principles to pick out a particular dynamical evolution. I do not see what in string theory picks out a specific dynamics. Perhaps the answer to the unanswered questions of what the initial condition is or what the final condition is will also specify the dynamics. All of this gets even harder to imagine as everyone says, with few objections, that space is emergent. We all accept this vision. Although it is not easy to do away with space as a basic concept we have all become accustomed to accepting this as a possibility. We have many examples in which part or all of space is emergent. But what really emerges from these dualities is space-time, in very simple geometries where time just goes along for the ride. I find it impossible to imagine a rigorous and well defined analytic description of an emergent time. If there is time for discussion, which there is not, but we will make time if somebody knows how to imagine a theory with no space, no time from which a bulk space-time would emerge. This is a good place to end because we can start the discussion with that question: Is time emergent and how could it be emergent?

G. 't Hooft May I just object to the statement that you said that everybody agrees that space is emergent. I do not.

D. Gross I will amend: everyone except Gerard. Any other comments that are urgent and that prevent us from coming to a conclusion?

E. Witten I kind of like to make one very narrow point, which is that I think when Juan Maldacena said there was an exact string theory, I took it to mean that the asymptotic expansion, for example in ten-dimensional Minkowski spacetime, is asymptotic to an exact theory there, an exact unitary quantum theory.

D. Gross Right, I understood it that way as well, but that would be an exact string state if you wish, not an exact string theory analogous to a specific Hamiltonian, or Lagrangian, or non-Lagrangian, whose solutions would be not just ten-dimensional superstring theory in flat space but also, say, quantum chromodynamics.

J. Malcadena I meant it in the narrow sense that Ed Witten said, in any situation where you have good observables, that you can perturbatively compute them using string perturbation theory, there is something exact you can compute. Let me make a comment here. You could imagine that maybe all orders in perturbation theory are well defined but the whole theory may not be well defined. For example, three-dimensional pure gravity is a renormalizable theory, it seems to make sense at all orders in perturbation theory, but when you consider it on AdS3 with toroidal boundary conditions, there seems to be some inconsistencies. It might be that the theory does not make sense due to these inconsistencies. So string theory could have been like that, and maybe it is like that. It might be we encounter some inconsistencies and there is something else, but if you believe in all these dualities and the fact that any theory is dual to string theory in this more generalized sense that you were imagining, it is more ...

D. Gross It is a major disappointment that, as Ed Witten knows very well as it was one of his dreams as well as mine, that it didn't turn out that ten-dimensional flat space superstring theory was not be an exact quantum state, so that we would find a much more restricted class of exact quantum states. The strange thing about string theory, although it appears to be a unique framework with no dynamical parameters, it does not appear to be as powerful as a specific Hamiltonian theory. We do not have any such a formulation of "string theory"; what we have are rules that are consistent from a variety of points of view for constructing quantum states. That is all. It is a framework. That is why I feel something is missing. It is like quantum field theory, which is a framework to which we have to add something to make predictions. Something is still missing in string theory before we can make predictions.

J. Malcadena Let me offer a wild speculation. In AdS/CFT what defines the theory are the boundary conditions. So maybe in our world what defines the theory is the question we are asking. So we are asking questions, and bad questions give zero, they are bad questions, and we can ask good questions, and all good questions have some answer.

D. Gross Does it depend on whose is asking?

J. Malcadena Yes, probably. Surely, I mean different observers in different vacua of the landscape get different answers.

E. Rabinovici For historical record, you mentioned maybe the lack of scales and so on. We never mentioned neutrinos here and we should remember neutrinos seem to have mass and that could give us an indication of a scale.

J. Hartle I am not sure if this is the right moment but I would like to propose a round of applause for our organizers.

Closing Session

Address by the Chair of the Conference

David Gross

I like to thank all of you, all of the contributors, speakers and discussants and, especially, the chairs of the sessions, who have been subject to extreme pressure over the last few months. You see – it paid off! We have had a wonderful conference, spanning all of quantum physics, with fascinating discussions. To have some measure of spontaneity requires extreme organization and I thank you for your efforts and hard work.

I also thank the staff of the Solvay Institutes for making all the arrangements that produced such a comfortable and enjoyable week in Brussels for all of us; I thank the people in the back of the room who created this wonderful environment for discussing physics; and I thank all the transcribers who, under the watchful eye of Alex Sevrin, will produce a written record of all of our proceedings.

We all have a deep sense of gratitude to the Solvay family and especially to Jean-Marie Solvay, for continuing the great tradition whose centenary we have helped to celebrate.

Finally, I would like to thank Marc, because this conference would not have been possible without his efforts and dedication. So thank you Marc.

Address by the Director of the International Solvay Institutes

Marc Henneaux

I started the conference with thanking David. This is the concluding session and I would like to conclude with the same words.

Indeed, it was a challenge to organize a successful scientific meeting with such a broad spectrum of fields and areas being covered. That meant a lot of work. It clearly paid off.

I am confident that all the participants will leave like in 1911 thinking that the conference was too short. Physics is as vivid now as it was one hundred years ago. We are extremely grateful to its chair for the remarkable scientific success of the Solvay centenary conference.

This is also the occasion for me to thank Alexander Sevrin, as well as the staff of scientific secretaries who have been working hard and will work even harder, now that they have to transcribe what has been recorded.

So thank you very much to all... and see you in 2111!

I.Antoniadis, X-G Wen, A.S

J.Maldacena, G.Giudice, D.Awschalom, N.Beisert, S.Dimopoulos, G.Veneziano, M.Sh

G.Horowitz,

L.Randall, W.Zurek, D.Kleppner, J.Hartle, J.Polchinski,   V.Mukhanov, G.Dvali, Y.Aharonov

A.Aspect, I.Cirac, J.Preskill, P.Zoller,   R.Dij

A.Sevrin, F.Englert, N.Seiberg, A.Leggett, E.Witten, M.Henneaux, F.Wilczek, D.Gr